CMOS Analog Integrated Circuits

Data Converters, Phase-Locked Loops, and Their Applications

T0225352

CMOS Analog Integrated Circuits

Data Converters, Phase-Locked Loops, and Their Applications

Tertulien Ndjountche

CRC Press
Taylor & Francis Group
Boca Raton London New York

CRC Press is an imprint of the
Taylor & Francis Group, an **informa** business

CRC Press
Taylor & Francis Group
6000 Broken Sound Parkway NW, Suite 300
Boca Raton, FL 33487-2742

First issued in paperback 2020

© 2019 by Taylor & Francis Group, LLC
CRC Press is an imprint of Taylor & Francis Group, an Informa business

No claim to original U.S. Government works

ISBN-13: 978-1-138-59973-4 (hbk)
ISBN-13: 978-0-367-73311-7 (pbk)

**Visit the Taylor & Francis Web site at
http://www.taylorandfrancis.com**

**and the CRC Press Web site at
http://www.crcpress.com**

Contents

Preface

The second edition of the book, *CMOS Analog Integrated Circuits*, is motivated by the scaling-down trend of the complementary metal-oxide semiconductor (CMOS) process used for the fabrication of integrated circuits. The shrinking of transistors is accompanied by the increase of chip density and circuit speed, and a reduction of the power supply voltage. However, important challenges (leakage currents, variability of technological parameters) for the analog circuit design can be associated to the deployment of nanometer CMOS process, especially below 65 nm. Approaches to overcome them rely on the use of appropriate analog synthesis techniques and computer-aided design tools at the circuit and physical levels.

The miniaturized silicon technology enabled the advent of various wireless and wearable devices. The significant impact of the electrical energy consumption on emerging applications has driven the need for low-power design methodologies. The book describes the important trends of designing high-speed and power-efficient front-end analog circuits, which can be used alone or to interface modern digital signal processors and micro-controllers in various applications such as multimedia, communication, instrumentation, and control systems.

The book contains resources to allow the reader to design CMOS analog integrated circuits with improved electrical performance. It offers a complete understanding of architectural- and transistor-level design issues of analog integrated circuits. It provides a comprehensive, self-contained, up-to-date, and in-depth treatment of design techniques, with an emphasis on practical aspects relevant to integrated circuit implementations.

Starting from an understanding of the basic physical behavior and modeling of MOS transistors, we review design techniques for more complex components such as amplifiers, comparators, and multipliers. The book details all aspects from specifications to the final chip related to the development and implementation process of filters, analog-to-digital converters (ADCs) and digital-to-analog converters (DACs), phase-locked loops (PLLs) and delay locked loops (DLLs). It provides the analysis of architectures and performance limitation issues affecting the circuit operation. The focus is on designing and verifying analog integrated circuits.

The book is intended to serve as a text for the core courses in analog integrated circuits and as a valuable guide and reference resource for analog circuit designers and graduate students in electrical engineering programs. It provides balanced coverage of both theoretical and practical issues in hierarchically organized format. With easy-to-follow mathematical derivations of all equations and formulas, the book also contains graphical plots, and a number of open-ended design problems to help determine the most suitable circuit architecture satisfying a given set of performance specifications. To appreciate the material in this book, it is expected that the reader has a rudimentary understanding of semiconductor physics, electronics, and signal processing.

New to this edition

Every chapter in the second edition has been revised to reflect the evolution of modern CMOS process technology. Furthermore, the text emphasizes paradigms that needed to be mastered and covers new materials such as:

1. DAC switching schemes

2. Element mismatch compensation techniques (optimal switching sequences, data-weighted averaging) for DACs

3. Generalized design method for continuous-time delta-sigma modulators

4. Review of phase detectors

5. Architectures and circuits for clock and data recovery

6. High-speed input/output link design

7. Phase-locked loop based on time-to-digital converter

8. Relaxation oscillator

9. Class-D power amplifier

Content overview

The book contains six chapters and two appendices.

Chapter 1
Mixed-Signal Integrated Systems: Limitations and Challenges
The use of CMOS technologies with a low device geometry and new architectures has accelerated the trend toward the system on a chip design, which merges analog, digital, and radio-frequency (RF) sections on a single integrated structure. While the manufacturing technology appears to be fundamentally limited by the material characteristics, the computer-aided design

tools have to face the computational intractability of design optimizations. In this context, design techniques should be concerned with the automated conception, synthesis, and testing of microelectronic systems.

Chapter 2
Data Converter Principles

The interface between real-world signals and digital-signal processors can be realized by data converters (analog-to-digital converters and digital-to-analog converters). An insight into the mathematical definitions of characteristics (quantization noise, component imperfections), which can affect the performance of data converters is provided. Depending on the sampling frequency, Nyquist and oversampling data converters can be distinguished. Generally, Nyquist converters are based on a parallel operation and can exhibit a high speed. On the other hand, digital filtering is combined with oversampling, which relies on using a sampling rate which is several times higher than two times the signal bandwidth, to improve the converter resolution. For a given dynamic range, the reduced sensitivity of oversampling structures to component imperfections is the result of a trade-off between speed and accuracy.

Chapter 3
Nyquist Digital-to-Analog Converters

Digital-to-analog converters can be designed using various architectures, each with its distinctive advantages and limitations. A review of various Nyquist converter architectures illustrates the system-level trade-offs and performance issues associated with the circuit design. For a given resolution, the difference between converter architectures can be an important factor for the selection of a specific application.

Chapter 4
Nyquist Analog-to-Digital Converters

A basic understanding of various ADC architectures is useful to meet the design challenges at the transistor level of high-resolution converters. There are various ADC architectures, each with its peculiar advantages and limitations. The description of Nyquist converters is presented along with their performance modeling. Applications are key in selecting a given ADC architecture even though some overlap can exist between the characteristics of various architectures.

Chapter 5
Oversampling Data Converters

Oversampling data converters generally consist of a delta-sigma modulator and digital (decimation or interpolation) filter. By combining the noise shaping and oversampling, which is similar to the sampling of a signal at a rate higher than twice the maximum frequency in the input signal, their quantization noise is removed from the signal band and spread over a larger range of frequencies. Various modulator architectures will be reviewed, the effects of circuit nonidealities on the converter performance are analyzed, and the digital filters used to remove the out-of-band noise will be presented. An evaluation and a comparison of the different delta-sigma modulation-based approaches used to improve the linearity of Nyquist converters are also provided. Another application area for delta-sigma modulators is the test and instrumentation where a precise test signal is required.

Chapter 6
Circuits for Signal Generation and Synchronization
Due to the increase of the IC clock frequency and data rate, circuits (phase-locked loop, delay-locked loop) for the clock signal generation and synchronization are generally included in electronic systems to avoid data read and transmission failures. They should be designed to operate with a low voltage, feature a low timing jitter, and be less sensitive to process and temperature variations. A tutorial survey of timing circuit and frequency synthesis architectures is presented. By combining timing and equalization circuits, power efficient input/output transceivers for high-speed serial links can be designed. Descriptions of relaxation oscillators, that generate non-sinusoidal signals, and class-D power amplifiers are provided.

Appendices
Two appendices cover the following topics:

Acknowledgments

Many of the changes in this edition were made in response to feedback received from some readers of the first edition. I would like to thank all those who took the time to send me messages.

I am grateful for the support of colleagues and students whose remarks helped refine the content of this book.

I would like to thank Prof. Dr.-Ing. h.c. R. Unbehauen (Erlangen-Nuremberg University, Germany). His continuing support, the discussions I had with him, and the comments he made have been very useful.

I express my sincere gratitude for all the support and spontaneous help I received from Dr. Fa-Long Luo (Element CXI, USA).

I wish to acknowledge the suggestions and comments provided by Prof. Avebe Zibi (UY-I, CM) and Prof. Emmanuel Tonye (ENSP, CM) during the early phase of this project.

While doing this work, I received much spontaneous help from some international experts: Prof. Ramesh Harjani (University of Minnesota, Minneapolis, Minnesota), Prof. Antonio Petraglia (Universidade Federal do Rio de Janeiro, Brazil), Dr. Schmid Hanspeter (Institute of Microelectronics, Windisch, Switzerland), Prof. Sanjit K. Mitra (University of California, Santa Babara, California), and Prof. August Kaelin (Siemens Schweiz AG, Zurich, Switzerland). I would like to express my thanks to all of them.

I am also indebted to the publisher, Nora Konopka, the project coordinator, Kyra Lindholm, the production editor, Michele Dimont, and the CRC Press editorial team of the previous edition, Jessica Vakili, Karen Simon, Brittany Gilbert, Stephany Wilken, Christian Munoz, and Shashi Kumar, for their valuable comments and reviews at various stages of the manuscript preparation, and their quality production of the book.

Finally, I would like to truly thank all members of my family and friends for the continual love and support they have given during the writing of this book.

1

Mixed-Signal Integrated Systems: Limitations and Challenges

CONTENTS

The objective of designing a complete system on a single chip has resulted in the complexity increase of application-specific integrated circuits (ASICs), application-specific standard parts (ASSPs), and very large-scale integrated circuits. The system on a chip (SoC), as shown in Figure 1.1, generally possesses complex signal paths through both analog devices and digital components (nonvolatile memory (NVM), random access memory (RAM), and digital signal processor (DSP)).

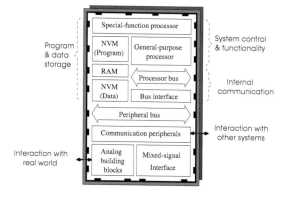

FIGURE 1.1
Example of an SoC floorplan.

Examples include multimedia devices, wireless transceivers, sensor and actuator controllers, instrumentation systems, and biomedical devices. The functions, which are realized in the analog domain, include:

1

☐ Biasing

☐ Sensor and actuator signal conditioning

☐ Driver and buffer

☐ Signal down-conversion and up-conversion

☐ Mixed-signal DSP interfaces

☐ Clock signal generation and frequency synthesis

They are implemented using basic building blocks such as:

- Voltage and current references
- Low-noise and power amplifiers
- Variable-gain amplifier and automatic gain control circuit
- Filter
- Oscillator
- Mixer
- Sample-and-hold circuit
- Analog-to-digital converter
- Digital-to-analog converter
- Phase-locked loop (PLL) and delay-locked loop (DLL)
- Input/output link transceiver

The SoC digital section essentially requires microprocessors, digital signal processors, memories, and control logics. The most important issues are then related to the integration of analog and digital sections. A fully monolithic chip appears to be limited, for instance, by the problematic isolation of analog sections with high-gain bandwidth from the noise generated by the substrate and digital circuits. Furthermore, the device-level simulation of mixed-signal integrated circuits in a realistic environment remains a challenge and testing chips with several complex functions is a difficult task.

1.1   Integrated circuit design flow

The specification partition into subsystems is illustrated in Figure 1.2. Tools such as SDL (Specification and Description Language), UML (Unified Modeling Language), and SystemC AMS are used to analyze the design at the higher level. At the system level, the design specifications are partitioned into hardware and software components. Note that SystemC AMS is particularly suited to provide functional modeling, architectural exploration, virtual prototyping, and integration validation for analog mixed-signal systems.

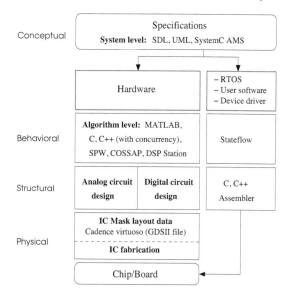

FIGURE 1.2
Specification partition into subsystems.

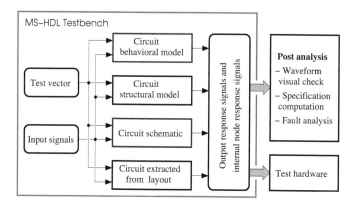

FIGURE 1.3
Circuit design verification.

The development of signal processing algorithms can be performed using MATLAB®, SPW (Signal Processing Workbench), COSSAP (Communication System Simulation and Application Processor), DSP Station, C, and C++ (extended to handle concurrency). To help manage the complexity, the design implementation in hardware is supported by providing a link to hardware synthesis tools.

The functional description is realized by an analog circuit and a digital circuit, which can be designed and verified using various computer-aided design programs such Cadence, Synopsys, and Mentor Graphics tools. When the data processing in the digital domain may require processors or microcontrollers, the design of a real-time operating system (RTOS), application software and device driver is necessary. Besides allowing a modular and scalable programming approach, the desirable features of an RTOS include the ability to provide basic support for task scheduling, resource management, inter-task and input-output communication such that the processor functionality is available to application software in an optimized and predictable way. Stateflow is an interactive design and simulation tool that can be used to describe complex logic, such as an RTOS, in a form that can easily be coded using C/C++ language or assembler.

The functional description can also be refined to analog and digital models, which can be analyzed and verified using a simulator that can interpret mixed-signal hardware description languages (MS-HDLs). Verilog-AMS and VHDL-AMS are two examples in this category (VHDL stands for very high-speed integrated-circuit HDL). MS-HDLs are particularly well suited for the verification of very large and complex mixed-signal integrated circuit designs. An MS-HDL testbench, as shown in Figure 1.3, provides the stimulus required to drive various representations of a circuit while the response signals at nodes of interest are monitored. Specifications are checked by comparing the behavioral and structural models, while the implementation is verified by emphasizing the similarities between the circuit schematic and the circuit extracted from the layout.

The design flow of an integrated system is illustrated in Figure 1.4. The top-down synthesis process consists of the topology selection, specification translation or circuit sizing, and design verification (design rule check (DRC), electrical rule check (ERC), and layout versus schematic (LVS)). It is then followed by a bottom-up generation and verification of the circuit layout. The performance specifications are required at each step. Throughout the design flow, any change should be taken into account by propagating the associated constraints down the hierarchy, thus ensuring that the top-level block meets the target specifications.

Nowadays, the methodologies of top-down design and bottom-up verification are well-accepted standards in the digital domain. From bit true models of signal processing algorithms, C, Verilog, or VHDL code is generated or written for custom hardware or a DSP-based software solution. By defining a digital circuit at an architectural or behavioral level rather than at the gate level, hardware description languages, such as Verilog or VHDL, can help manage more large designs than tools based on schematic entry. An automated design flow is then adopted to convert the high-level description of the circuit into industry-standard output formats, such as GDSII, that can be integrated into chip layout tools.

In the analog domain, the current design approach — design, simulate, optimize circuit specifications taking into account parasitic effects and process

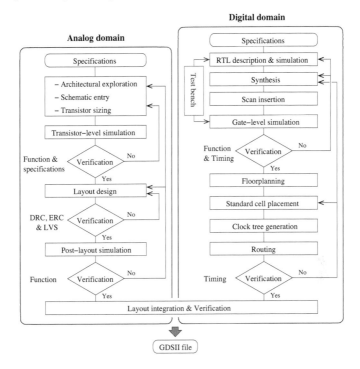

FIGURE 1.4
Design flow for an integrated system.

variations, repeat — can be very time consuming for large circuits and relies mostly on designer experience. This is due to the fact that second-order effects in analog circuits are difficult to model as the design evolves and automated tools are actually not available. Furthermore, the use of deep submicrometer CMOS processes contributes to making the verification of analog circuits substantially more difficult, as self-calibration or error cancelation schemes are often required to overcome the limitations of the components, thereby increasing the number of operating modes, behavioral complexity, and size of analog circuits.

The integration of circuit components takes place either during the physical design phase or after the fabrication. Physical design involves the floorplanning, timing optimization, placement and routing, or layout. The system timing and signal integrity verification is achieved by checking the electrical and design rules and comparing the layout and schematic. Detailed physical design information is required for accurate resistance, inductance, and capacitance parasitic extraction, and delay estimation. Note that the final system should include both hardware and software. The software platform binds the programmable cores and memories via the RTOS, the input-output interface via the device drivers, and the network connection via the communication subsystem.

1.2   Design technique issues

At the system level, an efficient solution is required for managing concurrency or assuring a real-time data flow. This is important for complex chips, which handle multiple tasks at the same time and in cases where the latency due to the interconnect delay dominates the signal bandwidth. The design reuse results in a reduction of the cost and development time. A higher level of abstraction is necessary to create a library of subsystems that can be used for different designs. The design of a complex chip should also include an adequate strategy for the verification of the functional blocks. The logic simulation is a suitable method for the functional verification. Here, the system is tested over a wide variety of operating conditions using simulated input patterns. However, the complexity of the chip can reduce the effectiveness of finding possible design errors. This is related to the higher number of likely events and the difficulty of determining whether the simulated behavior is correct. SoC testing suffers also from the lack of effective coverage metrics, that is, it is not always clear whether enough verification has been completed to confirm the reliability of a chip.

During the design, the circuit optimization is made with respect to timing, power, and area specifications for given values of interconnect load capacitances which can be different from the ones extracted from the final layout, specifically in deep submicrometer technologies. In addition, metal resistance effects are topology dependent and increase with the routing length, and the prediction of the delay propagation is not simple. That is, one-pass synthesis success becomes unlikely due to the requirement of physical design information. A possible solution can consist of using synthesis methods based on the delay equalization of all subsystems and the wire planning among blocks. The speed-power performance of a design based on a submicrometer integrated circuit (IC) process appears to be affected by the substrate and crosstalk-induced noises, signal delay, and parasitic inductance. The coupling effects can be controlled using low-swing differential pair structures, shield wires and repeater insertions, upper and lower bounding slew times, and increased spacing between wires. The increase in functionalities and operating frequencies results in more power dissipation. However, in addition to the supply voltage scaling, power consumption can be reduced by switching off unused subsystems via gated clocking modes.

It can be predicted that the use of an IC process with low geometries will increase the impact of fabrication techniques on the design and verification. The top-down design methodology provides a system-level model that can be used for the chip testing. But the mixed-signal nature of SoCs makes different test strategies suitable for each particular type of component, resulting in the requirement of a design for test and manufacturability across all abstraction levels.

1.3   Integrated system perspectives

The realization of integrated systems is influenced by several factors (IC process, circuit structure, package, software). Common goals such as performance optimization and development time reduction must be included in the suitable design framework.

Mixed-signal building blocks should be designed for reuse. Such a design is based on accurate high-level models, which can be used to evaluate the block suitability for a new design. The performance achievable in the hardware design reuse methodology seems to be limited in situations where the loading rules are highly complex or the circuit models exhibit interrelated features. Due to the chip complexity, power and performance can be lost by using a single clocked-synchronous approach to manage the on-chip concurrency. An optimal implementation and verification of reliable communication among a collection of components may then be necessary. The programmable platform-based design emerges as a viable approach for the SoC implementation, and the optimal power/performance is dependent on the trade-off between hardware and software. Methodologies for re-mapping and redesigning blocks based on physical information will be inessential only if the design approaches are either based on an improved nonideality (delay, noise, distortion) prediction or able to remove the requirement of predictability.

The performance of high-density ICs is mainly limited by noise and timing faults. For instance, the simultaneous switching of more devices increases the power supply noise. This can enlarge the timing delay by reducing the actual voltage that is applied to a device. The effect of capacitive coupling in submicrometer designs is also important and affects the signal integrity. Since the complexity of SoC makes a unified testing scheme difficult to implement, a self-test mechanism is required at the component and system levels. It can be developed based on new fault models with links to the layout and implemented as a program executed by the processor core. This approach has the benefit of eliminating any additional built-in self-test (BIST) hardware, such as linear feedback shift registers. Another important issue in critical applications is related to self-repair techniques, which take advantage of the reconfigurability provided by adding coprocessors, appropriate instruction sets, and peripheral units to the embedded processor core.

To eliminate the need for external testers, the implementation of BIST for analog blocks should preferably exploit the capability of the digital processor for the signal generation and analysis and down-sampling techniques for the specific case of RF circuits.

1.4   Built-in self-test structures

With the increase in the density and complexity of mixed-signal integrated circuits (ICs), more complex measuring devices are required to meet ever more severe test specifications. The built-in self-test (BIST) appears as a suitable approach to resolve the problem related to the fact that mixed-signal circuits are verified by functionality, the number of which can be high in a single chip. Furthermore, a BIST section can facilitate the initialization and observation of the circuit nodes. BIST structures for digital circuits have reached a good level of maturity, and it can be expected that testing solutions for the analog section in mixed-signal systems will exploit the computation capability of logic gates and digital signal processors.

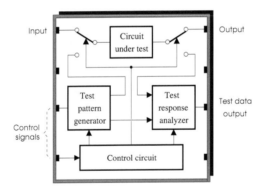

FIGURE 1.5
A chip including a built-in self-test structure.

The architecture of a chip including a BIST section is shown in Figure 1.5. It consists of a test pattern generator, a test response analyzer, and a control circuit, in addition to the circuit under test, which can be reconfigured by the control signals to support the test mode and the normal operation mode. Depending on the BIST flexibility, testing can be carried out while the circuit is in an idle state or during normal operation. A generator is used to provide the required test input signals and the analyzer features the capability of detecting various faults in the output response of the circuit under test.

1.5   Concluding remarks

The use of a submicrometer process for the IC implementation extends the operating frequency range, but also results in the enhancement of nonideal effects related to the interconnect crosstalk and latency. The density of test data is growing as the complexity and the number of intellectual property cores required in a single chip is increasing. Thus, the viable development of SoCs should take into account aspects of the design, manufacturing, and testing at all abstraction levels.

1.6   To probe further

- Special issue on limits of semiconductor technology, *Proc. of the IEEE*, vol. 89, no. 3, March 2001.

- W. Müller, W. Rosenstiel, and J. Ruf, Eds., *SystemC Methodologies and Applications*, Dordrecht, The Netherlands: Kluwer Academic Publishers, 2003.

- D. Jansen et al., Eds., *The Electronic Design Automation Handbook*, Dordrecht, The Netherlands: Kluwer Academic Publishers, 2003.

- M. D. Birnbaum, *Electronic Design Automation*, Upper Saddle River, NJ: Prentice Hall, 2004.

- Special issue on system on chip: Design and integration, *Proc. of the IEEE*, vol. 94, no. 6, June 2006.

- Special issue on leading-edge computer-aided design solutions for advanced digital and mixed-signal systems-on-chips, *Proc. of the IEEE*, vol. 59, no. 3, March 2007.

2

Data Converter Principles

CONTENTS

Data converters, or specifically analog-to-digital converters (ADCs) and digital-to-analog converters (DACs), play an important role in the design of data acquisition units in communication and microprocessor-based instrumentation systems. They include analog and digital building blocks and form the main interface component in mixed-signal processing systems (for some typical applications, see Table 2.1).

TABLE 2.1

Data Converter Specifications for Some Applications

Applications	Resolution (bits)	Sampling Frequency
Audio device	14–24	< 200 kHz
Digital oscilloscope	8	150 MHz
Magnetic read channel equalizer	6–8	(50–200) MHz
Wireless local area network	6–10	(1–50) MHz
Digital video camera	8–12	20 MHz
TV baseband processor	8–10	20 MHz
Modem	8–10	(10–20) MHz

• The process of converting an analog signal into a digital sequence, as illustrated in Figure 2.1, involves three operations: sampling, quantization, and coding. After the filtering operation, a sample-and-hold circuit first picks up the signal representative, which is maintained constant for the duration re-

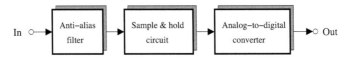

FIGURE 2.1
Typical ADC system.

quired by the converter to provide a digital word. Note that a continuous-time
signal with the maximum frequency, f_m, can adequately be represented by its
samples acquired at the rate $f_s \geq f_N = 2f_m$, where f_N is termed the Nyquist
rate or frequency. To maintain the frequency content of the input signal within
the Nyquist bandwidth, an analog lowpass filter, also known as an anti-aliasing
filter, is placed before the sample-and-hold circuit. Ideally, this filter should
attenuate signal components with a frequency above $f_s/2$. The sampled signal
is then transformed into one of a finite set of prescribed values by a quantizer,
the levels of which can be uniformly or nonuniformly spaced. The transfer
characteristic and error, e, of a uniform quantizer, whose implementation is
the simplest of both, is shown in Figure 2.2. The error caused by the quanti-

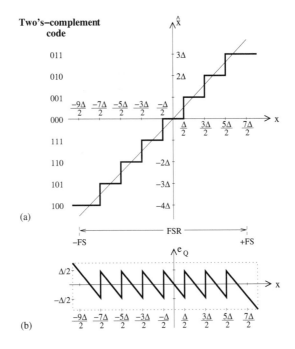

FIGURE 2.2
(a) Transfer characteristic and (b) quantization error of a 3-bit ADC.

zation is defined as the difference between the discrete output level and the actual analog input, $e_Q = \hat{x} - x$. It is in the range $\pm\Delta/2$ as long as the quantizer does not saturate. The transfer characteristic of Figure 2.2 belongs to a quantizer of the mid-tread type because it follows the input axis about zero. As a result, the output is insensitive to small input variation in the absence of signal in contrast to the mid-riser characteristic, which is supported by the output axis about zero. The coding then consists of assigning a unique binary number to each quantization level. Assuming that the number of bits, N, is 3, the 2^N quantization levels can be coded in an N-bit number representation. The step size of the converter, Δ, represents the least significant bit (LSB) of the digital number and is given by $\Delta = FSR/2^N$, where FSR is the quantizer range or full-scale range (FSR). By using the two's complement code, the sign of the input sample is determined by the most significant bit (MSB). A real number, X, which can be represented as

$$b_1 b_2 b_3 \cdots b_N \tag{2.1}$$

corresponds to the value

$$-b_1 2^0 + b_2 2^{-1} + b_3 2^{-2} + \cdots + b_N 2^{-(N-1)} \tag{2.2}$$

where b_1 is the MSB. Other codes can also be used depending on signal characteristics and the desired application; however, the two's complement representation is a convenient way of representing signed numbers and the most suitable for addition and subtraction operations.

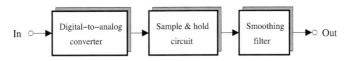

FIGURE 2.3
Typical DAC system.

• A digital-to-analog conversion stage, as shown in Figure 2.3, generally contains a DAC, a sample-and-hold (S/H) circuit, and a lowpass filter (LPF). The DAC is used to transform a finite number of digital codes into the corresponding analog discrete-time signal. Its transfer characteristic is shown in Figure 2.4 for a bipolar input code. In contrast to the ADC, which exhibits a quantization error due to the fact that any voltage within a given step size is mapped to the same output code, the DAC uniquely assigns each input code to an output level without an inherent error. Therefore, DACs do not directly realize the inverse function of ADCs.

In general, the DAC output signal, X_0, can ideally be put in the form

$$X_0 = G \cdot X_{REF} \left(K_1 \frac{D}{2^N} + K_2 \right) \tag{2.3}$$

where G is the gain, X_{REF} is the reference signal, D is the decimal equivalent

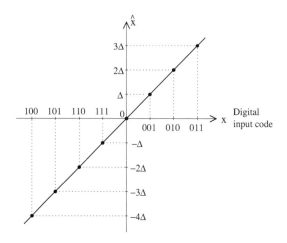

FIGURE 2.4
Transfer characteristic of a 3-bit DAC.

of the binary input code, and K_1 and K_2 are the gain and offset constants, respectively. In the case of unipolar conversion, $K_1 = 1$ and $K_2 = 0$, and the output range is from 0 to $G \cdot X_{REF}$. For bipolar DACs based on the offset binary input coding, the constants are chosen as $K_1 = 2$ and $K_2 = -1$ to produce an output swing between $-G \cdot X_{REF}$ and $G \cdot X_{REF}$. Note that the two's complement representation is converted into offset binary code only by inverting the MSB.

The digital-to-analog conversion process should be realized with the highest fidelity and minimal lag in time. The S/H introduces a delay in the output of the DAC to allow the current sample at the DAC output to reach the steady state. An LPF, often referred to as a smoothing (or reconstruction) filter, is used to remove the frequency components above $f_s/2$ from the converter output. It also smooths the signal provided by the S/H by removing all sharp discontinuities.

Various architectures are known for the implementation of data converters. They can be divided into two main groups: Nyquist and oversampling converters. Nyquist converters operate at a sampling rate close to the Nyquist frequency or slightly higher than twice the bandwidth of the input signal; therefore their output data rate can be very high. On the other hand, the operation of oversampling converters, which can achieve a higher resolution even with low-precision components, requires a sampling rate that is several times higher than the Nyquist frequency. By relying on the averaging of multiple samples performed by a digital filter for each conversion, oversampling converters feature a longer acquisition time than the one of Nyquist converters, which process each input sample independently.

2.1 Binary codes

One feature of data converters such as DAC and ADC is that either the input or output is in digital form and can be represented using binary codes.

Digital systems generally use a binary number coding rather than the most familiar decimal number representation. For binary codes, the base or radix is 2 and the digits are called bits and take the values 0 or 1. Positive integers with the value in the range 0 to $2^N - 1$ can be encoded using N bits. Starting with the LSB, which has a weight of 2^0 or equivalently, 1, the weight is increased by a factor of 2 from one bit to the next and up to the value of the MSB weight, 2^{N-1}. The binary representation is said to be positional because the location of a bit in the resulting sequence determines its weight. The value of a binary number corresponds to the sum of the weights of all nonzero bits.

In practice, the operating range of a data converter is bounded by the full scale (FS). Scaling down the full-scale range requires representing all of the numbers as fractions [1]. Any binary integer can be set into a fractional format by dividing its value by 2^N. With the binary point assumed to be at the left of the MSB, it is possible to encode the numbers in the range from 0 to $1 - 2^{-N}$ of the converter full-scale. The bit weight, which is inversely proportional to a power of 2, varies from $1/2$ for the MSB down to $1/2^N$ for the LSB.

2.1.1 Unipolar codes

Unipolar codes are used to represent signals with a predetermined sign. The most common unipolar codes, as shown in Table 2.2 for 4-bit converters, are natural binary, Gray, and binary-coded decimal (BCD) representations.

The natural binary representation is a positional system of numeration that uses the digits 0 and 1 and a radix of 2. Each successive digit of the binary code, which is generally adopted for a data converter, represents an inverse power of 2. The all-zero code corresponds to the zero-scale signal, while the all-one code is associated to the signal value, which is one LSB below the full scale.

Gray codes are used in the design of encoders for data converters in order to prevent errors due to the fact that all the required bit transitions between any two neighboring numbers may not occur at the same time. In general, the Gray code can be obtained by rearranging a binary sequence such that there is only one bit change between two adjacent numbers. The possibility to obtain multiple Gray code representations for a given number then increases with the resolution. The 4-bit example shown in Table 2.2 is known as binary-reflected Gray codes. From an initial binary sequence where all bits are set to zero, consecutive binary-reflected Gray codes are formed by changing the state of only one bit while starting from the rightmost bit side. Let b_k and g_k be the bits of the natural binary code and Gray code, respectively. The conversion

TABLE 2.2
Common Unipolar Codes for 4-Bit Converters

Decimal Number	Fraction of the FS	Natural Binary	Gray Code	8421 BCD
15	15/16	1111	1000	1001 0011
14	14/16	1110	1001	1000 0111
13	13/16	1101	1011	1000 0101
12	12/16	1100	1010	0111 0010
11	11/16	1011	1110	0110 1000
10	10/16	1010	1111	0110 0010
9	9/16	1001	1101	0101 0110
8	8/16	1000	1100	0101 0000
7	7/16	0111	0100	0100 0011
6	6/16	0110	0101	0011 0111
5	5/16	0101	0111	0011 0001
4	4/16	0100	0110	0010 0101
3	3/16	0011	0010	0001 1000
2	2/16	0010	0011	0001 0010
1	1/16	0001	0001	0000 0110
0	0	0000	0000	0000 0000

from the binary to Gray code is based on the next algorithm:

$$g_N = b_N \tag{2.4}$$
$$g_k = b_{k+1} \oplus b_k \quad \text{for} \quad k = N - 1 \quad \text{down to} \quad 1 \tag{2.5}$$

while the Gray-to-binary conversion is obtained as follows:

$$b_N = g_N \tag{2.6}$$
$$b_k = b_{k+1} \oplus g_k \quad \text{for} \quad k = N - 1 \quad \text{down to} \quad 1 \tag{2.7}$$

where $\oplus$ denotes the exclusive or (XOR) logic operation and N is the number of bits of the code.

In the BCD code, each decimal digit, D_k, with a value of 0 through 9 is replaced by its 4-bit binary equivalent, $b_{k,j}$, where $j = 1, 2, 3, 4$. Table 2.2 shows 8421 BCD codes of the first two fractional digits of FS fractions. The designation 8421 indicates the binary weights of the four bits representing each digit. Note that different versions of BCD codes can be obtained using other weight combinations. A further digit can be appended to a BCD code just by adding another 4-bit sequence. One advantage of BCD over binary representations is that the range of numbers that can be represented is not limited. The BCD code is particularly useful for interfacing to printing or display devices, which can process individual decimal digits (e.g., digital multimeter, digital instruments for physical measurements). On the other hand, the natural binary code for a given number requires fewer bits than the corresponding BCD code. The BCD representation is a relatively inefficient coding due to the fact that only 10 of the 2^4 or 16 combinations allowed by any 4-bit sequence are exploited.

Note that additional bit combinations are sometimes used in the BCD code in order to take into account the data sign or other meaningful indications.

2.1.2 Bipolar codes

TABLE 2.3

Common Bipolar Codes for 4-Bit Converters

Decimal Number	Fraction of the FS	Sign Magnitude	One's Complement	Offset Binary	Two's Complement
7	7/8	0111	0111	1111	0111
6	6/8	0110	0110	1110	0110
5	5/8	0101	0101	1101	0101
4	4/8	0100	0100	1100	0100
3	3/8	0011	0011	1011	0011
2	2/8	0010	0010	1010	0010
1	1/8	0001	0001	1001	0001
0	0	0000 1000	0000 1111	1000	0000
−1	−1/8	1001	1110	0111	1111
−2	−2/8	1010	1101	0110	1110
−3	−3/8	1011	1100	0101	1101
−4	−4/8	1100	1011	0100	1100
−5	−5/8	1101	1010	0011	1011
−6	−6/8	1110	1001	0010	1010
−7	−7/8	1111	1000	0001	1001
−8	−8/8	—	—	0000	1000

Bipolar codes are used to represent signals that can be either positive or negative. The most common bipolar codes, which are sign magnitude, one's complement, offset binary, and two's complement representations, are shown in Table 2.3 for a resolution of 4 bits.

In the sign-magnitude representation, the bit in the MSB position is reserved for the number sign and the remaining bits indicate the number magnitude. The sign bit can be either 0 for positive numbers or 1 for negative numbers. The sign-magnitude representation has the drawback of having two different codes for zero and requiring a rather complex hardware for the realization of arithmetic operations.

The one's complement representation is formed by inverting each bit of the natural binary code for the number to be converted. It can also be obtained by subtracting each bit of the natural binary code from one. The bit in the MSB position can be set either to 0 for positive numbers, or 1 for negative numbers. Even if the one's complement representation can help reduce the algorithm complexity for some arithmetic operations, it still leads to ambiguity because there are two different codes for zero.

The offset binary coding is a binary representation that is shifted so that a signal with the zero value corresponds to the mid-scale code, that is, the code consisting of a one at the MSB position followed by zeros at all the remaining bit positions. The all-zero code is then used for the negative full scale and the all-one code is assigned to the signal value that is one LSB below the positive full scale.

In the two's complement representation, zero and positive signal values have the same code as in the natural binary format while the negative signal values are represented by forming the two's complement of the corresponding positive number. The two's complement is formed by complementing each bit of the binary code and then adding one LSB without taking into account any carry-out. The MSB is either 0 for positive numbers, or 1 in the case of negative numbers. The two's-complement representation is an efficient coding approach for bipolar signal values in microprocessors because it allows the use of only an adder to implement both the addition and subtraction. Note that the conversion from the offset binary format to two's complement code only requires the inversion of the MSB logic state.

2.1.3   Remarks

The choice of a number representation system has repercussions on the complexity of algorithm implementations for arithmetic operations and the input or output interface of the data converter with other circuits. The two's complement representation is used in most digital systems, and hence is commonly considered for data converter implementations. Sign-magnitude and BCD codes are mainly used for instrumentation applications.

Number expressions and dynamic ranges in common binary representations are summarized in Table 2.4, where X_0 denotes the encoded signal value and X_{FS} designates the data converter full-scale. Note that the variable X can stand for a voltage or current signal.

In some data converter configurations, it is required to use the aforementioned codes with all bits inverted, also known as complementary codes. In differential structures, the required inversion is carried out simply by permuting the input or output nodes.

TABLE 2.4

Number Expressions and Dynamic Range in Common Binary Representations

Representation	Range
Natural binary	

$$\frac{X_0}{X_{FS}} = \sum_{k=1}^{N} b_k 2^{-k} \qquad\qquad 0 \le \frac{X_0}{X_{FS}} \le 1 - 2^{-N}$$

8421 BCD

$$\frac{X_0}{X_{FS}} = \sum_{k=1}^{N} D_k 10^{-k}$$

$$\text{where } D_k = \sum_{j=1}^{4} b_{k,j} 2^{-j+1} \qquad\qquad 0 \le \frac{X_0}{X_{FS}} \le 1 - 10^{-N}$$

Offset binary

$$\frac{X_0}{X_{FS}} = -1 + \sum_{k=1}^{N} b_k 2^{-k+1} \qquad\qquad -1 \le \frac{X_0}{X_{FS}} \le 1 - 2^{-N+1}$$

Sign-magnitude

$$\frac{X_0}{X_{FS}} = (-1)^{b_1} \sum_{k=2}^{N} b_k 2^{-k+1} \qquad\qquad 2^{-N+1} - 1 \le \frac{X_0}{X_{FS}} \le 1 - 2^{-N+1}$$

One's complement

$$\frac{X_0}{X_{FS}} = \left(2^{-N+1} - 1\right) b_1 + \sum_{k=2}^{N} b_k 2^{-k+1} \qquad 2^{-N+1} - 1 \le \frac{X_0}{X_{FS}} \le 1 - 2^{-N+1}$$

Two's complement

$$\frac{X_0}{X_{FS}} = -b_1 + \sum_{k=2}^{N} b_k 2^{-k+1} \qquad\qquad -1 \le \frac{X_0}{X_{FS}} \le 1 - 2^{-N+1}$$

2.2 Data converter characterization

In addition to the errors introduced by the quantization process [2], the performance of data converters can be affected by device nonlinearities.

2.2.1 Quantization errors

Let x and $\hat{x}$ be the input and output samples of the quantizer, respectively. According to the rounding quantizer model depicted in Figure 2.5, the quantization error is defined as

$$e_Q(k) = \hat{x}(k) - x(k) \tag{2.8}$$
$$\hat{x}(k) = Q(x(k)) \tag{2.9}$$

where Q denotes the quantizer operation. Its value should not exceed half of the quantization level,

$$-\frac{\Delta}{2} < e_Q(n) \le \frac{\Delta}{2} \tag{2.10}$$

where Δ is the quantizer step size. However, for input signals with a high dynamic range, the samples that go over the quantizer limit are clipped and e_Q can be greater than $\Delta/2$. Note that, for the rounding quantizer, the signal values that are below an integer multiple of Δ located between two adjacent transitions are quantized to the lower level; otherwise they should be mapped to a higher level.

The converter performance can be described by the signal-to-noise ratio (SNR) given by

$$\text{SNR} = 10 \log_{10} \left(\frac{P_x}{P_Q} \right) \quad \text{in dB}, \tag{2.11}$$

where $P_x = \sigma_x^2 = \text{E}[x^2(k)]$ is the input signal power or variance, and $P_Q = \sigma_Q^2 = \text{E}[e_Q^2(k)]$ is the variance or power of the quantization noise.

The SNR can also be expressed in terms of root-mean square (rms) amplitudes. By exploiting the fact that the average power is proportional to the square of the signal rms amplitude, we can write that

$$\text{SNR} = 10 \log_{10} \left(\frac{A_x^2}{A_Q^2} \right) = 20 \log_{10} \left(\frac{A_x}{A_Q} \right) \quad \text{in dB}, \tag{2.12}$$

where A is the rms amplitude. The logarithmic decibel scale helps describe signal-level ratios, that span many orders of magnitude, with numbers of modest size without losing information.

(a)   (b)   (c)

FIGURE 2.5
(a) Quantizer; (b) linear quantizer model; (c) probability density function of the quantization error.

It is convenient to deal with an input signal, x, that is zero mean, stationary, and uncorrelated with e_Q. Furthermore, by assuming that e_Q is a uniformly distributed white noise sequence over the interval $-\Delta/2$ to $\Delta/2$, the probability function p (see Figure 2.5(c)) is given by

$$p(e_Q) = \begin{cases} \dfrac{1}{\Delta} & \text{for} \quad |e_Q| \leq \dfrac{\Delta}{2} \\ 0 & \text{otherwise.} \end{cases} \tag{2.13}$$

The power of the quantization noise can be obtained as

$$P_Q = \sigma_Q^2 = \int_{-\Delta/2}^{\Delta/2} e_Q^2 p(e_Q) de_Q = \frac{1}{\Delta} \int_{-\Delta/2}^{\Delta/2} e_Q^2 de_Q = \frac{\Delta^2}{12} \tag{2.14}$$

where Δ is the quantizer step size.

Compute the power of the discrete-time sinusoidal signal given by

$$x(k) = A \sin(\omega k) \tag{2.15}$$

where $\omega = 2\pi(f/f_s)$, f is the frequency of the sinusoid, f_s represents the sampling frequency, and A is the amplitude. The power is defined as

$$P_x = \sigma_x^2 = \mathrm{E}[x^2(k)] = \frac{1}{N} \sum_{k=0}^{N-1} x^2(k) \tag{2.16}$$

With the substitution of $x(k)$ and using the fact that the cosine has a zero mean, we obtain

$$P_x = \frac{A^2}{N} \sum_{k=0}^{N-1} \sin^2(\omega k) = \frac{A^2}{N} \sum_{k=0}^{N-1} \frac{1}{2}[1 - \cos(2\omega k)] = \frac{A^2}{2} \tag{2.17}$$

The power is proportional to the square of the sinusoidal signal amplitude.

Assuming a sinusoidal input signal with an amplitude equal to half of the quantizer full-scale range, the power can be written as

$$P_x = \frac{(FSR/2)^2}{2} \tag{2.18}$$

where FSR denotes the full-scale range. With the full-scale range of the N-bit quantizer given by $FSR = 2^N \Delta$, the SNR can take the next form

$$\text{SNR} = 10 \log_{10} \left(\frac{3}{2} 2^{2N} \right)$$

$$= 6.02N + 1.76 \quad \text{(dB)}. \tag{2.19}$$

The SNR increases by about 6 dB for each additional bit of the quantizer. However, practical implementations rarely achieve the theoretical SNR due to various imperfections associated to circuit components.

An approach to improve the SNR can consist of using the oversampling technique, which distributes the power of the quantization noise over a wider frequency band. A digital filter is then required to reduce the quantization noise to a great extent without affecting the signal of interest, thereby increasing the number of bits.

By increasing the value of the sampling rate, the initial power of the quantization noise, which remains unchanged, can be expressed as a function of the quantization noise spectral density, E_Q. Hence,

$$P_Q = \int_{-f_s/2}^{f_s/2} |E_Q(f)|^2 df = |E_Q(f)|^2 \int_{-f_s/2}^{f_s/2} df = |E_Q(f)|^2 f_s \tag{2.20}$$

Consequently, we have

$$|E_Q(f)|^2 = \frac{\Delta^2}{12} \frac{1}{f_s} \tag{2.21}$$

The quantized signal is then processed by an ideal lowpass filter with the frequency response

$$H(f) = \begin{cases} 1 & |f| \leq f_B \\ 0 & \text{otherwise,} \end{cases} \tag{2.22}$$

where f_B represents the cutoff frequency. The resulting quantization noise power is now due to the spectral contributions of E_Q, which are confined between $-f_B$ and f_B. That is,

$$P_Q' = \int_{-f_s/2}^{f_s/2} [E_Q(f)H(f)]^2 df = \int_{-f_B}^{f_B} [E_Q(f)]^2 df$$

$$= \frac{\Delta^2}{12} \frac{2f_B}{f_s} = \frac{\Delta^2}{12} \frac{1}{OSR} \tag{2.23}$$

where OSR $= f_s/2f_B$ is the oversampling ratio. Recalling the signal-to-noise ratio definition, we have

$$\text{SNR'} = 10 \log_{10} \left(\frac{P_x}{P_Q'} \right) = 10 \log_{10} \left(\frac{3}{2} 2^{2N} \right) + 10 \log_{10}(OSR) \tag{2.24}$$

Thus

$$\text{SNR'} = 6.02N + 1.76 + 10 \log_{10}(OSR) \quad \text{(dB)}. \tag{2.25}$$

It should be emphasized that for every doubling of the OSR, the signal-to-noise ratio is increased by 3 dB or equivalently 0.5 bit of resolution. With the oversampling technique, the sampling frequency must be multiplied by a factor of 2^{2N} to yield an increase of N bits. It should be noted that oversampling has the advantage of relaxing the requirements of the anti-aliasing or smoothing analog filter.

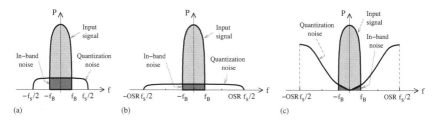

FIGURE 2.6
Power spectrum of the quantization noise: (a) Nyquist ADC, (b) oversampling ADC, (c) delta-sigma ADC.

An efficient architecture used to improve the resolution without requiring an excessive high oversampling is the delta-sigma ADC [3], which modulates the quantization noise so that its magnitude is attenuated in the signal band and increased for out-of-band frequencies. The quantization noise outside the signal bandwidth is reduced using a digital filter, which can also adjust the rate of the output data, if necessary. The delta-sigma converter can achieve a resolution up to 24 bits without the need for high-precision components. For its implementation, the choice of an architecture depends on the characteristics required for the quantization noise shaping.

The frequency spectrum of the digitized signal, which includes the contributions of the input signal and the quantization noise, is shown in Figure 2.6 for Nyquist, oversampling, and delta-sigma ADCs. Due to the oversampling, the quantization noise inherent in the conversion process is spread over a large band of frequencies, including the signal bandwidth. For delta-sigma converters, only a small fraction of the quantization noise falls in the frequency range of interest.

2.2.2 Errors related to circuit components

In addition to the quantization errors, the overall noise available at the converter output is related to the component noise and the nonuniform allocation of the sampling instants or clock jitter. In the particular case of ADCs, the contribution due to the comparator ambiguity should also be taken into account. Furthermore, the dynamic performance of data converters is limited by the frequency characteristics of passive and active components.

• The noise caused by the different components of a converter can be analyzed using equivalent models. It is dominated by the thermal and flicker noise contributions.

• For the jitter analysis, let us consider an input sinusoid

$$x(t) = A \sin(2\pi f t + \phi) \tag{2.26}$$

where A is the amplitude, and f and ϕ are the frequency and phase of the signal, respectively. Ideally, the sampling instants should be given by

$$t_k = kT, \quad k = 0, 1, 2, \cdots, K - 1. \tag{2.27}$$

However, they are affected by errors, δ_k, due to the clock jitter and can be written as

$$t_k = kT + \delta_k \tag{2.28}$$

The samples of the input signal can be obtained using

$$\bar{x}(k) = A \sin(\Omega k + J_k + \phi) \tag{2.29}$$

where the digital angular frequency, Ω, and the phase jitter, J_k, are given by $\Omega = 2\pi f T$ and $J_k = 2\pi f \delta_k$, respectively. The error due to the jitter can be computed as

$$e_J(k) = \bar{x}(k) - x(k) \tag{2.30}$$

where $x(k)$ is the uniformly sampled version of the input signal, especially, $x(k) = A \sin(\Omega k + \phi)$. The SNR contribution of the jitter is

$$\text{SNR} = 10 \log_{10} \left(\frac{\sigma_x^2}{\sigma_Q^2} \right) \quad \text{in dB,} \tag{2.31}$$

where $\sigma_x^2 = \text{E}[x^2(k)]$ is the input signal variance and $\sigma_Q^2 = \text{E}[e_J^2(k)]$ is the jitter noise variance. The value of σ_Q^2 is dependent on the jitter statistical model and ADCs generally exhibit a clock jitter in the range of 0.5 to 2 ps. In practice, high-precision oscillators are used in conjunction with phase-locked loop or delay-locked loop circuits to minimize the effects of clock and timing errors.

• Input signals around the decision level of the comparator, which often consists of an amplifier stage followed by a latch, may result in ambiguous output codes. Let v_Q and $v_{\overline{Q}}$ be the voltages related to the positive and negative outputs of the comparator. The difference , $v_d = v_Q - v_{\overline{Q}}$, can be obtained by solving a differential equation of the form

$$\frac{dv_d(t)}{dt} - \frac{1}{\tau} v_d(t) = 0 \tag{2.32}$$

where the time constant, τ, characterizes the latch ability to resolve intermediate voltage levels and depends on the loading conditions, transistor parameters, and the IC process. That is,

$$v_d(t) = v_d(0) \exp(t/\tau) \tag{2.33}$$

with $v_d(0)$ being the initial condition at the beginning of the metastable region. Ideally, the latch requires an extra time, T, to generate a valid output, which can be processed by the next circuit section. This requires that v_d reaches a given value, say $V_{FS}/2$, within the time period, T. Otherwise, an inaccurate decision will be made if $v_d(T) < V_{FS}/2$. Taking into account the fact that v_d is equal to the amplifier output at $t = 0$, we have

$$v_d(0) = A(|v_i(0) - V_{th}|) = A\Delta v_i(0) \tag{2.34}$$

where v_i is the input voltage, and V_{th} and A denote the threshold level and gain of the comparator, respectively. Hence, the failure condition reads

$$\Delta v_i(0) < \frac{V_{FS}}{2A} \exp(-T/\tau) = \Delta V_i \tag{2.35}$$

Assuming uniformly distributed input sample over the comparator range, the probability, P_e, to produce an uncertain output voltage is given by

$$P_e = P[\Delta v_i(0) < \Delta V_i] = \frac{2\Delta V_i}{V_L} \tag{2.36}$$

where V_L is the effective LSB voltage. The error likelihood P_e can be considered an additive contribution to the quantization noise, that is,

$$\sigma_Q^2 = \frac{\Delta^2}{12}(1 + P_e) \tag{2.37}$$

Note that P_e should include the contribution of all comparators used in the ADC.

Data converters are limited by various error sources. Static errors affect the accuracy of converters during the conversion of dc signals, whereas dynamic errors essentially degrade the high-speed performance. Offset, gain, differential nonlinearity, and integral nonlinearity errors are generally associated with static performance of data converters. Their impact on the signal level can also be characterized in the frequency domain by estimating dynamic characteristics such as the signal-to-noise ratio, total harmonic distortion, and spurious-free dynamic range. On the other hand, dynamic errors can also be related to the limitations (acquisition time, settling time, glitches) of the transient response.

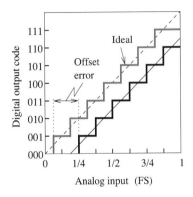

FIGURE 2.7
ADC offset error.

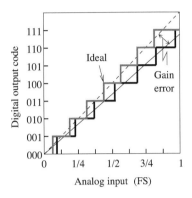

FIGURE 2.8
ADC gain error.

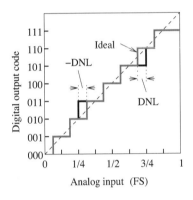

FIGURE 2.9
ADC differential nonlinearity.

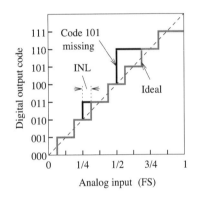

FIGURE 2.10
ADC integral nonlinearity.

2.2.3   Static errors

The converter can be affected by the following static errors [2, 4] (see Figures. 2.7 through 2.14) when it transforms a signal. These errors are most commonly expressed in LSB units, or as a percentage of the converter FSR.

- *Offset error* — The offset error corresponds to the converter output deviation obtained by applying an input signal with a zero-scale to the converter. It can be either positive or negative, and affects all the output data in the same way (see Figures 2.7 and 2.11).

- *Gain error* — The gain of the transfer characteristic is given by the slope of the straight line joining the two endpoints. The gain error results in a slope

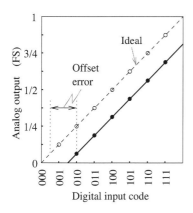

FIGURE 2.11
DAC offset error.

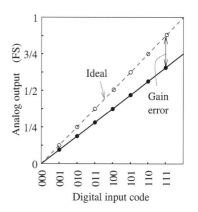

FIGURE 2.12
DAC gain error.

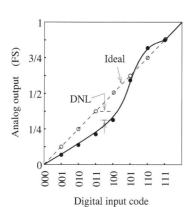

FIGURE 2.13
DAC differential nonlinearity.

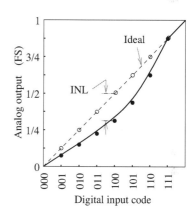

FIGURE 2.14
DAC integral nonlinearity.

difference between the ideal and real converters. All the codes exhibit the same percentage of deviation.

- *Differential nonlinearity (DNL) error* — The DNL is the deviation of either the step width for the ADC (see Figure 2.15) or the step height for the DAC from the ideal value of 1 LSB. In the specific case of a data converter, where a quantization step size can be associated with each code k, it is defined as

$$\text{DNL}_k = \frac{\triangle_k}{V_{LSB}} - 1 \qquad (2.38)$$

where $\triangle_k$ represents the actual quantization step size for the code k and

V_{LSB} is the ideal quantization step size. Generally, the highest value of $|DNL_k|$ is considered the DNL of the data converter. The missing code is the result of a DNL equal to or less than -1 LSB. In this case, the corresponding step does not appear in the transfer characteristic. It is possible that the converter becomes nonmonotonic for DNL values greater than 1 LSB. The magnitude of the converter output diminishes as the input data increases.

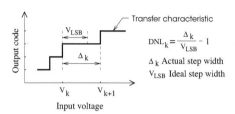

FIGURE 2.15
Illustration of the DNL error.

- *Integral nonlinearity (INL) error* — The INL denotes the deviation at any point of the transfer characteristic of output data from a reference straight line drawn through the zero and full scale. Its value can depend on the definition of the two endpoints. Assuming that the computation of quantization levels is possible, we can obtain

$$INL_k = \left| \frac{V_k - V_0}{V_{LSB}} - k \right| \qquad (2.39)$$

where V_k is threshold level associated with the code k, V_0 is the threshold level corresponding to the lowest transition code, and V_{LSB} is the ideal quantization step size. It can also be shown that

$$INL_k = \sum_{j=0}^{k} DNL_j \qquad (2.40)$$

where the first output code index is supposed to be zero.

Linearity errors are more important than the ones due to offset and gain deviations, which can be adjusted using a suitable calibration technique.

2.2.4 Dynamic errors

In addition to static errors, the converter performance is also affected by errors whose root is related to the time-varying nature of the input signals. The signal-to-noise ratio (SNR), total harmonic distortion (THD), signal-to-noise and distortion (SINAD) ratio, effective number of bits (ENOB), and spurious-free dynamic range (SFDR), together with specifications in the time domain such as the settling time, slew rate, and glitch impulse, are generally used to specify the converter dynamic performance. Parameters that are normally specified in dB, can also be given in units of dBc (decibels to carrier) when the absolute power of the fundamental is used as the reference, or dBFS (decibels to full-scale) when the power of the fundamental is extrapolated to the converter full-scale range.

In general, all the spectral components available at the converter output and that are different from the one of the input signal are considered to be noise. However, a better insight into the conversion process is provided by separately estimating the noise floor and harmonic distortion levels. Hence, the harmonic distortion components are excluded in the measurement of the SNR, but are taken into account in the determination of the SINAD.

• The SNR is the ratio of the power of the fundamental input signal to the noise power, which is caused by all spectral components from dc to half of the sampling frequency, excluding noise at dc and harmonic distortion components. By considering a full-scale input signal and only the noise due to quantization errors, the SNR in decibels (dB) of a Nyquist converter is given by

$$SNR = 10 \log_{10} \left(\frac{P_1}{P_Q} \right) = 6.02N + 1.76 \tag{2.41}$$

where P_1 is the power of the first harmonic component or fundamental signal, P_Q denotes the noise power, and N is the number of bits. In the case of oversampling converters, the SNR includes additional terms depending on the OSR and modulator feedback structure.

• The dynamic range (DR) is the ratio of the power of a full-scale sinusoidal input signal to the power of the noise delivered by the converter with inputs shorted together.

• The THD is the ratio of the rms sum of the powers of the harmonic components above the fundamental frequency to the power of the fundamental signal at the converter output. The THD can be derived as

$$THD = 10 \log_{10} \left(\frac{P_D}{P_1} \right) \tag{2.42}$$

where P_D is the sum of the powers, P_j ($j \geq 2$), of all distortion spectral components. The distortion measurement is realized with an input signal whose amplitude is generally 0.5 dB to 1 dB below the full scale to avoid clipping, and only takes into account harmonic components within the Nyquist bandwidth. In practice, the distortion effects are directly observable on time-domain waveforms for a THD value of about -30 dB.

• The SINAD is the ratio of the power of the fundamental signal to the power of all the remaining spectral (except dc) components below half of the sampling frequency. It can be expressed as

$$SINAD = 10 \log_{10}\left(\frac{P_1}{P_Q + P_D}\right) \tag{2.43}$$

where P_D is the sum of all distortion spectral component powers. The SINAD is also known as the signal-to-noise and distortion ratio (SNDR) and is measured in dB at a specified input frequency and sampling rate. In the case of an ideal converter, the quantization error is the only source of noise and the SINAD is reduced to the SNR.

• The ENOB specifies the resolution that an ideal converter would realize in order to exhibit the same SINAD as the one measured on the real converter. It can be derived for Nyquist converters as

$$ENOB = (SINAD - 1.76)/6.02 \tag{2.44}$$

The difference between the ENOB and the nominal number of bits indicates the impact of circuit imperfections on the conversion process.

Assuming that an input signal with the same amplitude and frequency is used for all measurements, determine the relationship between the SINAD, SNR, and THD.

From the definitions of the SNR, THD, and SINAD, we obtain

$$\frac{P_Q}{P_1} = 10^{-SNR/10} \tag{2.45}$$

$$\frac{P_D}{P_1} = 10^{THD/10} \tag{2.46}$$

and

$$\frac{P_Q + P_D}{P_1} = 10^{-SINAD/10} \tag{2.47}$$

It can then be shown that

$$\frac{P_1}{P_Q + P_D} = \left(\frac{P_Q}{P_1} + \frac{P_D}{P_1}\right)^{-1} = \left(10^{-SNR/10} + 10^{THD/10}\right)^{-1} \tag{2.48}$$

Hence,

$$SINAD = 10 \log_{10}\left(\frac{P_1}{P_Q + P_D}\right)$$
$$= -10 \log_{10}\left(10^{-SNR/10} + 10^{THD/10}\right) \tag{2.49}$$

The degradation of the SINAD for high frequencies is primarily due to the fact that the importance of distortion effects increases with the input signal frequency.

• The SFDR can be obtained as the ratio of the power of the fundamental signal to the power of the highest spurious component in the converter spectrum (excluding dc). It is generally plotted as a function of the test signal amplitude. The SFDR is commonly used in communication applications, as an indication of the usable dynamic range.

• The two-tone intermodulation distortion (IMD) is measured by connecting the converter input to the sum of two sinusoidal signals with the same magnitude, but having slightly different frequencies. It is computed as the ratio of the power of the worst third-order intermodulation product to the power of either input tone. The IMD is a key specification used in the selection of building blocks for multi-carrier communication systems. With the use of test signals at frequencies f_1 and f_2, the determination of the third-order IMD can rely only on the distortion products occurring at $2f_2 - f_1$ and $2f_1 - f_2$, or relatively near the fundamental signals, as the distortion components at the higher frequencies, or say at $2f_2 + f_1$ and $2f_1 + f_2$, usually fall outside the passband, where they can be filtered out easily.

It should be noted that the specifications of different converters can be fairly compared only if the measurements are realized with the same fundamental input frequency or are valid over the same bandwidth, which is defined as the frequency range over which the input signal can be converted with an amplitude attenuation less than or equal to 3 dB.

• The glitch impulse represents the undesired signal transients that can appear at the DAC output. It is characterized by measuring its area at the mid-scale code transition where the logic states of the maximum number of bits are changed. The unit of measurement can be chosen as $nV \cdot s$, for instance.

• The settling time is the time elapsed from the application of the input signal until the converter output reaches and remains within a given error range about the final value. It can include components such as a short delay time related to the propagation delay, the slew time required for the output to reach its highest value, and the ring time needed by the output to recover from the overload condition associated with the slewing and to settle to within the specified error band, which can be a certain percent of the final value or $\pm 1/4$ LSB. The settling time is usually specified for a full-scale transition.

• The latency time denotes the time elapsed from the initiation of one input conversion to the next. It includes the conversion time, which is defined as the time required for a single conversion, and data retrieval time. For ADCs, the latency time is measured by assuming that the signal is already sampled.

2.3   Summary

Data converters are generally designed to meet the requirements of various applications such as imaging, video, instrumentation, control, and communication systems. For most commonly used architectures, the specifications for the resolution and sampling frequency generally extend beyond the range from 8 bits, 250 Msps (digital oscilloscope) to 24 bits, 2.5 Msps (seismic monitoring systems) for ADCs, and from 8 bits, 330 Msps (digital radio modulation) to 24 bits, 200 ksps (DVD systems) for DACs. The distribution of resolution versus sampling frequency can furnish insight into the performance limitations of data converters.

The resolution appears to be limited at low frequencies by the component mismatches and thermal noise. It tends to be reduced on the order of one bit when the sampling frequency is increased by two times. This is due to the enhanced effect related to the comparator ambiguity and clock jitter at high frequencies. Furthermore, the maximum sampling frequency of the converter cannot exceed the transition frequency of the transistor, which is determined by the IC process.

Data converters used in portable equipment should meet the requirement of low supply voltage and low power consumption. In this case, the important limitation, which is introduced on the dynamic range and speed, depends on the architecture and IC technology.

Bibliography

[1] B. M. Gordon, "Linear electronic analog/digital conversion architectures, theirs origins, parameters, limitations, and applications," *IEEE Trans. on Circuits and Systems*, vol. 25, no. 7, pp. 391–418, July 1978.

[2] Understanding data converters, Application notes, Texas Instruments, 1999.

[3] S. R. Norsworthy, R. Schreier, and G. C. Temes, Eds., *Delta-Sigma Data Converters: Theory, Design, and Simulation,* New York, NY: Wiley-IEEE Press, 1996.

[4] J. R. Naylor, "Testing digital/analog and analog/digital converters," *IEEE Trans. on Circuits and Systems*, vol. 25, no. 7, pp. 526–538, July 1978.

3

Nyquist Digital-to-Analog Converters

CONTENTS

Digital-to-analog converters (DACs) enable the interfacing of digital systems with the real world. They are used to transform binary code into an analog output signal, either in the form of a voltage or current.

Ideally, signals from *dc* up to the Nyquist frequency, which is defined as half of the DAC sampling frequency, can be generated by a converter. In practice, in addition to the increased difficulty to meet the specifications of the reconstruction filter, the converter performance tends to degrade significantly when approaching the Nyquist frequency. Nyquist DACs are based on various architectures (binary-weighted, thermometer-coded, and segmented structures) and conversion techniques such as voltage scaling, current scaling, charge scaling or redistribution, and hybrid methods. In each of these types of DACs, the conversion is achieved by summing all the output signal contributions associated with the different bits of the input digital code. The resolution of high-speed DACs based on each of the aforementioned techniques is generally limited by the circuit size and complexity. One approach adopted to reduce the number of elements (transistors, resistors, and capacitors) is to use a segmented DAC. In a segmented converter, a first stage decodes the most significant bit (MSB) part of the input digital code and a second stage is driven by the remaining least significant bits (LSBs). A further issue with high-resolution DACs is that the linearity and monotonicity of the conversion characteristic is limited by component matching. The effect of mismatches, which are generally caused by IC process gradients, can be alleviated either by adding a calibration stage to the DAC or by laser trimming components to adjust their values after wafer fabrication.

3.1 Digital-to-analog converter (DAC)

The transformation of a digital code by an N-bit digital-to-analog converter (DAC) can rely on the use of switches to select the appropriate reference voltages, which are then summed to provide a discrete time signal. The format of the analog output generated by a DAC is dependent on the switching scheme.

According to the digital input code, a DAC ideally operates by updating its analog output, which is then held for a certain time. The timing of this operation is controlled by a clock signal with a frequency $f_s = 1/T$. Assuming a zero-order hold, the analog output, $x(t)$, reconstructed from the discrete-time sample, $x(n)$, available at time intervals spaced by T, can be written as,

$$x(t) = \sum_{n=0}^{N} x(n) \cdot \mathrm{rect}\left(\frac{t - (n + 1/2)T}{T}\right) \tag{3.1}$$

where $\mathrm{rect}(t)$ is the rectangular function. The impulse response is of the form,

$$h(t) = \frac{1}{T} \cdot \mathrm{rect}\left(\frac{t - T/2}{T}\right) = \begin{cases} \dfrac{1}{T} & \text{if } 0 \leq t < T \\ 0 & \text{otherwise.} \end{cases} \tag{3.2}$$

In the frequency domain, the Fourier transform of the impulse response, $h(t)$, is given by

$$H(f) = \mathcal{F}\{h(t)\} = \frac{1 - e^{-j2\pi fT}}{j2\pi fT} = e^{-j\pi fT}\mathrm{sinc}(fT) \tag{3.3}$$

where the sinc function is defined as, $\mathrm{sinc}(fT) = \sin(\pi fT)/(\pi fT)$.

The desired signal frequency in the first Nyquist zone, that extends from 0 to $f_s/2$, is reflected as mirror images, but with attenuated amplitudes due to the sinc frequency response, into higher Nyquist zones. A lowpass or bandpass filter is often used to attenuate these image signals. Increasing the clock frequency helps reduce but does not eliminate the sinc-frequency roll-off effect. In applications that require a response remaining flat over the entire frequency range, a pre-equalization digital filter (or post-equalization analog filter), whose frequency response is the inverse of the sinc function, should be placed in front of (or after) the DAC.

Here, the hold time is equal to the clock signal period, and the signal power is essentially delivered over the first Nyquist zone that encompasses the DAC operating range.

Alternatively, by choosing the hold time that is equal to a fraction of the clock signal period, the magnitude of the DAC response is much lower and flatter in the first Nyquist zone and significantly higher in the second Nyquist zone (or other Nyquist zones) than that of the previous case. As a result, the DAC operation now becomes possible in other Nyquist zones.

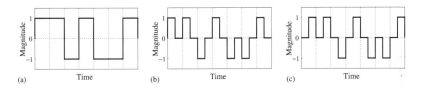

FIGURE 3.1
DAC output for the digital code 1101001: (a) NRZ, (b) RZ, (c) HRZ.

A 1-bit DAC output can be represented, as shown in Figure 3.1, for nonreturn-to-zero (NRZ), return-to-zero (RZ), and half-clock-period delayed return-to-zero (HRZ) pulses. The output of a DAC with NRZ pulse remains constant over a full clock period, T, while the output of a DAC with RZ pulse is held constant only between 0 and $T/2$, and the output of a DAC with HRZ pulse is a half-clock-period delayed version of the one of a DAC with RZ pulse. In general, a DAC generates a stepwise output signal that can then be transformed into an analog signal using a lowpass smoothing filter.

Intuitively, DACs with NRZ pulse are less affected by clock signal jitter (or clock edge fluctuation) due to the fact that jitter errors can occur only during the clock edges at which the data sample is changing. On the other hand, DACs with RZ or HRZ pulse are the most sensitive to clock signal jitter because random variations can affect the rising and falling edges of the waveform at every clock cycle. However, DACs with NRZ pulse can be severely limited by the inter-symbol interference (ISI) that is caused by mismatches of the clock signal rise and fall times than other DACs. In the presence of ISI, the DAC output is no longer dependent on only the current data sample, but also on the previous data sample.

The performance of the converter is related to the choice of the IC technology and switching scheme.

3.1.1 Binary-weighted structure

The simplest and area-efficient way to control the switches consists of using the digital input directly. However, the resulting structure has some drawbacks. The components, which determine the reference voltages, are required to be binary weighted. They should be matched to within $\pm 1/2$ LSB to meet the specifications of a high-precision DAC. This is equivalent to achieving a relative accuracy of less than $1/2^N$ LSB for the MSB. The statistical spread appears to be a limiting factor for the realization of this objective, and the monotonicity of the DAC is not guaranteed. Furthermore, the output data can exhibit a large spike or glitch at the mid-code transition, such as from 0111 to 1000 in a 4-bit example, due to the required asymmetrical switching. Hence, the state of all bits is modified. The LSBs all can switch faster than the MSB and the DAC output will first attempt to change toward zero before returning to the right state. A glitch can be characterized by measuring its energy, which is expressed in units of picovolt-seconds for high-performance DACs. A sample-and-hold circuit can be used to maintain the DAC output constant during the code transition, reducing in this way the glitch effect.

3.1.2 Thermometer-coded structure

An N-bit thermometer-coded DAC consists of 2^N unit elements, which are connected to switches, whose control signals are generated by a binary-to-thermometer decoder. The switching of only one element is needed when the input code changes by 1 LSB. That is, the relative accuracy to be realized is $1/2$, corresponding to a relaxed matching constraint. Furthermore, the glitch energy is considerably reduced. The monotonicity is guaranteed because the output remains constant or follows the variation of the input code. The primary inconvenience of the thermometer-coded DAC is the need for a large chip area.

3.1.3 Segmented architecture

A segmented DAC is based on an array of binary-weighted elements directly controlled by the L least significant bits and an array of unit elements steered by the remaining $(N - L)$ bits, which are thermometer encoded. It is then a compromise between both aforementioned structures and can combine the high accuracy and conversion rate.

Note that the glitch problem at the mid-code can also be solved by never turning off elements as the digital code is increased. This results in the interpolated architecture. However, the drawback in this case is related to the requirement of a complicated switching scheme.

3.2 Voltage-scaling DACs

In response to a digital input code, voltage-scaling DACs produce an output voltage by exploiting the voltage-divider principle. As simple in operation as a digitally controlled potentiometer, voltage-scaling DACs can easily be designed to meet the high-speed and low-power specifications without affecting their inherent monotonicity. While such DAC architectures are generally widely used for applications requiring a resolution not exceeding 8 bits, they are not suitable in cases where the resolution is high. This is due to the fact that the required number of resistors and switches increases exponentially with the resolution, making the resulting chip area very large.

3.2.1 Basic resistor-string DAC

The simplest architecture used to design a DAC based on the voltage divider principle consists of a series of resistors connected between two supply voltages, switches controlled by the digital input code, and a buffer to drive low impedance loads. Each digital code is converted by selecting a node between two resistors so that the appropriate voltage level is transferred to the converter output.

A 3-bit resistor-string DAC based on a binary-tree structure is shown in Figure 3.2. For an N-bit resolution, such a converter requires 2^N resistors and $2(2^N - 1)$ switches. Using the voltage divider rule, the voltage at node j of the resistor string can be expressed as

$$V_j = \frac{j-1}{2^N} V_{REF} \quad j = 1, 2, \cdots, 2^N \tag{3.4}$$

where V_{REF} is the reference voltage. The input and output characteristics of the unipolar DAC of Figure 3.2 are shown in Figure 3.3(a). Ideally, the DAC establishes a unique correspondence between an input digital code and a reference voltage. Because the reference levels are separated by an LSB or equivalently $V_{REF}/2^N$, the characteristic can be shifted upward by LSB/2, as shown in Figure 3.3(b), where the output level corresponding to the zero digital code is now LSB/2. The resistor-string DAC based on the output characteristic of Figure 3.3(b) is depicted in Figure 3.4. In comparison with the

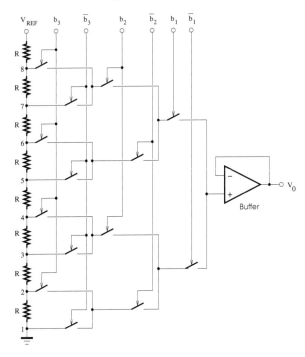

FIGURE 3.2
Block diagram of a 3-bit binary-decoded resistor-string DAC.

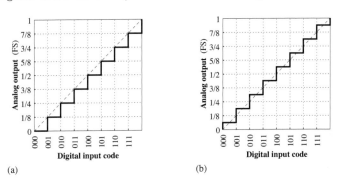

(a)   (b)

FIGURE 3.3
Transfer characteristics of a 3-bit DAC.

previous structure, the overall number of resistors increases by one. Using the topmost and bottommost resistors with a value of $R/2$, the voltage at the node j of the resistor string is given by

$$V_j = \frac{j-1/2}{2^N} V_{REF} \quad j = 1, 2, \cdots, 2^N \tag{3.5}$$

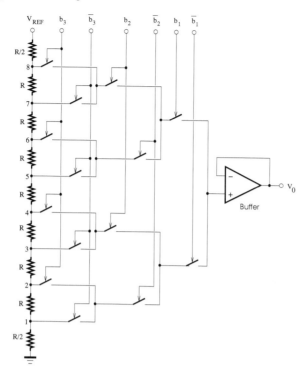

FIGURE 3.4
Block diagram of a 3-bit binary-decoded resistor-string DAC with code transitions at multiples of LSB/2.

In addition to its simple structure, the resistor-string DAC exhibits an inherent monotonicity [1, 2]. For increasing input values, a DAC with a monotonic characteristic generates strictly increasing output values. The linearity of the resistor-string DAC then appears to be less affected by resistor mismatches. However, the resistor-string DAC architecture is limited by the number of resistors and switches, which grows exponentially as the resolution increases.

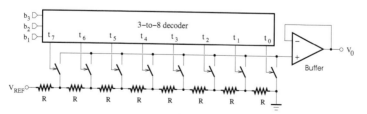

FIGURE 3.5
Block diagram of a 3-bit resistor-string DAC with a 3-to-8 decoder.

The number of switches can be reduced using an N-to-2^N decoder, as shown in Figure 3.5 for a resolution of 3 bits. Because only one switch is included in the signal path for a given digital code conversion, the effects of switch resistances and parasitic capacitance are reduced, yielding an improved speed performance.

TABLE 3.1

Truth Table and Logic Equations of the 3-to-8 Decoder

	b_1	b_2	b_3	t_0	t_1	t_2	t_3	t_4	t_5	t_6	t_7	
		Binary Code					1-out-of-8 Code					
0	0	0	0	1	0	0	0	0	0	0	0	$t_0 = \bar{b}_1 \cdot \bar{b}_2 \cdot \bar{b}_3$
1	0	0	1	0	1	0	0	0	0	0	0	$t_1 = \bar{b}_1 \cdot \bar{b}_2 \cdot b_3$
2	0	1	0	0	0	1	0	0	0	0	0	$t_2 = \bar{b}_1 \cdot b_2 \cdot \bar{b}_3$
3	0	1	1	0	0	0	1	0	0	0	0	$t_3 = \bar{b}_1 \cdot b_2 \cdot b_3$
4	1	0	0	0	0	0	0	1	0	0	0	$t_4 = b_1 \cdot \bar{b}_2 \cdot \bar{b}_3$
5	1	0	1	0	0	0	0	0	1	0	0	$t_5 = b_1 \cdot \bar{b}_2 \cdot b_3$
6	1	1	0	0	0	0	0	0	0	1	0	$t_6 = b_1 \cdot b_2 \cdot \bar{b}_3$
7	1	1	1	0	0	0	0	0	0	0	1	$t_7 = b_1 \cdot b_2 \cdot b_3$

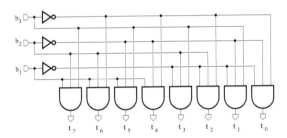

FIGURE 3.6

Circuit diagram of a 3-to-8 decoder.

The digital input code is translated into a format suitable for the switch control by the 3-out-of-8 decoder, which can be implemented using various architectures.

The 3-out-of-8 decoder detects the occurrence of a specific bit combination in the input code, as illustrated by the truth table in Table 3.2. The decoding leads to the activation of only the output with the subscript that is the same as the decimal equivalent of the input code. Figure 3.6 shows the circuit diagram of a 3-out-of-8 decoder using inverters and AND gates. This implementation is directly related to the decoder logic equations and is simple. However, the

TABLE 3.2

Truth Table of the 3-Bit Binary-to-Thermometer Decoder and 1-out-of-8 Encoder

	Binary Code			Thermometer Code							1-out-of-8 Code							
	b_1	b_2	b_3	T_0	T_1	T_2	T_3	T_4	T_5	T_6	t_0	t_1	t_2	t_3	t_4	t_5	t_6	t_7
0	0	0	0	0	0	0	0	0	0	0	1	0	0	0	0	0	0	0
1	0	0	1	1	0	0	0	0	0	0	0	1	0	0	0	0	0	0
2	0	1	0	1	1	0	0	0	0	0	0	0	1	0	0	0	0	0
3	0	1	1	1	1	1	0	0	0	0	0	0	0	1	0	0	0	0
4	1	0	0	1	1	1	1	0	0	0	0	0	0	0	1	0	0	0
5	1	0	1	1	1	1	1	1	0	0	0	0	0	0	0	1	0	0
6	1	1	0	1	1	1	1	1	1	0	0	0	0	0	0	0	1	0
7	1	1	1	1	1	1	1	1	1	1	0	0	0	0	0	0	0	1

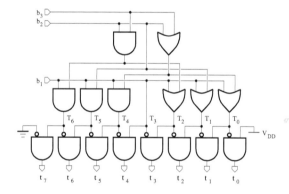

FIGURE 3.7

Circuit diagram of a modular 3-out-of-8 decoder.

maximum number of digital inputs that a single AND gate can support can be quickly reached as the input code resolution is increased.

For high resolutions, an implementation solution of the N-out-of-2^N decoder is to use an N-bit binary-to-thermometer decoder, followed by a 1-out-of-2^N encoder. This results in a structure that is modular and realizable with two-input gates. In the specific case of $N = 3$, let T_p, $p = 0, 1, 2 \cdots, 6$, and t_q, $q = 0, 1, 2 \cdots, 7$, denote, respectively, the thermometer code and the 1-out-of-8 code for the binary code, b_k, $k = 1, 2, 3$. Table 3.2 shows the correspondence between the logic states of b_k and T_p, and all possible combinations of T_p and

t_q. The logic equations derived from the truth table are given by

$$
\begin{aligned}
&T_0 = b_1 + b_2 + b_3 \qquad T_1 = b_1 + b_2 \quad T_2 = b_1 + b_2 \cdot b_3 \quad T_3 = b_1 \\
&T_4 = b_1 \cdot b_2 + b_1 \cdot b_3 \quad T_5 = b_1 \cdot b_2 \qquad T_6 = b_1 \cdot b_2 \cdot b_3
\end{aligned}
\tag{3.6}
$$

and

$$
t_j = \begin{cases}
\overline{T_0} & \text{for} \quad j = 0 \\
T_{j-1} \cdot \overline{T_j} & \text{for} \quad j = 1, 2, \cdots, 6 \\
T_6 & \text{for} \quad j = 7
\end{cases}
\tag{3.7}
$$

The circuit realization of the 3-out-of-8 decoder is shown in Figure 3.7. It should be noted that the required number of gates can be high as the converter resolution increases.

Design a 4-bit binary-to-thermometer decoder based on a 3-bit binary-to-thermometer decoder.

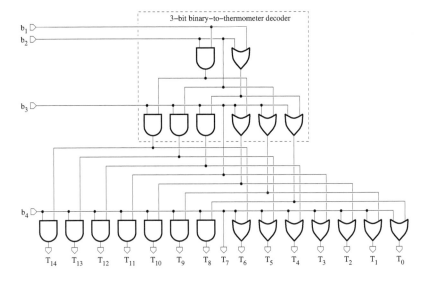

FIGURE 3.8
Circuit diagram of a 4-bit binary-to-thermometer decoder.

The circuit diagram of a 4-bit binary-to-thermometer decoder is shown in Figure 3.8. The encoding of a 4-bit binary code produces a thermometer code with j consecutive zeros followed by $15 - j$ consecutive ones. The number of ones in the thermometer code is equal to the decimal equivalent of the binary input code. The logic equations of the 4-bit binary-to-thermometer decoder are given by

$$
\begin{array}{ll}
T_0 = b_1 + b_2 + b_3 + b_4 & T_1 = b_2 + b_3 + b_4 \\
T_2 = b_1 \cdot b_2 + b_3 + b_4 & T_3 = b_3 + b_4 \\
T_4 = b_1 \cdot b_3 + b_2 \cdot b_3 + b_4 & T_5 = b_2 \cdot b_3 + b_4 \\
T_6 = b_1 \cdot b_2 \cdot b_3 + b_4 & T_7 = b_4 \\
T_8 = (b_1 + b_2 + b_3) \cdot b_4 & T_9 = (b_2 + b_3) \cdot b_4 \\
T_{10} = (b_1 \cdot b_2 + b_3) \cdot b_4 & T_{11} = b_3 \cdot b_4 \\
T_{12} = (b_1 \cdot b_3 + b_2 \cdot b_3) \cdot b_4 & T_{13} = b_2 \cdot b_3 \cdot b_4 \\
T_{14} = b_1 \cdot b_2 \cdot b_3 \cdot b_4 &
\end{array}
\tag{3.8}
$$

In general, the required number of AND and OR gates for an N-bit binary-to-thermometer decoder is equal to $2^N - (N + 1)$.

Switches are used to connect the different reference voltages provided by the resistor string to the output. Fully decoded architectures have the advantage of providing a monotonic output, even if the resistances drift from their nominal values. The output resistance of the DAC is dependent on the digital code. This results in a code-dependent settling time when charging a capacitive output load. The switches must be appropriately sized or compensated to reduce the effect of this nonideality on the converter performance.

3.2.2 Intermeshed resistor-string DAC

The complexity of the digital decoder can be reduced by using a DAC with intermeshed resistor strings [3, 4]. The reference voltage is subdivided into 2^P coarse voltage segments using a P-bit coarse resistor string with a low impedance. Each coarse resistor is related in parallel to a Q-bit fine resistor string with a relatively high impedance. The conversion of an input digital code then consists of selecting the nodes of the fine resistor string connected to the appropriate coarse voltage segments. With an overall resolution of N bits, where $N = P + Q$, the resulting intermeshed resistor-string DAC required a P-out-of-2^P decoder and a Q-out-of-2^Q decoder. Because the integers P and Q are less than N, the complexity of the required decoders remains low in comparison with a single N-out-of-2^N decoder. Furthermore, by connecting in parallel each coarse resistor with a fine resistor string, the converter output impedance value and its variations with the input code are reduced, resulting in an improved settling time.

The circuit diagram of a 5-bit intermeshed resistor-string DAC is shown in Figure 3.9. The proper switches are selected by splitting the 5-bit input code in the three MSBs and the two LSBs, that are applied to a 3-out-of-8 decoder and a 2-out-of-4 decoder, respectively. Latches can be inserted between the decoders and switches, if necessary, to reduce glitch errors during a code transition. In general, an N-bit intermeshed resistor-string DAC requires 2^P

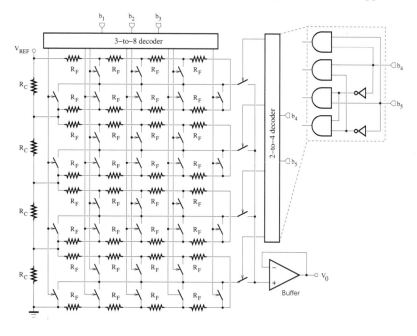

FIGURE 3.9
Circuit diagram of a 5-bit intermeshed resistor-string DAC.

resistors R_C in addition to 2^N resistors R_F. For resolutions greater than 8 bits, the overall number of resistors can be too high to be practical.

3.2.3   Two-stage resistor-string DAC

The number of passive elements in the DAC can be further reduced by adopting a design technique based on a segmented architecture [5], that consists of a cascade of P-bit and Q-bit resistor-string DACs through buffer amplifiers to realize a resolution of N bits, where $N = P + Q$. In a segmented DAC, the input code is partitioned so that the P MSBs are processed by the first stage of the converter, while the remaining Q LSBs are converted by the second stage. The switch control signals are then obtained using P-to-2^P and Q-to-2^Q decoders. Figure 3.10 shows a segmented converter based on two 3-bit resistor-string DACs. Buffer amplifiers are required to prevent the second-stage resistor string from loading the first-stage resistor string.

    With the switch configuration defined by the P MSBs of a given input code, the first resistor string can generate two reference voltage levels of the form

$$V_p = p\frac{V_{REF}}{2^P} \tag{3.9}$$

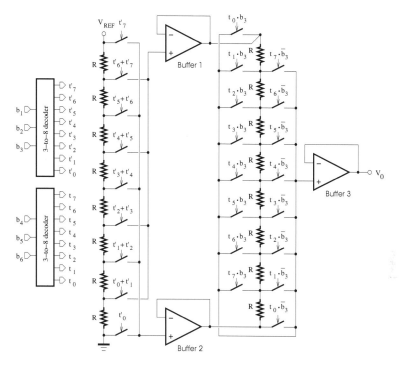

FIGURE 3.10
Circuit diagram of a 6-bit two-stage resistor-string DAC.

and

$$V_{p+1} = (p+1)\frac{V_{REF}}{2^P} \tag{3.10}$$

where p is the decimal equivalent of the P MSBs and $p = 0, 1, 2, \cdots, 2^P - 1$. These voltages are applied at the inputs of the buffer amplifiers, whose outputs are connected to the top and bottom of the second resistor string. Using the superposition theorem, the converter output voltage generated in accordance with the remaining Q LSBs of the input code can be expressed as

$$V_0 = q\frac{V_{01}}{2^Q} + (2^Q - q)\frac{V_{02}}{2^Q} = V_{02} + q\frac{V_{01} - V_{02}}{2^Q} \tag{3.11}$$

where V_{01} and V_{02} denote the output voltages of the first and second buffer amplifiers, respectively; q is the decimal equivalent of the Q LSBs; and $q = 0, 1, 2, \cdots, 2^Q - 1$. The output voltage is then generated by linearly inter-polating between V_{01} and V_{02}. The step size by which the output voltage of the second resistor-string DAC can change has a value of $(V_{01} - V_{02})/2^Q$. If the buffer amplifiers are assumed to be ideal, $V_{01} = V_{p+1}$ and $V_{02} = V_p$, the voltage V_0 will be given by

$$V_0 = \left(p \cdot 2^Q + q\right) \frac{V_{REF}}{2^N} \tag{3.12}$$

With k being the decimal representation of the input code, the converter output voltage is reduced to

$$V_0 = k \frac{V_{REF}}{2^N} \tag{3.13}$$

where $k = p \cdot 2^Q + q$. For an N-bit unipolar DAC, the value of k can vary from 0 to $2^N - 1$.

A resistor-string DAC is inherently monotonic. However, the monotonicity in a segmented string DAC can be affected by the offset voltage of practical buffer amplifiers. By taking into account the offset voltages V_{off1} and V_{off2} of the first and second buffer amplifiers, we have $V_{01} = V_{p+1} + V_{off1}$ and $V_{02} = V_p + V_{off2}$. The converter output voltage is then given by

$$V_0 = k \frac{V_{REF}}{2^N} + V_{off} \tag{3.14}$$

where

$$V_{off} = V_{off2} + q \frac{V_{off1} - V_{off2}}{2^Q} \tag{3.15}$$

Assuming that the buffer offset voltages are identical, the output voltage for each input code is shifted by the same amount and the monotonicity of the converter characteristic is preserved. However, this last requirement is rarely met in practice.

When q reaches its highest value, $2^Q - 1$, the voltage V_0 becomes

$$V_0 = V_{02} + (2^Q - 1) \frac{V_{01} - V_{02}}{2^Q} = V_{01} - \frac{V_{01} - V_{02}}{2^Q} \tag{3.16}$$

where $V_{01} = V_{p+1} + V_{off1}$ and $V_{02} = V_p + V_{off2}$. Hence, the output voltage is one step size below $V_{p+1} + V_{off1}$. The next increase of the converter output occurs when the input codes of the second and first resistor-string DACs are, respectively, reset and incremented by one.

With a switching scheme designed such that only the value of V_{02} is changed to $V_{p+2} + V_{off2}$ as a result of the digital code decoding, the digital-to-analog conversion can preserve the numerical order of the input codes. Because V_{p+1} and V_{p+2} are two adjacent voltage levels of the first resistor-string DAC, we have

$$V_{p+2} - V_{p+1} = V_{REF}/2^P \tag{3.17}$$

In the worst case, $V_{off2} = -V_{off1}$, and a monotonic conversion is guaranteed provided the magnitude of the buffer offset voltage remains lower than $V_{REF}/2^{P+1}$, or equivalently, half of the LSB of the first resistor-string DAC.

The polarity of the voltage supplied to the second resistor string is reversed whenever the input code of the first resistor-string DAC is incremented by one. This polarity inversion should be compensated by appropriately changing the order used in the selection of the tap voltages of the second resistor-string DAC. The switches responsive to the Q-bit LSBs are then configured such that the tap voltages can be selected in either a top-down or a bottom-up fashion.

The circuit diagram of a 6-bit two-stage resistor-string DAC using the foregoing reversal switching scheme is depicted in Figure 3.10. The decoder responsive to the three LSBs selects one of the eight tap voltages of the second resistor string, starting with the node connected to either V_{01} or V_{02}, depending on the state of b_3.

The two-stage architecture can be used to implement a DAC with a resolution on the order of 16 bits. The required number of resistors and switches is reduced to $2^P + 2^{N-P}$ for a resolution of N bits. However, the main disadvantages are the offset voltage and settling-time requirements placed on the design of buffer amplifiers to meet the monotonicity and conversion speed specifications.

In general, even with the use of improved IC process technology, resistor-string DACs can feature nonlinearity errors only in the 4 LSB range, and are then essentially used in closed-loop applications such as motor control or process control, where the INL and DNL specifications are undemanding.

3.3 Current-scaling DACs

Current-scaling DACs generally consist of a number of current sources that are selectively switched into a summing node in response to a digital input code. Various structures can be used in the implementation of a switchable current source, while the sum of currents is generally formed at the inverting node of an amplifier.

3.3.1 Binary-weighted resistor DAC

A binary-weighted resistor DAC is realized by summing the current contributions associated to the different bits of a digital input code. It then consists of weighted current sources, which can simply be a set of resistors with power-of-two values and connected in parallel to either the reference voltage or the ground, depending on each bit state. Figure 3.11 shows the circuit diagram of a binary-weighted resistor DAC. This architecture allows only unipolar digital-to-analog conversion. The output voltage can be computed as,

$$V_0 = -R_F(I_1 + I_2 + I_3 + \cdots + I_N) \qquad (3.18)$$

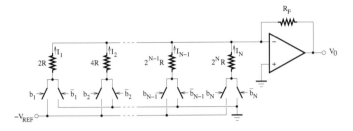

FIGURE 3.11
Circuit diagram of a binary-weighted resistor DAC.

where N represents the DAC resolution. Because various binary-weighted resistors are used to convert the reference voltage V_{REF} into the currents I_k, the voltage V_0 is found to be

$$V_0 = V_{REF}\frac{R_F}{R}\left(\frac{b_1}{2} + \frac{b_2}{2^2} + \frac{b_3}{2^3} + \cdots + \frac{b_N}{2^N}\right) = V_{REF}\frac{R_F}{R}\sum_{k=1}^{N}\frac{b_k}{2^k} \qquad (3.19)$$

The converter full scale, which is the difference between the output voltages for the highest and smallest input codes, is obtained as

$$FS = \frac{(2^N - 1)R_F}{2^N R}V_{REF} \qquad (3.20)$$

The resolution of the binary-weighted DAC depends on the precision achieved for the resistor values. The ratio between the resistor values associated with the MSB, b_1, and the LSB, b_N, is given by

$$\frac{R_1}{R_N} = \frac{2R}{2^N R} = \frac{1}{2^{N-1}} \qquad (3.21)$$

In the specific case of an 8-bit binary-weighted resistor DAC, the range of resistor values can vary by a factor of 128. The spread of resistances can then be very large for high resolutions, resulting in a difficulty to design resistors with accurate values using classical integrated-circuit fabrication methods. Furthermore, by increasing the DAC resolution, the current related to the MSB can be slightly smaller than the sum of all currents due to the other remaining bits, leading to a nonmonotonic conversion characteristic. In general, the binary-weighted resistor architecture is used with a resolution not exceeding 4 bits as a building block in the design of large systems.

3.3.2   R-2R ladder DAC

One of the most common DAC structures is based on the R-2R resistor ladder network, which is symmetric and uses only two values of resistors, thus greatly simplifying the matching requirements. The block diagram of a DAC using the R-2R ladder network is shown in Figure 3.12. The switches are controlled by digital logic. A 2R resistor can be connected either to the amplifier inverting node or to the ground. In the first case, the corresponding bit, b_k, assumes the high state, while it is in the low state in the latter. Ideally, the load of

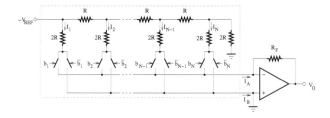

FIGURE 3.12
Block diagram of R-2R ladder DAC.

each 2R resistor has a resistance of 2R. That is, the currents can be obtained according to

$$I_1 = 2I_2 = 4I_3 = \cdots = 2^{N-1}I_N \qquad (3.22)$$

where $I_1 = -V_{REF}/(2R)$. The DAC output is given by

$$V_0 = -R_F \sum_{k=1}^{N} b_k I_k \qquad (3.23)$$

$$= \frac{R_F V_{REF}}{R} \sum_{k=1}^{N} \frac{b_k}{2^k} \qquad (3.24)$$

where $I_k = -V_{REF}/(2^k R)$ and b_k is either 1 or 0. The voltage V_0 is proportional to the switched-on bits. The R-2R DAC can exhibit a lower noise and nonlinearity errors of only ±1 LSB. However, without involving some amount of resistor trimming to reduce the matching errors, the resolution of an R-2R DAC is limited to 12 bits. Furthermore, the glitch caused by switch timing differences has a more significant effect on R-2R structures than on resistor-string architectures. Hence, the R-2R DAC is less attractive for glitch-sensitive applications such as waveform generation.

3.3.3 Switched-current DAC

In general, switched-current DACs or current-steering DACs are preferred for applications requiring a resolution on the order of 10 bits and several megahertz of signal bandwidth. By driving a resistive load directly without the need for a voltage buffer, it can exhibit a very high power efficiency because almost all power is directed to the output load. Furthermore, the switched-current DAC has the advantage of featuring a low power consumption and requiring a small chip area for low to medium resolutions. Switched-current DACs can be designed using a binary-weighted, or thermometer-coded current array, as shown in Figure 3.13, where R_L is the load resistor. The number of current sources required to achieve N bits of resolution is, respectively, N and $2^N - 1$ for the binary-weighted and thermometer-coded architectures because the converter state without any switched-current contribution represents the zero input code.

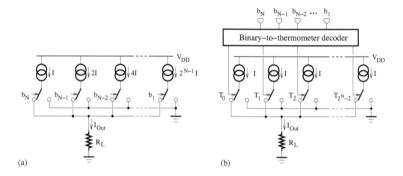

FIGURE 3.13
Block diagram of the (a) binary-weighted and (b) thermometer-coded switched-current DACs.

A binary-weighted DAC consists of current sources that are sized to have power-of-two values and are associated with switches directly controlled by the input code. Even though it has a simple structure, its operation can be limited by mismatches between the current sources. The worst-case error affecting the conversion transfer characteristic can be observed at mid-code transitions, or when the MSB current source is enabled, and all other current sources are disconnected from the output. As it is difficult to match the value of the MSB current source and the sum of remaining current sources to within 0.5 LSB, the DAC monotonicity cannot be guaranteed. Furthermore, glitches can arise in the DAC output as a result of the effects of asynchronous current switching and parasitic capacitive coupling.

For a thermometer-coded DAC, $2^N - 1$ current sources of equal weight are required to achieve a resolution of N bits. Each unit current source is connected to a switch controlled by a signal generated by the binary-to-thermometer decoder. For each increase in the input digital code by one, only one additional current source is switched to the converter output, and the connection configuration of the other current sources is maintained unchanged. This greatly reduces the effect of glitches and the monotonicity of the DAC characteristic is guaranteed. However, the main drawback of the thermometer-coded DAC is the large chip area required as the resolution is increased.

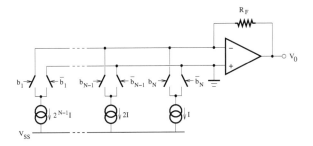

FIGURE 3.14
Block diagram of a binary-weighted switched-current DAC.

Switched-current DACs can also be designed to generate an output voltage, as shown in Figure 3.14, for a converter of the binary-weighted type. The DAC operation principle relies on summing the digitally selected outputs of a set of current sources. Each bit is assigned to a switch, which directs the current flow either to the common summing node or to the ground, depending on the bit state. The current direction adopted here allows the inverting amplifier of the DAC to generate a positive output voltage of the form

$$V_0 = R_F I (b_N + 2b_{N-1} + \cdots + 2^{N-1} b_1)$$

$$= 2^N R_F I \sum_{k=1}^{N} \frac{b_k}{2^k} \tag{3.25}$$

where I denotes the LSB current and N is the converter resolution. Note that the operation of the switched-current DAC with an output current can be affected by the finite output impedance.

The circuit diagram of a binary-weighted switched-current DAC is depicted in Figure 3.15. Each current source is formed by superposing a power-of-two multiple of unit currents obtained by duplicating the current I. This is realized by a current mirror with the number of output transistors associated with a current source equal to the weight of the corresponding bit.

The major drawbacks of the thermometer-coded switched-current DAC are the complexity of the decoder and the high number of the required current sources. On the other hand, the binary-weighted switched-current DAC architecture is limited by its large element spread. A common practice is to divide a high-resolution DAC into subconverters with a small number of bits, and then either combine their output currents with a weighting network or scale the weights of the current sources used in the subconverters [9]. The MSBs, which require a high level of accuracy, are thermometer coded and the LSBs can either be implemented using the binary-weighted or thermometer-coded design technique.

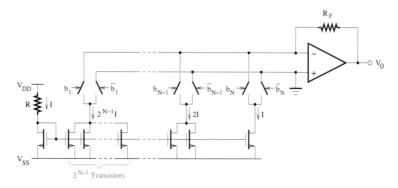

FIGURE 3.15

Circuit diagram of a binary-weighted switched-current DAC.

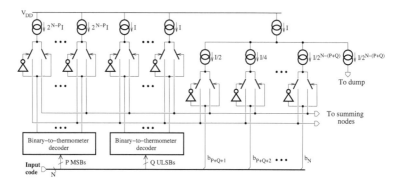

FIGURE 3.16

Block diagram of a segmented switched-current DAC.

The block diagram of a segmented switched-current DAC is shown in Figure 3.16, and its transistor implementation is depicted in Figure 3.17. Current sources with weight values determined according to the subconverter resolutions are used to ensure the monotonicity of the conversion characteristic. They are implemented using p-channel MOS transistors, which can allow an output swing from the ground level. The segmented DAC combines thermometer-decoded current cells, which are associated with the P most significant bits (MSBs) and the Q upper least significant bits (ULLSBs), and binary-decoded current sources, which determine the $N - (P + Q)$ lower least significant bits (LLSBs). With the LLSB transistor array being driven by a ULLSB transistor, an extra LLSB transistor connected to the ground is required to make the sum of the LSBs equal to 1 ULLSB.

Note that, by eliminating the intermediate ULSB segment, the segmented DAC can be reduced to an upper array of thermometer-decoded MSB current sources and an additional MSB current source loaded by a lower array of

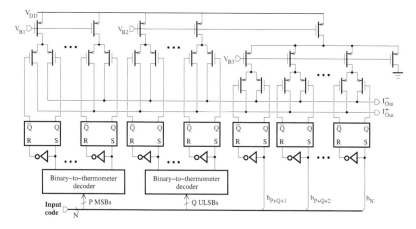

FIGURE 3.17
Transistor implementation of a segmented switched-current DAC.

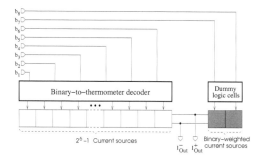

FIGURE 3.18
Block diagram of a $(6+2)$-bit segmented DAC.

binary-decoded LSB current sources. In this way, the use of multiple reference currents, which can be difficult to match in practice to ensure the linearity between the DAC segments, is no longer necessary. Fig 3.18 shows a DAC architecture including a thermometer-encoded segment controlled by the 6 MSBs and a binary-weighted segment connected to the 2 LSBs.

With the number of components required in the thermometer-coded segment of the DAC remaining large for high resolutions, the current sources and switches can be arranged in a two-dimensional pattern to enable efficient routing of switch control signals.

In the structure of Figure 3.19 [10, 11], the output signal is generated by summing the contributions of the thermometer-coded and binary-weighted DAC segments, which are, respectively, controlled by the 6 MSBs and 2 LSBs. The number of thermometer-coded current sources, which are sized to have a value of 4 LSBs and arranged in a matrix form, is 63 since no current source

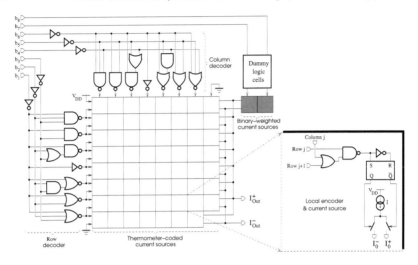

FIGURE 3.19
Block diagram of a $(2 \times 3 + 2)$-bit segmented DAC.

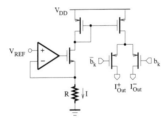

FIGURE 3.20
Circuit diagram of a single switched-current cell with a bias circuit.

is enabled for the input code 0. The values of the two binary-weighted current sources are 2 LSBs and 1 LSB, respectively. The first 3 MSBs steer the column decoder, the next 3 bits are applied to the row decoder, and the remaining 2 bits are equalized to have equal delay with the MSBs and used to select the binary-weighted current sources.

By using a two-stage decoding process for the 6 MSBs, the decoder can be implemented with a minimum gate delay. The first stage is carried out by the row and column decoders, and the second stage is completed by a local decoder provided for each current cell. A given current source is enabled according to the output state of the local decoder in response to the selection signals generated by the row and column decoders. The output signal of each local decoder can be derived simply by combining two of the row selection signals *Row j* and *Row* $(j + 1)$ and one of the column selection signals *Column j* $(j = 1, 2, \cdots, 7)$. This is due to the fact that, for the conversion of a given input code, the matrix of current sources is configured to exhibit, at most, three types of rows. The first rows in which all the current sources are enabled are followed in succession by a row in which only the first current sources are enabled, and the remaining last rows in which all the current sources are disabled. As the digital input code is increased, current cells are enabled along the rows.

The decoders are implemented with inverters, NAND and NOR gates. The skew between the signals of the row and column decoders is eliminated by the latch connected to the switching transistors of each current cell. Figure 3.20 shows the detailed circuit diagram of a switched-current source, which is necessary to build the DAC. The current source is implemented by applying the reference voltage, V_{REF}, to the noninverting input of an amplifier with a high input impedance, which helps maintain the voltage at the inverting input also equal to V_{REF}. The current flowing through the resistor R as a result of the conduction of the MOSFET included in the negative feedback of the amplifier is given by

$$I = V_{REF}/R \tag{3.26}$$

The duplication or scaling of the current I can then be achieved using a current mirror.

The matching errors at the edge of the array can be eliminated by surrounding the active current cells with layers of dummy cells. Because the smallest difference between the values of the current sources is 1 LSB, a DNL error of 0.5 LSB can still be achieved with a relative current mismatch as large as $\pm 12.5\%$. The segmented DAC architecture has the advantage of reducing the linearity error caused by random errors in the current source array. However, the INL characteristic is especially sensitive to graded and symmetrical errors caused, respectively, by a voltage drop along the power supply lines and thermal distribution inside the DAC chip. In practice, for converters with resolutions greater than 8 bits, the linearity error due to nonuniform current sources can be reduced using improved switching schemes or calibration circuits [12, 13].

The performance of switched-current DACs can also be affected by the limited output impedance of current sources. A different number of current cells is connected to the output, depending on the digital input code. This results in a nonlinearity caused by the variation of the output impedance, which can be reduced by using current sources with a cascode configuration as shown in Figure 3.21, where each switch is implemented using a pMOS transistor with the gate connected to the inverted input code.

A cascode current source has the advantage of featuring an output impedance greater than the one of a single transistor. To provide the highest possible voltage swing at the output terminal, the current-source bias circuit is designed to maintain the transistors in the cascode configuration in the saturation region.

For the normal operation of the structure illustrated in Figure 3.21(a), two reference currents are required to establish the dc bias levels. The output

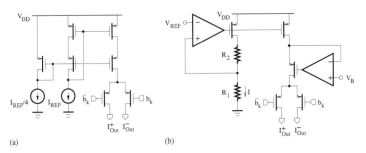

FIGURE 3.21
(a) Circuit diagram of a cascode current source; (b) circuit diagram of an active cascode current source.

current is approximately equal to the reference current I_{REF}, provided the sizes of transistors are identical.

In the current source of Figure 3.21(b), a pair of active cascode transistors is used to further improve the output impedance. The output of the operational amplifier having the noninverting input connected to the reference voltage, V_{REF}, and the inverting input connected to the voltage at the node between resistors R_1 and R_2, drives the transistor gate to set the delivered current at the value $I = V_{REF}/R_1$. The voltage divider including R_1 and R_2 helps maintain the transistor drain at the voltage $V_{REF}(1 + R_2/R_1)$. The duplication of the current I is realized by applying the amplifier output to the gate of other transistors. An additional amplifier uses the bias voltage V_B applied to its noninverting input to regulate the voltage at its inverting node, which is connected between the drain and source of the transistor. In this way, the output impedance of the current source is enhanced by the gain of the amplifier in the negative feedback loop. For high values of the output current, the performance of the active gain enhancement technique may be limited by stability requirements, which can become critically dependent on the parasitic capacitances.

3.3.3.1   Static nonlinearity errors

Given a digital input code k, the differential nonlinearity (DNL) (in LSBs) can be expressed as

$$DNL_k = \frac{I_{Out}(k) - I_{Out}(k-1)}{I_{LSB}} - 1 \tag{3.27}$$

where I_{Out} is the analog output current. The DAC DNL is the maximum value of DNL_k. For monotonic operation, the DAC should exhibit a DNL value that is less than 1 LSB.

The integral nonlinearity (INL) (in LSBs) is given by

$$INL_k = \frac{I_{Out}(k) - I_{Out}(0)}{I_{LSB}} - k \tag{3.28}$$

where $k = 1, 2, \cdots, 2^N - 1$ for an N-bit DAC. The DAC INL is taken as the maximum value of DNL_k. The DAC exhibits a monotonic characteristic if the INL value remains less than 0.5 LSB.

It can be remarked that

$$DNL_k = INL_k - INL_{k-1} \tag{3.29}$$

and

$$INL_k = \sum_{j=1}^{k} DNL_j \tag{3.30}$$

Generally, precise analog levels required in the DAC implementation are generated by using an array of identical unit current sources. Due to transistor mismatches, each unit current source, I_j, of an N-bit current-switched DAC can be modeled by adding a current variation $\triangle I_j$ to the nominal current, $\overline{I}$. Hence,

$$I_j = \overline{I} + \triangle I_j \tag{3.31}$$

where $\triangle I_j$ is the unit current error and the average current $\overline{I}$, that is almost equal to the ideal LSB current, can be expressed as

$$\overline{I} = \frac{1}{2^N - 1} \sum_{j=1}^{2^N - 1} I_j \tag{3.32}$$

The errors $\triangle I_j$ are normally distributed with zero mean and variance σ_I^2, and are assumed to be uncorrelated with each other.

For a thermometer-coded DAC, the output current related to the input digital code k can be written as

$$I_{Out}(k) = k\overline{I} + \sum_{j=1}^{k} \triangle I_j \tag{3.33}$$

where $k = 0, 1, 2, 3, \cdots, 2^N - 1$. For large numbers of unit current sources, the INL at the digital code k can be computed as

$$INL_k = \frac{\displaystyle\sum_{j=1}^{k} \triangle I_j - \frac{k}{2^N - 1} \sum_{l=1}^{2^N - 1} \triangle I_l}{\overline{I} + \frac{1}{2^N - 1} \sum_{l=1}^{2^N - 1} \triangle I_l} \simeq \frac{\displaystyle\sum_{j=1}^{k} \triangle I_j - \frac{k}{2^N - 1} \sum_{l=1}^{2^N - 1} \triangle I_l}{\overline{I}} \tag{3.34}$$

The variance of INL_k is given by

$$\sigma_{INL_k}^2 = k\left(1 - \frac{k}{2^N - 1}\right)\frac{\sigma_I^2}{\overline{I}^2} \tag{3.35}$$

Assuming that the maximum value of the INL_k variance occurs at midcode, or for $k = 2^{N-1}$, we can obtain

$$\sigma_{INL}^2 = 2^{N-1}\left(1 - \frac{2^{N-1}}{2^N - 1}\right)\frac{\sigma_I^2}{\overline{I}^2} \tag{3.36}$$

Considering the DNL at the digital code k, it can be shown that

$$DNL_k = \frac{\triangle I_{Out}(k) - \overline{I} - \dfrac{1}{2^N - 1}\displaystyle\sum_{l=1}^{2^N - 1} \triangle I_l}{\overline{I} + \dfrac{1}{2^N - 1}\displaystyle\sum_{l=1}^{2^N - 1} \triangle I_l} \simeq \frac{\triangle I_{Out}(k) - \overline{I}}{\overline{I}} \tag{3.37}$$

where

$$\triangle I_{Out}(k) = I_{Out}(k) - I_{Out}(k-1) = \overline{I} + \triangle I_k \tag{3.38}$$

The variance of DNL_k can be expressed as

$$\sigma_{DNL_k}^2 = \frac{\sigma_I^2}{\overline{I}^2} \tag{3.39}$$

The variance $\sigma_{DNL_k}^2$ is essentially dependent on the standard deviation of a single unit current source and may appear to be rather small.

The performance difference between thermometer-coded and binary-weighted DACs is related to the way the current sources are combined to form the output current. In the thermometer-coded DAC, the unit current sources are individually switched on and off, while in the binary-weighted DAC, the unit current sources are first gathered into groups of 2^j, with j being an integer, one of which is selected by closing or opening just one of the switches.

The output current of a binary-weighted DAC can be put into the form,

$$I_{Out}(k) = k\overline{I} + \sum_{j=1}^{N}\left[b_{k+1,j}\left(\sum_{i=2^{j-1}}^{2^j-1} \triangle I_i\right)\right] \tag{3.40}$$

where $k = 0, 1, 2, 3, \cdots, 2^N - 1$ and $b_{k+1,j}$ are the elements of the $2^N \times N$ switching matrix given by

$$\mathbf{B} = (b_{k,j}) = \begin{pmatrix} 0 & 0 & 0 & 0 & \cdots & 0 \\ 1 & 0 & 0 & 0 & \cdots & 0 \\ 0 & 1 & 0 & 0 & \cdots & 0 \\ 1 & 1 & 0 & 0 & \cdots & 0 \\ 0 & 0 & 1 & 0 & \cdots & 0 \\ \vdots & \vdots & \vdots & \vdots & \ddots & \vdots \\ 1 & 1 & 1 & 1 & \cdots & 1 \end{pmatrix} \tag{3.41}$$

It should be noted that $b_{k+1,j}$ refers to the element in row $k+1$ and column j of matrix $\mathbf{B}$. The rows of matrix $\mathbf{B}$ are determined by the binary representation of the first $2^N - 1$ integers, the LSB being in the left-most column. The 1's and 0's represent the current sources that are switched on and off, respectively, for the conversion of a given digital code.

The INL_k is approximately given by

$$
INL_k \simeq \frac{\sum_{j=1}^{N} \left[b_{k+1,j} \left(\sum_{i=2^{j-1}}^{2^j-1} \triangle I_i \right) \right] - \frac{k}{2^N - 1} \sum_{l=1}^{2^N-1} \triangle I_l}{\overline{I}} \tag{3.42}
$$

The maximum INL variance of the binary-weighted DAC is almost identical to that of the thermometer-coded DAC.

The DNL_k can be roughly estimated as

$$
DNL_k \simeq \frac{\triangle I_{Out}(k) - \overline{I}}{\overline{I}} \tag{3.43}
$$

where

$$
\triangle I_{Out}(k) = I_{Out}(k) - I_{Out}(k-1) \tag{3.44}
$$

$$
= \overline{I} + \sum_{j=1}^{N} \left[b_{k+1,j} \left(\sum_{i=2^{j-1}}^{2^j-1} \triangle I_i \right) \right] - \sum_{j=1}^{N} \left[b_{k,j} \left(\sum_{i=2^{j-1}}^{2^j-1} \triangle I_i \right) \right] \tag{3.45}
$$

The worst-case transition occurs at mid-scale, or between the digital code $k = 2^{N-1}$ (associated to the row $[1111 \cdots 10]$ of $\mathbf{B}$) and $k = 2^{N-1} + 1$ (associated to the row $[0000 \cdots 01]$ of $\mathbf{B}$) and involves all $2^N - 1$ current sources. The maximum INL variance can then be computed as,

$$
\sigma_{DNL}^2 \simeq (2^N - 1) \frac{\sigma_I^2}{\overline{I}^2} \tag{3.46}
$$

where N is the converter's resolution in number of bits.

Due to transistor mismatches, the non-linearity characteristic of DACs implemented using the same IC process technology can vary randomly. A yield-analysis technique for predicting the values of this characteristic within certain boundaries is necessary to improve the sizing of the unit current sources [14–17]. Especially, analytical methods or Monte Carlo simulations can be used to relate the current matching error variance to the non-linearity characteristic and yield that is defined, for instance, as the percentage of functional devices that have an INL less than 1/2 LSB in a set of fabricated devices.

3.3.3.2 Current source sizing

Typical DAC current sources are implemented using transistors in a cascode configuration. Assuming that the cascode transistor has a high gain, the generated current is reduced to

$$I = I_D = K(V_{GS} - V_{T_n})^2 \tag{3.47}$$

where I_D is the drain current, K is the transconductance parameter, V_{T_n} is the threshold voltage, and V_{GS} is the gate-source voltage. Due to variations of the transistor parameters, the current can deviate from its nominal value. By differentiating the current, it can be shown that

$$\delta I = \frac{\partial I}{\partial K}\delta K + \frac{\partial I}{\partial V_{T_n}}\delta V_{T_n} \tag{3.48}$$

$$= (V_{GS} - V_{T_n})^2 \delta K - 2K(V_{GS} - V_{T_n})\delta V_{T_n} \tag{3.49}$$

and

$$\frac{\delta I}{I} = \frac{\delta K}{K} - \frac{2\delta V_{T_n}}{V_{GS} - V_{T_n}} \tag{3.50}$$

The relative error of the current is given by

$$\frac{\Delta I}{I} = \frac{\Delta K}{K} + \frac{2\Delta V_{T_n}}{V_{GS} - V_{T_n}} \tag{3.51}$$

The variance of the current error can be obtained as

$$\frac{\sigma_I^2}{I^2} = \frac{\sigma_K^2}{K^2} + \frac{4\sigma_{V_{T_n}}^2}{(V_{GS} - V_{T_n})^2} \tag{3.52}$$

The transconductance parameter and threshold voltage differences between two identically drawn and closely spaced transistors can be approximately characterized by the following variances:

$$\frac{\sigma_K^2}{K^2} \simeq \frac{A_K^2}{WL} \tag{3.53}$$

$$\sigma_{V_{T_n}}^2 \simeq \frac{A_{V_{T_n}}^2}{WL} \tag{3.54}$$

where A_K and $A_{V_{T_n}}$ are technology-dependent constants.

The LSB current exhibits the smallest magnitude and can then be the most critical in terms of accuracy. The use of the transistor mismatch model and square law equation to size the transistor of the LSB current source leads to

$$WL = \frac{1}{2}\frac{A_K^2 + \dfrac{4A_{V_{T_n}}^2}{(V_{GS} - V_{T_n})^2}}{\dfrac{\sigma_{I_{LSB}}^2}{I_{LSB}^2}} \tag{3.55}$$

and

$$\frac{W}{L} = \frac{I_{LSB}}{K'(V_{GS} - V_{T_n})^2} \tag{3.56}$$

where I_{LSB} was assumed to be equal to the transistor drain current, and $K' = \mu_n C_{ox}/2$. For a given overdrive voltage, $V_{GS} - V_{T_n}$, the transistor sizes can be estimated as follows:

$$W = \left[\frac{I_{LSB}}{2K'(\sigma^2_{I_{LSB}}/I^2_{LSB})}\left(\frac{A_K^2}{(V_{GS} - V_{T_n})^2} + \frac{4A^2_{V_{T_n}}}{(V_{GS} - V_{T_n})^4}\right)\right]^{1/2} \tag{3.57}$$

$$L = \left[\frac{K'}{2I_{LSB}(\sigma^2_{I_{LSB}}/I^2_{LSB})}[A_K^2(V_{GS} - V_{T_n})^2 + 4A^2_{V_{T_n}}]\right]^{1/2} \tag{3.58}$$

Hence, the transistor sizes of the current source are dependent on matching errors.

3.3.3.3 Switching scheme

High-resolution current-switched DACs are generally based on segmented structures consisting of an LSB binary-weighted array and an MSB thermometer-coded unary array.

In contrast to the thermometer-coded DAC, the binary-weighted DAC exhibits a low overhead because it does not require a decoding logic. The advantage of the thermometer-coded DAC over the binary-weighted DAC is its inherent monotonicity and relaxed matching requirements to meet the same DNL specification. In addition, the worst-case glitch that occurs at mid-code in the binary-weighted DAC is eliminated because only one current source can be switched at once. However, the large area required by the thermometer-coded DAC, especially when a resolution greater than 10 bits is of interest, places a limitation on the achievable current source matching.

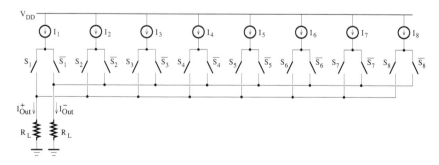

FIGURE 3.22
Block diagram of a 3-bit current-switched DAC.

In practice, transistor mismatch errors attributable to process variations include both stochastic (or random) and gradient errors. The reduction of

TABLE 3.3

Comparison of 4 switching sequences for a 3-bit DAC

	I_1 1.86	I_2 1.90	I_3 1.94	I_4 1.98	I_5 2.02	I_6 2.06	I_7 2.10	I_8 2.14	(μA)
Sequence 1	1	2	3	4	5	6	7	8	
Error (%)	−7	−5	−3	−1	1	3	5	7	
INL_k (%)	−7	−12	−15	−16	−15	−12	−7	0	$INL = 16$
Sequence 2	7	5	3	1	2	4	6	8	
Error (%)	−1	1	−3	3	−5	5	−7	7	
INL_k (%)	−1	0	−3	0	−5	0	−7	0	$INL = 7$
Sequence 3	2	6	4	8	5	1	7	3	
Error (%)	3	−7	7	−3	1	−5	5	−1	
INL_k (%)	3	−4	3	0	1	−4	1	0	$INL = 4$
Sequence 4	5	3	7	1	8	2	6	4	
Error (%)	−1	3	−5	7	−7	5	−3	1	
INL_k (%)	−1	2	−3	4	−3	2	−1	0	$INL = 4$

stochastic errors leads to the chip area increase that in turn induces systematic (or gradient) errors in oxide thickness and along wires or voltage drops across supply-voltage interconnection lines resulting in linear matching errors that are primarily dependent on the DAC layout techniques or the current source location in the DAC array. Thermometer-coded DACs can be implemented with various switching schemes that assign a selection order to each of the current sources as the digital code changes. Appropriate DAC switching schemes [10, 18–20] can help reduce the effect of gradient matching errors. They determine the interconnection between the thermometer decoder/latch outputs and the switch control terminals, and then affect the DAC layout drawing.

Consider the block diagram of a 3-bit current-switched DAC with differential outputs is shown in Figure 3.22. Each unit element can be realized as a cascoded pMOS transistor current source in order to increase the current source output resistance and reduce the effect of parasitic capacitance between the current source and switches. The switches can be implemented using MOS transistors or transmission gates. The output load resistances are 50 Ω.

The nominal value of the unit current is defined as the average current generated by all current sources in the DAC array. Hence,

$$\bar{I} = \frac{1}{8} \sum_{j=1}^{8} I_j \tag{3.59}$$

The INL (in LSBs) for the digital code k is approximately computed as

$$INL_k \simeq \sum_{j=1}^{k} \epsilon_j \tag{3.60}$$

where $\epsilon_j = \triangle I_j/\overline{I}$. The DAC INL is the maximum value of INL_k.

Table 3.3 presents the comparison of 4 switching sequences for a 3-bit DAC. The actual values of unit currents and the corresponding relative errors are listed in the first and second rows, respectively.

The conventional sequential switching sequence (sequence 1) leads to an INL of 16%. The use of the alternating switching sequence (sequence 2)—here, the current sources are switched in alternating fashion relative to matching errors with identical absolute value, starting with the current source whose negative error is the highest -1% up to the one whose positive error is the highest 7%—helps reduce the INL to 7%. For the optimal switching sequences (sequence 3 and 4), the resulting INL takes the minimum value of 4%.

The switching sequence 3 starts with a current source, whose relative error is equal to or less than the INL lower bound. The remaining current sources are then switched so that each of the INL_k values remains within the negative and positive lower bounds of the INL as the digital code is increased. Note that, when the INL_k maximum and minimum are symmetrical about zero, their identical absolute value is equal to the minimum INL.

Considering the switching sequence 4, the first and last current sources to be switched are the ones with the highest negative error and lowest positive error, respectively. The error sign changes from one current source to the next and the remaining current sources are switched so that there is a symmetry between current sources with negative and positive errors of the same absolute value. As a result, a further reduction of the error variance can be achieved.

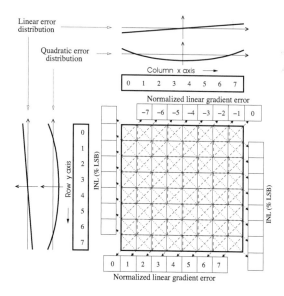

FIGURE 3.23
Structure of a 8 × 8 DAC array showing error distributions.

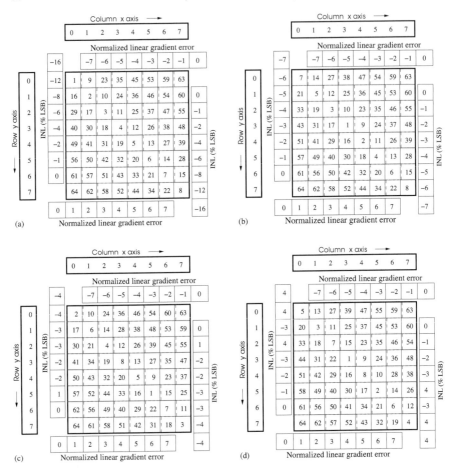

FIGURE 3.24
Four possible switching sequences for a 8 × 8 DAC array.

The aforementioned switching schemes can be applied to current sources arranged in a two-dimensional or matrix array. Figure 3.23 shows the structure of a 8 × 8 DAC array that can provide a resolution of 6 bits. Each current source is identified by the coordinates (x, y), where x and y are the column and row positions, respectively.

Gradient error distributions across a thermometer-coded DAC matrix can be reduced to linear (first-order) and quadratic (second-order) terms in x and y. Their representations use the center of the DAC array as the origin. The normalized linear gradient error is determined for the set of current sources along each dashed line. Its maximum absolute value is associated to the current sources at $(0, 0)$ and $(7, 7)$, while its minimum value is related to current sources that are located along the diagonal dashed line from $(0, 7)$ to $(7, 0)$.

A switching sequence, together with the corresponding digital code INL, is associated to each dotted line along the direction indicated by the arrow.

Figure 3.24 shows 8×8 DAC arrays making use of each of the four possible switching sequences, as well as the corresponding digital code INL. The maximum absolute value of INL for each of the four switching sequences is 16, 7, 4, and 4, respectively. Optimal switching sequences help reduce significantly the effects of nonlinearities due to gradient errors.

3.3.4 NRZ and RZ SC DAC

A switched-current DAC is suitable for high-speed applications (with clock signal frequency up to the GHz range) requiring a moderate resolution. It consists of an array of current sources and switches controlled by the digital code to be converted.

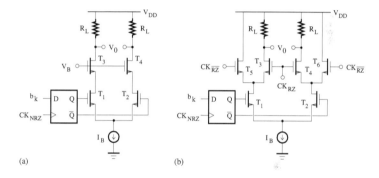

(a)   (b)

FIGURE 3.25
Switched-current DAC cell with (a) NRZ and (b) RZ switching schemes.

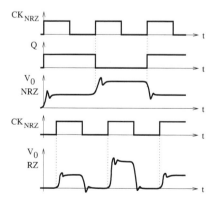

FIGURE 3.26
Waveform representation for NRZ and RZ DACs.

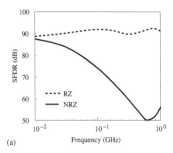 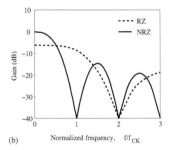

(a)   (b)

FIGURE 3.27
SFDR plots (a) and gain response (b) for NRZ and RZ DACs.

Various switching schemes can be adopted to minimize the correlation between the input code and the resulting DAC error. Figure 3.25(a) shows a switched-current DAC cell with a non-return-to-zero (NRZ) switching scheme. The NRZ output response is obtained by holding constant the code b_k during each clock signal period. In a DAC cell with a return-to-zero (RZ) switching scheme, as depicted in Figure 3.25(b), an additional clock signal is used to periodically return the output to a known state in order to reduce the data dependencies of switching glitches. To ensure that cell data path delay errors do not propagate to the output, transitions of the flip-flop outputs should occur during the reset phase (signal CK_{RZ} in the low state). Figure 3.26 shows the waveforms illustrating the operation of NRZ and RZ DACs.

For NRZ and RZ DACs with a resolution of 12 bits, Figure 3.27(a) presents spurious-free dynamic range (SFDR) plots. The SFDR corresponds to the ratio between the output signal power and the largest magnitude of any spectral component (excluding the DC component). RZ DACs exhibit the best SFDR performance because they are less sensitive to switching errors.

However, RZ DACs are two times more sensitive to random clock jitter due to output transitions that can occur on both (rising and falling) edges of the clock signal. Furthermore, they can exhibit a 6-dB reduction of the output signal power at *dc* and a flatter frequency response with the first null shifted at two times the clock frequency, as illustrated in the output gain response of Figure 3.27(b).

3.4   Charge-scaling DAC

A typical charge-scaling DAC exploits the principle of charge transfer between capacitors for the conversion of the digital input code. Charge-scaling DACs [6] that are capable of achieving a high linearity can be fabricated using CMOS technology.

The circuit diagram of a binary-weighted capacitor-array DAC is shown in Figure 3.28, where b_1 and b_N are the MSB and LSB, respectively. With the assumption that all capacitors are initially discharged, the converter operation requires a two-phase, nonoverlapping clock signal. The converter is based on the charge redistribution principle. The amplifier is assumed to have an offset voltage V_{off} and an infinite dc gain.

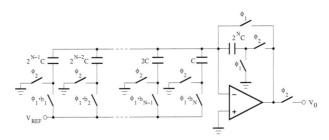

FIGURE 3.28
Circuit diagram of a binary-weighted capacitor-array DAC.

During the first clock phase, the amplifier operates as a unity-gain voltage follower. Each input capacitor $2^{N-k}C$ is connected between the amplifier inverting node and either the reference voltage or the ground, depending on whether the corresponding bit is high or low, while the capacitor 2^NC is connected between the amplifier inverting node and the ground. The charge stored on the input capacitors is of the form

$$\sum_{k=1}^{N}\left(2^{N-k}b_k CV_{REF} - 2^{N-k}CV_{off}\right) = CV_{REF}\sum_{k=1}^{N} b_k 2^{N-k} - (2^N - 1)CV_{off}$$

$$(3.61)$$

and the capacitor 2^NC is charged to V_{off}. Note that the capacitor array has a total capacitance of $(2^N - 1)C$. During the second clock phase, the input capacitors are connected between the ground and the amplifier inverting node and are then charged to V_{off}, while the voltage $V_{off} - V_0$ is applied across the capacitor 2^NC, which is now included in the amplifier feedback path. Applying the charge conservation rule at the amplifier inverting node, we obtain

$$2^NC(V_{off} - V_0) - 2^NCV_{off}$$

$$= -(2^N - 1)CV_{off} - CV_{REF}\sum_{k=1}^{N} b_k 2^{N-k} + (2^N - 1)CV_{off} \qquad (3.62)$$

Hence, the resulting output voltage can be expressed as

$$V_0 = V_{REF}\sum_{k=1}^{N} \frac{b_k}{2^k} \qquad (3.63)$$

where V_{REF} is the reference voltage. For resolutions greater than 8 bits, the capacitance spread, which increases exponentially with the number of bits, can become too high, thereby greatly limiting the achievable matching accuracy.

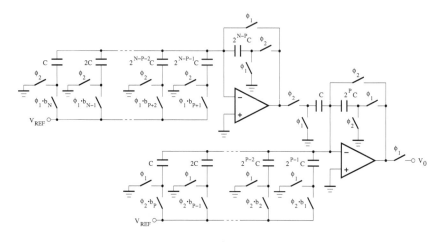

FIGURE 3.29
Circuit diagram of an N-bit cascaded binary-weighted capacitor-array DAC.

One technique to reduce the capacitance spread can consist of using an architecture with two conversion stages as shown in Figure 3.29. The capacitor array of the first conversion stage is responsive to the $(N - P)$ LSBs of the N-bit input code, while the P MSBs are applied to the capacitor array of the second conversion stage. The final result is produced by exploiting the inverting node of the amplifier used in the second conversion stage to combine the output of the first stage with the signal due to the input code MSBs. The output voltage, V_0, can then be computed as

$$V_0 = V_{REF} \sum_{k=1}^{P} \frac{b_k}{2^k} + \frac{1}{2^P} V_{REF} \sum_{k=P+1}^{N} \frac{b_k}{2^{k-P}} = V_{REF} \sum_{k=1}^{N} \frac{b_k}{2^k} \qquad (3.64)$$

Note that the reference signal sampling and the charge transfer phase take place during opposite clock phases in the first and second stage of the DAC. Using the above two-stage architecture, the capacitance spread for an N-bit DAC can be reduced from $1/2^N$ to $\max(1/2^P, 1/2^{N-P})$.

Because the charge is simply redistributed between the capacitors of a charge-scaling DAC, the power consumption is significantly reduced compared with other DAC structures. However, precise clock signals are indispensable for the control of the charge transfer; otherwise glitches may appear at the converter output. In addition, because of the charge slowly leaking from the capacitors over time, the accuracy of charge-scaling DACs starts to decrease within a few milliseconds after the beginning of the conversion. Charge-sharing DACs then appear to be unsuitable for general-purpose DAC applications, and are preferably used in successive-approximation ADCs, whose conversion cycle, by lasting only about a few microseconds, is generally ended well before the effect of the leakage current can become significant.

3.5 Hybrid DAC

An effective method to reduce the component spread and still achieve a high resolution is to use hybrid DAC architectures. By dividing the input code into two subwords to be processed by different sections of the DAC, it is possible to realize the conversion using a reduced number of components.

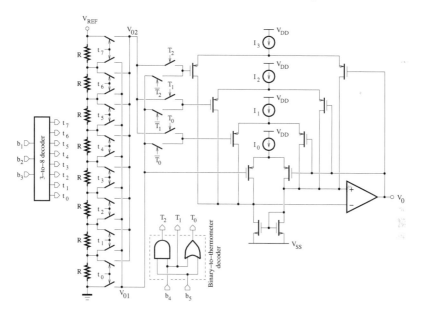

FIGURE 3.30
Circuit diagram of a 5-bit unipolar hybrid DAC including a resistor-string network and an interpolation amplifier stage ($P = 3, Q = 2$).

The unipolar hybrid DAC, as shown in Figure 3.30, includes a 3-bit resistor string and a 2-bit interpolation amplifier stage. This architecture has the advantages of improving the conversion speed and output settling characteristics [7].

With an N-bit input code assumed to be of the form $N = P + Q$, the resistor string is composed of 2^P resistors and the interpolation amplifier requires 2^Q differential transistor pairs, each having their source nodes connected to the corresponding current source. Two consecutive voltages, V_{01} and V_{02}, of the resistor string are selected according to the decoding of the P MSBs and connected to the interpolation amplifier by switches controlled by the Q LSBs. All the inputs of the interpolation amplifier, except the first one which is driven by V_{01}, can be switched to either V_{01} or V_{02}.

In general, the output voltage representing the DAC input code can be obtained as

$$V_0 = V_{01} + q\frac{V_{02} - V_{01}}{2^Q} \tag{3.65}$$

where V_{02} is greater than V_{01} and q is the decimal equivalent of the Q LSBs.

Both the resistor string and the interpolation sections are inherently monotonic, and the monotonicity of the entire DAC is related to the operation principle consisting of adding the value interpolated from V_{01} and V_{02} to the voltage V_{01}.

Let us assume that all the inputs of the interpolation amplifier are initially connected to V_{01}. The negative feedback from the amplifier output forces V_0 to be equal to V_{01} and all the differential transistor pairs become balanced. Furthermore, all differential transistor pairs are actively loaded with the same current mirror, whose output current should be the same as the input current.

By incrementing the Q LSBs by one, one of the interpolation amplifier inputs is switched from V_{01} to V_{02}. The drain current flowing through the corresponding input transistor is then decreased by $\triangle I_D$. The negative feedback provided by the amplifier responds to this imbalance by inducing an increase in $\triangle V_0$ in the output voltage, where $\triangle V_0$ is equal to $(V_{02} - V_{01})/2^Q$. The drain current through each of the transistors with the gates connected to V_0, except the transistor forming a differential stage with the transistor connected to V_{02} and whose drain current is increased by $\triangle I_D$, is augmented by $\triangle I_D/3$ due to the variations in their gate-source voltages. Because the input and output currents of the current mirror are to be maintained equal, the drain current through each of the input transistors connected to V_{01} is increased by $\triangle I_D/3$.

In general, for a given value of the P MSBs, the interpolation controlled by the Q LSBs is realized to allow variation in the output voltage from V_{01} to $V_{01} + (2^Q - 1)(V_{02} - V_{01})/2^Q$ in a step of $(V_{02} - V_{01})/2^Q$.

Due to the transistor sizes required to determine the transconductances in the differential stages of the interpolation amplifier, the effect of parasitic capacitances becomes critical as the number P of LSBs exceeds 8. Taking into account the fact that the converter linearity is also limited by mismatches in the resistor string, the maximum resolution of a DAC including a resistor-string network followed by an interpolating amplifier stage should be on the order of 16 bits.

The aforementioned hybrid DAC uses an amplifier in the interpolation stage. To achieve a high resolution, the amplifier should have a large common-mode rejection ratio to maintain the accuracy over the entire input range and a low offset voltage. An alternative design solution can consist of combining a resistor-string network with a capacitor array DAC. The resulting DACs can exploit the operation principles of switched-capacitor circuits to reduce the effect of the amplifier offset voltage and to achieve a bipolar conversion with a single reference voltage. In addition, the operation of the required amplifier is not affected by common-mode input signals.

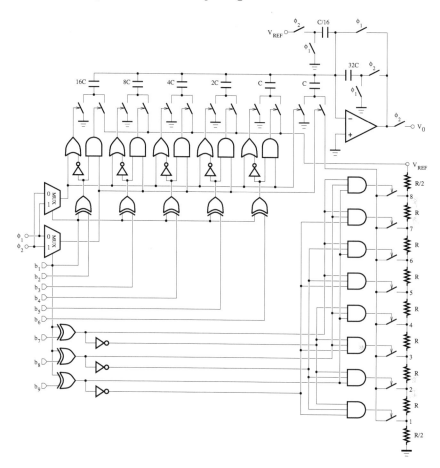

FIGURE 3.31
Circuit diagram of a bipolar hybrid DAC with a 3-bit resistor-string LSB network and a 5-bit binary-weighted capacitor MSB array.

The circuit diagram of a bipolar hybrid DAC is depicted in Figure 3.31. It is based on the voltage-scaling and charge-scaling principles [8]. This DAC structure features a resolution of 8 bits plus the sign bit and has the advantage of performing a bipolar conversion with only a single reference voltage. The input code is supposed to be in two's complement format, which is commonly found in DSP applications.

The absolute value of the two's complement representation equivalent to a negative value is obtained by first inverting all the bits, and then adding one to the result. For the realization of the first step, it is usual to perform the exclusive OR operation on the sign bit, b_1, and each of the remaining bits, $b_2 - b_9$. To implement the second step, the DAC is designed such that the output swings associated with positive and negative input codes can appear to be, respectively, shifted upward and downward by an offset voltage corresponding to 1/2 LSB. The MSBs of the input code are applied to a 5-bit binary-weighted capacitor array, while the remaining bits are used to control a 3-bit resistor-string network with the topmost and bottommost resistors chosen to have the value of $R/2$ so that the aforementioned offset voltages can be produced. The operation of adding a value of $-1/2$ LSB to the output voltage of the DAC is required to set the converter output for the zero input code to 0 V.

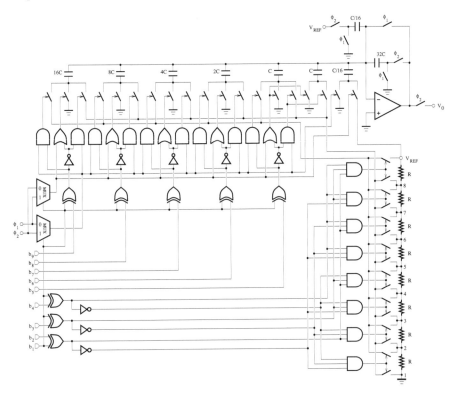

FIGURE 3.32
Circuit diagram of a bipolar hybrid DAC with a 3-bit resistor-string MSB network and a 5-bit binary-weighted capacitor LSB array.

The DAC operation requires two nonoverlapping clock phases. In the case of a positive input code, the multiplexer configuration defined by the state of the sign bit, b_1, which is a logic low, allows the reference voltage sampling and charge transfer to occur, respectively, during the clock phases ϕ_1 and ϕ_2, and the DAC operates as a noninverting gain stage. For a negative input code, the state of b_1 is a logic high, leading to the role interchange between ϕ_1 and ϕ_2 in the input branch, and the DAC is now equivalent to an inverting gain stage.

The output voltage of the DAC will be expressed as

$$V_0 = V_{REF} \left[\frac{b_2}{2} + \frac{b_3}{2^2} + \cdots + \frac{b_6}{2^5} + \frac{1}{2^5} \left(\frac{b_7}{2} + \frac{b_8}{2^2} + \frac{b_9}{2^3} \right) + \frac{1}{2} \frac{1}{2^8} \right] - \frac{1}{2} \frac{V_{REF}}{2^8}$$

(3.66)

if $b_1 = 0$, and

$$V_0 = -V_{REF} \left[\frac{\bar{b}_2}{2} + \frac{\bar{b}_3}{2^2} + \cdots + \frac{\bar{b}_6}{2^5} + \frac{1}{2^5} \left(\frac{\bar{b}_7}{2} + \frac{\bar{b}_8}{2^2} + \frac{\bar{b}_9}{2^3} \right) + \frac{1}{2} \frac{1}{2^8} \right] - \frac{1}{2} \frac{V_{REF}}{2^8}$$

(3.67)

if $b_1 = 1$. Hence,

$$V_0 = \begin{cases} \dfrac{V_{REF}}{2^8} \left(2^7 b_2 + 2^6 b_3 + \cdots + 2^1 b_8 + 2^0 b_9 \right) & \text{if } b_1 = 0 \\[2mm] -\dfrac{V_{REF}}{2^8} \left(2^7 \bar{b}_2 + 2^6 \bar{b}_3 + \cdots + 2^1 \bar{b}_8 + 2^0 \bar{b}_9 + 1 \right) & \text{if } b_1 = 1 \end{cases}$$

(3.68)

Note that the conversion is not affected by the dc offset voltage of the amplifier. The DAC accuracy is mainly determined by the achievable component matching.

Another approach for the hybrid-DAC design consists of using the MSBs to control a resistor-string network and the LSBs to drive a binary-weighted capacitor array. Figure 3.32 shows the circuit diagram of a bipolar hybrid DAC with a 3-bit resistor-string MSB network and a 5-bit binary-weighted capacitor LSB array. The input digital code is supposed to be in two's complement representation.

A set of node voltages is defined on the resistor string as the result of the reference voltage division by resistors with equal values. After the MSB decoding, two adjacent nodes of the resistor string are selected and connected to the binary-weighted capacitors, depending on the decoded state of the remaining LSBs. If the decoded state of an LSB is a high logic, the corresponding capacitor will be switched between the node with the highest voltage and ground; and if it is a low logic, the capacitor switching will take place between the lowest voltage and ground. In both cases, the initial position of the switches and the polarity of the output voltage are determined by the sign bit, b_1.

3.6 Configuring a unipolar DAC for the bipolar conversion

In general, DACs can be designed to operate in either a unipolar or bipolar mode. But, unipolar DACs can be associated with a differential amplifier stage to achieve conversions with a bipolar output-voltage range.

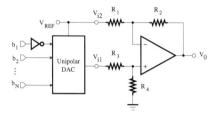

FIGURE 3.33
Circuit diagram of a unipolar DAC with a bipolar output-voltage range.

The circuit diagram of a unipolar DAC with a bipolar output-voltage range is shown in Figure 3.33. Here, the DAC uses an input code in the offset binary representation and an incoming two's complement code is converted to the offset binary format by inverting the MSB, b_1. Using the voltage divider principle, we have

$$V^+ = \frac{R_4}{R_3 + R_4} V_{i1} \tag{3.69}$$

and

$$V^- = \frac{R_2}{R_1 + R_2} V_{i2} + \frac{R_1}{R_1 + R_2} V_0 \tag{3.70}$$

where $V_{i1} = \sum_{k=1}^{N}(b_k/2^k)$ and $V_{i2} = V_{REF}$. For the usual case of an amplifier with a high dc gain, the relation $V^+ = V^-$ is exploited to express the output voltage in the form

$$V_0 = \frac{R_4}{R_3 + R_4}\left(1 + \frac{R_2}{R_1}\right)\sum_{k=1}^{N}\frac{b_k}{2^k} - \frac{R_2}{R_1}V_{REF} \tag{3.71}$$

When $R_2/R_1 = R_4/R_3$, the DAC output stage operates as a differential amplifier, and

$$V_0 = \frac{R_2}{R_1}\left(\sum_{k=1}^{N}\frac{b_k}{2^k} - V_{REF}\right) \tag{3.72}$$

The DAC output voltage depends on the resistor ratios. However, due to mismatches in the resistor ratios, the effect of common-mode signals on the converter characteristics can become critical. Furthermore, the amplifier should be designed to exhibit a low offset voltage.

An alternative design solution consists of using a differential stage with more than one amplifier to improve the common mode rejection. In the special case of DACs based on R-2R resistor network, a converter with a bipolar output range can be realized as shown in Figure 3.34. From the analysis of an R-2R network, the currents I_A and I_B can be expressed as

$$I_A = -\frac{V_{REF}}{2R}\sum_{k=1}^{N}\frac{b_k}{2^k} \tag{3.73}$$

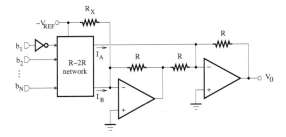

FIGURE 3.34
Circuit diagram of a bipolar R-2R ladder DAC.

and

$$I_B = -\frac{V_{REF}}{2R} \sum_{k=1}^{N} \frac{\overline{b_k}}{2^k} \qquad (3.74)$$

The output voltage of the DAC is given by

$$V_0 = -R(I_A - I_B) - \frac{R}{R_X} V_{REF} \qquad (3.75)$$

Exploiting the sum formula for geometric series,[1] it can be deduced that

$$I_A + I_B = -\frac{V_{REF}}{2R} \sum_{k=1}^{N} \frac{1}{2^k} = -\frac{V_{REF}}{R} \left(1 - \frac{1}{2^N} \right) \qquad (3.76)$$

and

$$I_A - I_B = 2I_A - (I_A + I_B) = -\frac{V_{REF}}{R} \left(-1 + \sum_{k=1}^{N} \frac{b_k}{2^k} \right) - \frac{V_{REF}}{R} \frac{1}{2^N} \qquad (3.77)$$

This last expression without the term $V_{REF}/(2^N R)$ is proportional to the offset binary representation of the DAC input code. By choosing the value of the resistor R_X such that

$$R_X = 2^N R \qquad (3.78)$$

---

[1] For the geometric series with the common ratio r, the sum S is given by

$$S = \sum_{k=0}^{n} r^k = 1 + r + r^2 + \cdots + r^n = \frac{1 - r^{n+1}}{1 - r}$$

and if the sum is taken starting at $k = 1$, we have

$$\sum_{k=1}^{n} r^k = \frac{r(1 - r^n)}{1 - r}$$

the DAC output voltage is reduced to

$$V_0 = V_{REF} \left(-1 + \sum_{k=1}^{N} \frac{b_k}{2^k} \right) \tag{3.79}$$

The performance of this DAC architecture depends on the achievable matching between the resistors.

When an ac source is used as the reference voltage, it is amplified by a factor determined by the digital input code. In this case, the converter is then referred to as a multiplying DAC. Two-quadrant or four-quadrant multiplications are performed, depending on whether the multiplying DAC is designed for unipolar or bipolar conversion.

3.7   Algorithmic DAC

The algorithmic DAC operates according to a selection tree structure. It can be efficiently implemented with less power dissipation and fewer circuit components than other conversion architectures for a given resolution and bandwidth [21].

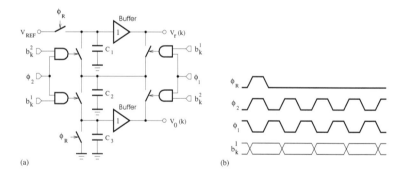

(a)   (b)

FIGURE 3.35
(a) Circuit diagram and (b) clock timing of an algorithmic DAC.

Starting from the MSB, a reference voltage, V_{REF}, is added or not to the previous output, depending on the state (low or high) of the present bit, b_k^l ($l = 1, 2$). During the k-th cycle, the output voltage, V_0, is computed as

$$V_0(k) = V_0(k-1) + b_k^l V_r(k) \tag{3.80}$$

where

$$V_0(0) = 0 \qquad (3.81)$$

$$V_r(k) = \begin{cases} V_{REF} & \text{if} \quad k = 0 \\ \dfrac{V_r(k-1) - V_0(k-1)}{2} & \text{otherwise.} \end{cases} \qquad (3.82)$$

At the end of N cycles, the analog output is obtained as the sum of the voltages associated to each bit of the digital input code. An implementation of the algorithmic DAC with its clock timing is shown in Figure 3.35. The converter consists of three identical capacitors C_1, C_2, and C_3 and two unity buffers. Initially, the pulse signal, ϕ_R, is used to allow the capacitors C_1 and C_3 to be charged by the initial voltages $V_r(0) = V_{REF}$ and $V_0(0) = 0$. The charge transfer is controlled by two nonoverlapping clock signals, ϕ_1 and ϕ_2, and the state of the bits, b_k^l. For the high state of the bits, $V_r(k-1)$ is held by C_1 and $V_0(k-1)$ is updated because the charge stored on C_2 during the clock phase ϕ_1 is redistributed between C_2 and C_3 during the clock phase ϕ_2. In the case of low-state bits, the charge produced by $V_0(k-1)$ on C_3 remains unchanged and the update of $V_r(k-1)$ is achieved as the result of the charge sharing between C_2, which was first connected to $V_0(k-1)$, and C_1.

However, it should be mentioned that component nonidealities (mismatch, charge injection, clock feed-through) limit the resolution of the algorithmic DAC to no more than 10 bits.

3.8 Direct digital synthesizer

Direct digital synthesis is a method that can be used to produce analog waveforms, such as square, triangular and sinusoidal signals, by generating a time-varying signal in digital form and then performing a digital-to-analog conversion. Direct digital synthesizers (DDSs) are also known as numerically controlled oscillators. They find applications in communication systems (quadrature synthesizers) and test equipment (arbitrary waveform generators), where it is necessary to produce and control waveforms of various frequencies and profiles.

The block diagram of a DDS is shown in figure 3.36. It consists of a phase accumulator, a phase-to-amplitude converter (conventionally a read-only memory (ROM) used as look-up table (LUT)), a digital-to-analog converter (DAC) and a lowpass filter.

The phase accumulator is a digital integrator that is realized using a register and an adder. The input binary number, M, represents the phase increment that is added, at each clock pulse, to the data previously held in the phase register. The N-bit phase accumulator steps through each of the 2^N possible

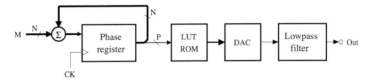

FIGURE 3.36
Block diagram of a direct digital synthesizer.

phase increment values before it overflows and the cycle begins again. The overflow rate corresponds to the DDS output frequency that is given by

$$f_0 = \frac{M \cdot f_{CK}}{2^N} \tag{3.83}$$

where M is the binary tuning word, f_{CK} is frequency of the clock signal, and N is the length (in bits) of the phase accumulator.

The smallest incremental change in frequency, or the frequency resolution, is of the form,

$$\triangle f = \frac{f_{CK}}{2^N} \tag{3.84}$$

The output frequency of a DDS can range from $f_{CK}/2^N$ (in the case where $M = 1$) up to the Nyquist frequency $f_{CK}/2$ imposed by the sampling theorem.

To reduce the power consumption and the size of the LUT ROM (or die area), the output of the phase accumulator is truncated and only P most significant bits are passed to the phase-to-amplitude converter or used to index the LUT ROM with 2^P entries. The phase truncation results in a small increase of the output phase noise, but it does not affect the frequency resolution.

A DDS uses an addressing scheme with an appropriate LUT ROM to produce samples of an arbitrary sinewave. The content of the LUT for a sinewave with an offset B and a peak amplitude A can be obtained as follows,

$$LUT(k) = \left\lfloor B + A \sin\left(\frac{2\pi k}{2^P}\right) + \frac{1}{2} \right\rfloor \tag{3.85}$$

where $\lfloor \ \rfloor$ denotes the floor operator and the range of the index k is from 0 to $2^P - 1$.

The DDS can be designed to generate two quadrature output signals by using the LUT ROM contents, especially, $LUT(k)$ and $LUT(k + 2^P/4)$ for the sine and cosine waves, respectively. A square wave can be generated with no computational overhead by exploiting the fact that the most significant bit of the phase accumulator toggles periodically at one-half of the phase represented by the accumulator. However, this square wave can be corrupted by a phase jitter of one clock signal period.

Quarter wave symmetry in the sine waveform can be exploited to design a DDS that uses a LUT ROM with reduced size. This is achieved by storing the sine wave, $\sin(\phi)$, only for $0 \leq \phi \leq \pi/2$, instead of $0 \leq \phi \leq 2\pi$, and then using the two most significant bits of the quantized phase angle to perform quadrant mapping (or to deduce the remaining waveform samples). As a result, only 2^{P-2} entries are required in the LUT ROM, leading to a size compression ratio of 4-to-1 for the LUT ROM.

The output of the LUT ROM is quantized to the number of bits of the DAC. However, this quantization results in an increase of the signal-to-noise ratio. The resolution of the DAC is then typically 1 to 4 bits less than the output word length of the LUT ROM so as not to significantly increase the noise level.

DDSs are generally designed with N ranging from 24 to 32 bits and can generate periodic waveforms at frequencies from less than 1 Hz up to 400 MHz with 1-GHz clock signal. The DDS performance for a particular application can be evaluated using the phase noise (in dBc/Hz), jitter (in degrees rms), and spurious-free dynamic range (in dB) specifications.

3.9   Summary

Depending on the trade-offs between the power consumption, resolution, conversion speed, and latency, Nyquist-rate DACs can be designed using parallel, pipeline, or serial architecture.

Parallel DAC structures can require a high power consumption to meet the speed and settling requirements of the amplifier, which is used to perform the charge transfer or current summing. Pipeline DACs can help reduce the power consumption and the spread of component values. However, this is achieved at the cost of an increase in latency time. Serial DAC structures often require a low circuit area, but they can be very slow. Hybrid or segmented DACs can be adopted to reduce the spread of component values, and thereby the effect of the amplifier loading on the converter performance.

3.10   Circuit design assessment

1.   Analysis of R-2R DACs

A multiplying DAC can be implemented as shown in Figure 3.37.

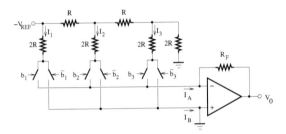

FIGURE 3.37

Multiplying DAC based on current-mode R-2R network.

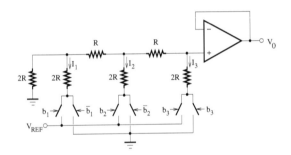

FIGURE 3.38

DAC based on voltage-mode R-2R network.

The R-2R network generates a current that is proportional to the input digital code and is converted by the feedback structure consisting of the amplifier and resistor R_F to the required voltage level.

Find the output voltage V_0 as a function of the input digital code.

Show that the output resistance of the R-2R network is given by

$$R_0 = \frac{3R}{(3/2)b_1 + (9/2^3)b_2 + (33/2^5)b_3 + (3/2^3)b_2 b_3} \qquad (3.86)$$

To reduce the effects of nonlinear errors on the converter performance, the DAC can be realized using the voltage-mode R-2R network, as illustrated in Figure 3.37.

Verify that the output resistance of the R-2R network is constant and equal to R.

Determine the output voltage V_0.

2. Implementation of the R-2R DAC using MOS transistors

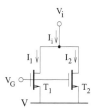

FIGURE 3.39
Current division principle based on two transistors.

The two-transistor circuit of Figure 3.39 can be used to divide an input current I_i into two components, I_1 and I_2. Assuming that V and V_G are chosen so that the transistors remain in the on-state, show that

$$\frac{I_1}{I_2} = \frac{W_1/L_1}{W_2/L_2} \qquad (3.87)$$

where W_i and L_i $(i = 1, 2)$ are the width and length of the corresponding transistors, respectively.

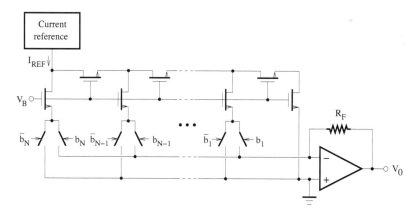

FIGURE 3.40
Transistor implementation of a R-2R DAC.

Using the current division principle [22], it is possible to implement an R-2R ladder using MOS transistors as shown in Figure 3.40. To improve the linearity, the transistors are assumed to operate in the triode region. The switches are realized using transistors driven either by b_k or $\bar{b}_k$.

Use SPICE simulations to verify that the small-signal equivalent resistance seen between the drain and source terminals of the MOS-FETs is not identical throughout the transistor network.

Ideally, each stage of the transistor network splits its input current into two equal parts, that is,

$$I_{0,k} = I_{i,k}/2 = b_k 2^{-k} I_{REF} \qquad (3.88)$$

Due to transistor mismatches, the current division is affected by an error of the form

$$\Delta I_{0,k} = \epsilon_k I_{REF} \qquad (3.89)$$

Assuming that the requirements INL < 0.5 LSB and DNL < 1 LSB are met, provided that $\max(\sum_k |b_k \epsilon_k|) < 1/2^{N+1}$, determine the worst-case error for a converter resolution, $N = 8$ bits.

3. Segmented DAC

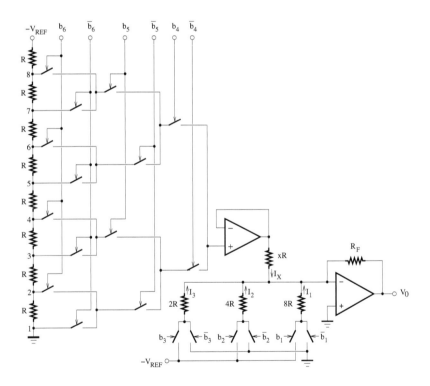

FIGURE 3.41
Segmented DAC.

Consider the segmented DAC depicted in Figure 3.41, which is based on resistor string and binary weighted resistors.

Assuming ideal amplifiers, determine x to realize a 6-bit segmented DAC.

Express the output voltage, V_0, as a function of the input digital code, b_k, $k = 1, 2, 3, 4, 5, 6$.

Put the output voltage into the form

$$V_0 = V_{FS} \sum_{k=1}^{6} \frac{b_k}{2^k}$$

where V_{FS} is the full-scale output voltage to be determined.

Assuming that the DAC is designed to feature a slew rate, SR, of 0.75 V/µs, use the equation, $f_{max} = 1/t_{max}$, where $t_{max} = \Delta V_0 / SR$ and $\Delta V_0 = V_{FS}$, to estimate the maximum data rate, f_{max}, for $V_{FS} = 5$ V.

4. **Binary-weighted charge-scaling DAC**

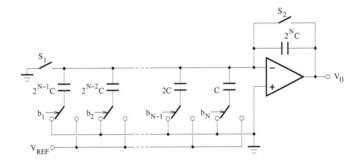

FIGURE 3.42
Binary-weighted charge-scaling DAC.

The circuit diagram of an N-bit binary-weighted charge-scaling DAC is shown in Figure 3.42. A digital-to-analog conversion starts with the discharge of all capacitors, and each switched capacitor is then connected either to the ground or the reference voltage, depending on the state of its corresponding bit.

With the assumption that the operational amplifier is ideal, verify the following charge conservation equation.

$$\Delta Q = CV_{REF} \sum_{i=1}^{N} 2^{N-k} b_k \qquad (3.90)$$

$$= 2^N CV_0 \qquad (3.91)$$

Deduce the expression of the output voltage, V_0.

In practical DAC implementations, the capacitors exhibit a variation of $\pm\Delta C$. Let $V_{LSB} = V_{REF}/2^N$ and use

$$|DNL_{max}| = V_0^{\Delta C}(100\cdots0) - V_0^{\Delta C}(011\cdots1) - V_{LSB} \qquad (3.92)$$

and

$$INL(k) = V_0^{\Delta C}(k) - V_0(k) \qquad (3.93)$$

where

$$V_0^{\Delta C}(100\cdots0) = 2^{N-1}\frac{C+\Delta C}{C}\frac{V_{REF}}{2^N} \qquad (3.94)$$

$$V_0^{\Delta C}(011\cdots1) = \frac{V_{REF}}{2^N}\frac{C-\Delta C}{C}\sum_{k=2}^{N}2^{N-k} \qquad (3.95)$$

$$V_0^{\Delta C}(k) = \frac{V_{REF}}{2^N}\left(1+\frac{\Delta C}{C}\right)2^{N-k} \qquad (3.96)$$

and

$$V_0(k) = \frac{V_{REF}}{2^N}2^{N-k} \qquad (3.97)$$

to show that

$$|DNL_{max}| = (2^N - 1)\frac{\Delta C}{C}\frac{V_{REF}}{2^N} \qquad (3.98)$$

and

$$INL(k) = 2^{N-k}\frac{\Delta C}{C}\frac{V_{REF}}{2^N} \qquad (3.99)$$

respectively. In the case of the DNL, it is assumed that the MSB capacitor takes its maximum value, while the remaining capacitors exhibit their minimum values.

Deduce the worst-case INL value given by $|INL_{max}| = INL(1)$.

Determine the number of bits, N, if the capacitor tolerance, $\Delta C/C$, is equal to $\pm0.5\%$ and $INL = \pm0.5$ LSB.

5. **Sensor signal conditioner**
   Consider the conditioner circuit of Figure 3.43 that is required to make the sensor output suitable for further processing. The voltage V_{DAC} is provided by a digital-to-analog converter.

   Assuming that the operational amplifiers are ideal, use

$$\frac{V_A - V_B}{R_2} = \frac{V_B - V_0^+}{R_1} \qquad (3.100)$$

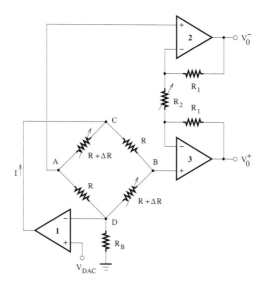

FIGURE 3.43
Sensor signal conditioner.

and

$$\frac{V_B - V_A}{R_2} = \frac{V_A - V_0^-}{R_1} \qquad (3.101)$$

to verify that

$$V_0 = V_0^+ - V_0^- = \left(1 + \frac{2R_1}{R_2}\right)(V_B - V_A) \qquad (3.102)$$

where $V_B - V_A = \triangle R \cdot I/2$ and $I = V_{DAC}/R_B$.

Bibliography

[1] A. R. Hamadé, "A single chip all-MOS 8 bit A/D converter," *IEEE J. of Solid-State Circuits*, vol. 13, no. 6, pp. 785–791, Dec. 1978.

[2] P. E. Allen and D. R. Holberg, *CMOS Analog Circuit Design*, 2nd ed., New York, NY: Oxford University Press, 2002.

[3] A. Abrial, J. Bouvier, J.-M. Fournier, P. Senn, and M. Veillard, "A 27-MHz digital-to-analog video processor," *IEEE J. of Solid-State Circuits*, vol. 23, no. 6, pp. 1358–1369, Dec. 1988.

[4] M. J. M. Pelgrom, "A 10-b 50-MHz CMOS D/A converter with 75-Ω buffer," *IEEE J. of Solid-State Circuits*, vol. 25, no. 6, pp. 1347–1352, Dec. 1990.

[5] P. Holloway, "A trimless 16b digital potentiometer," *1984 IEEE ISSCC Digest of Technical Papers*, p. 66–67, 320–321, Feb. 1984.

[6] R. Gregorian and G. Amir, "A single chip speech synthesizer using a switched-capacitor multiplier," *IEEE J. of Solid-State Circuits*, vol. 18, no. 1, pp. 65–75, Feb. 1983.

[7] A. Yilmaz, "LSB interpolation circuit and method for segmented digital-to-analog converter," U.S. Patent 6,246,351, filed October 7, 1999; issued June 12, 2001.

[8] M. Kokubo, S. Nishita, and K. Yamakido, "Interpolative D/A converter," U.S. Patent 4,652,858, filed April 16, 1986; issued March 24, 1987.

[9] A. R. Bugeja, B.-S. Song, P. L. Rakers, and S. F. Gillig, "A 14-b, 100-MS/s CMOS DAC designed for spectral performance," *IEEE J. of Solid-State Circuits*, vol. 34, pp. 1719–1732, Dec. 1999.

[10] T. Miki, Y. Nakamura, M. Nakaya, S. Asai, Y. Akasaka, and Y. Horiba, "An 80-MHz 8-bit CMOS D/A converter," *IEEE J. of Solid-State Circuits*, vol. 21, pp. 983–988, Dec. 1986.

[11] J. M. Fournier and P. Senn, "A 130-MHz 8-b CMOS video DAC for HDTV applications," *IEEE J. of Solid-State Circuits*, vol. 26, pp. 1073–1077, July 1991.

[12] Y. Nakamura, T. Miki, A. Maeda, H. Kondoh, and N. Yazawa, "A 10-b 70-MS/s CMOS D/A converter," *IEEE J. of Solid-State Circuits*, vol. 26, pp. 637–642, April 1991.

[13] G. A. M. Van der Plas, J. Vandenbussche, W. Sansen, M. S. J. Steyaert, and G. G. E. Gielen, "A 14-bit intrinsic accuracy random walk CMOS DAC," *IEEE J. of Solid-State Circuits*, vol. 34, pp. 1708–1718, Dec. 1999.

[14] K. R. Lakshmikumar, R. A. Hadaway, and M. A. Copeland, "Characterization and modeling of mismatch in MOS transistors for precision analog design," *IEEE J. of Solid-State Circuits*, vol. 21, no. 6, pp. 1057–1066, Dec. 1986.

[15] G. I. Radulov, M. Heydenreich, R. W. van der Hofstad, J. A. Hegt, and A. H. M. van Roermund, "Brownian-bridge-based statistical analysis of the DAC INL caused by current mismatch," *IEEE Trans. Circuits Syst. II*, Exp. Briefs, vol. 54, no. 2, pp. 146–150, Feb. 2007.

[16] H. Park and C.-K. Ken Yang, "Nearly exact analytical formulation of the DNL yield of the digital-to-analog converter," *IEEE Trans. Circuits Systems–II*, vol. 59, no. 9, pp. 563–567, Sep. 2012.

[17] H. Park and C.-K. Ken Yang, "An INL yield model of the digital-to-analog converter," *IEEE Trans. Circuits Systems–I*, vol. 60, no. 3, pp. 582–592, Mar. 2013.

[18] Y. Cong and R. L. Geiger, "Switching sequence optimization for gradient error compensation in thermometer-decoded DAC arrays," *IEEE Trans. Circuits Syst. II*, vol. 47, no. 7, pp. 585–595, Jul. 2000.

[19] J. Deveugele, G. Van der Plas, M. Steyaert, G. Gielen, and W. Sansen, "A gradient-error and edge-effect tolerant switching scheme for a high accuracy DAC," *IEEE Trans. Circuits Syst. I*, vol. 51, no. 1, pp. 191–195, Jan. 2004.

[20] K.-C. Kuo, and C.-W. Wu, "A switching sequence for linear gradient error compensation in the DAC design," *IEEE Trans. Circuits Syst. II*, vol. 58, no. 8, pp. 502–506, Aug. 2011.

[21] K. Watanabe, G. C. Temes, and T. Tagami, "A new algorithm for cyclic and pipeline data conversion," *IEEE Trans. on Circuits and Systems*, vol. 37, no. 2, pp. 249–252, Feb. 1990.

[22] C. M. Hammerschmied and Q. Huang, "Design and implementation of an untrimmed MOSFET-only 10-bit A/D converter with −79-dB THD," *IEEE J. of Solid-State Circuits*, vol. 33, pp. 1148–1157, Aug. 1998.

4

Nyquist Analog-to-Digital Converters

CONTENTS

In general, analog-to-digital converters (ADCs) are required for any application where an analog or continuous-time signal must be processed by digital systems. A variety of architectures is available for the design of ADCs with different characteristics, such as resolution, bandwidth, sampling frequency, power consumption, latency, and chip area.

For the special case of Nyquist ADCs, the sampling frequency can be at least two times the maximum frequency of the input signal. This converter group includes, but is not limited to, successive approximation register (SAR) ADC, integrating ADC, flash ADC, pipelined ADCs, and cyclic ADC. All techniques for analog-to-digital conversion rely on at least one operation of comparison between the input signal and a reference level. The flash ADC, which uses one comparator for each comparison, exhibits a higher speed than the SAR ADC, whose operation involves various comparisons realized by the same comparator. Because of the parallelism of the flash ADC, the number of comparators grows exponentially with the resolution, leading to an increase in the power consumption and chip area, and a decrease in the bandwidth as a result of an augmentation in the input capacitance. Some variations of flash architecture, such as the folding and interpolating ADC, and pipelined ADC, have been proposed in order to reduce the effect of some of these limitations.

In general, the trade-offs between the converter characteristics (speed, resolution, power consumption, size, linearity) play an important role in determining the better ADC architecture for a given application. For instance, a significantly faster conversion is usually achieved at the price of a reduction in the initial resolution. Due to the component matching requirement, the resolution of a flash ADC is limited to 9 bits. SAR ADCs most commonly exhibit a resolution in the range from 8 to 16 bits and provide a low power consumption as well as a small IC size. They are ideal for many real-time applications. On the other hand, resolutions up to 18 bits can be achieved by integrating ADCs, which can work well also with low-level signals. However, this architecture generally has a low conversion speed and is only suitable for applications such as portable instruments, where the signal bandwidth remains low.

ADC architectures offer different compromises between the performance metrics. They are then chosen to be consistent with the specifications of the target application.

4.1 Analog-to-digital converter (ADC) architectures

Generally, trade-offs between resolution, power consumption, chip area size, conversion time, static performance, and dynamic performance play an important role in the choice of the proper ADC architecture for a given application (data acquisition, measurement, voiceband audio, image and video, wireless communication systems).

4.1.1 Successive approximation register ADC

Successive approximation register (SAR) ADCs can be designed using either a single-ended architecture or a differential architecture.

- **Single-ended architecture**

The block diagram of a SAR ADC is shown in Figure 4.1. It consists of a digital control logic, a clock generator, a comparator, a successive approximation register (SAR), a digital-to-analog converter (DAC), and an output buffer based on latches. The operations of resetting the SAR, enabling the clock signal, cycling each bit, and halting the clock signal at the end are conducted by the control logic.

For positive input signals, the operation principle of a SAR ADC can be based on the algorithm illustrated in Table 4.1. The SAR ADC exploits the concept of the binary search algorithm to find the nearest digital code to an input voltage.

The principle is to compare the analog version of various DAC digital codes with the input analog signal. Starting with the DAC most-significant bit (MSB), each bit is initially set to the logic high state. If the signal to be converted is higher than the one generated by the DAC and which actually represents one-half of the full scale (FS), the initial value of the MSB is main-

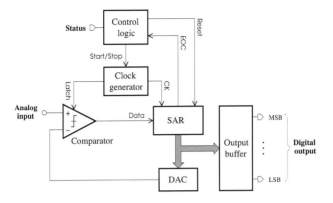

FIGURE 4.1
Block diagram of a SAR ADC.

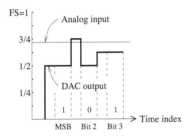

FIGURE 4.2
Three-bit successive approximation of the input signal.

tained and the comparison procedure continues with the next DAC output signal of $(3/4)$ FS. Otherwise, the MSB is reset to the logic low state. In this case, the output signal of the DAC required for the next comparison step corresponds to $(1/4)$ FS. This successive conversion (see Figure 4.2) continues until the least-significant bit (LSB) is reached, that is, the DAC output is within $\pm(1/2)$ LSB of the input voltage. For an N-bit ADC with 2^N levels of resolution, the conversion is carried out in N clock periods.

The SAR of Figure 4.3 is built with two sets of N-bit registers (control and data registers), the function of which is to store and guess the conversion result, respectively. Each flip-flop in the data register is sequentially set by the control register to a state such that on the next rising edge of the clock pulse, the current value of the input data is transferred to the output. This approach has the advantage of simplifying the layout design, which consists of reproducing each bit cell containing two D flip-flops. However, this approach can require a large chip area (19 D flip-flops for an 8-bit SAR). Furthermore, while the flip-flops of the control register use the same clock signal, the clock input of a given flip-flop of the data register is obtained from the next flip-

TABLE 4.1
The Computation Scheme of a SAR ADC

Begin

1. Initialization:
Acquire a sample of V_i
Specify the initial time index, $k \leftarrow 1$
Set the most significant bit, b_1, to the logic high and clear the remaining bits
Assign $V_r^1 = V_{FS}/2$

2. Repeat

(a) Compare V_i with V_r^k
   If $V_i \geq V_r^k$ then
      return the actual logic state of the bit under test
   else
      invert the logic state of the bit under test
   End If

(b) Adjust the time index, $k \leftarrow k + 1$

(c) Set the bit b_k to the logic high

(d) Update V_r^k
   If $V_i \geq V_r^{k-1}$ then
   $$V_r^k \leftarrow V_r^{k-1} + \frac{V_{FS}}{2^k}$$
   else
   $$V_r^k \leftarrow V_r^{k-1} - \frac{V_{FS}}{2^k}$$
   End If

until the DAC output is within $\pm(1/2)$ LSB of the input voltage, V_i

End

flop output. As a result, the SAR output can be affected by a three-flip-flop propagation delay in the worst case. If this delay is comparable to the DAC settling time or the comparator response time, its effect can become significant [1, 2].

The design of Figure 4.4 makes use of a single register with $N + 1$ JK flip-flops for the generation of the digital code and the end of conversion signal. The k-th cell consists of a JK flip-flop and two OR gates. The input terminals are labeled **Reset** and **Data**, and the N-bit code and **End** represent the output variables. The JK flip-flops are cleared by **Reset** $= 1$, and then **Reset** $= 0$ is maintained. A 1 at the **Data** input shows that the analog version of the SAR

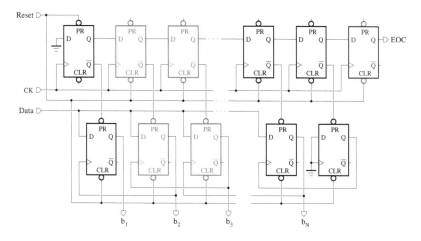

FIGURE 4.3
Circuit diagram of a SAR using two sets of registers (b_1 is the MSB and b_N is the LSB).

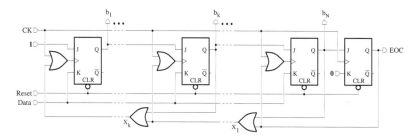

FIGURE 4.4
Circuit diagram of a SAR with a reduced number of registers.

code provided by the DAC is greater than the input signal and the current approximation must be reduced.

A given flip-flop is initialized to 0 and held in this state as long as $X_k = 1$. On the other hand, its behavior will be described by referring to the truth table of a JK flip-flop if $X_k = 0$ (see Appendix A). The output of the flip-flop is changed to 1 for $J = 1$. This state is held during the next clock pulse if **Data** $= 0$, otherwise the current output bit is modified to 0. As the next bit is set to 1, X_k will take the value 1 and the flip-flop output state will be maintained. At this step, a further change can only be initiated by the **Reset** signal.

Because capacitors are easily fabricated in CMOS technologies, the charge redistribution technique is generally adopted in the SAR ADC implementation [3]. Figure 4.5 shows the block diagram of a SAR ADC exploiting the charge redistribution on weighted capacitors. Note that the extra LSB ca-

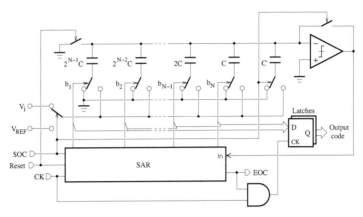

FIGURE 4.5
Block diagram of a charge-redistribution unipolar SAR ADC ($C' = C$).

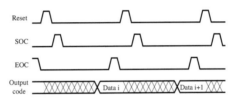

FIGURE 4.6
Timing diagram of the SAR ADC.

pacitor, C', of value C is required to make the total value of the capacitor array equal to $2^N C$, where N is the number of bits, so that a binary voltage division can be performed by switching any weighted capacitor of the DAC. The timing diagram of the SAR ADC is illustrated in Figure 4.6. A start-of-conversion (SOC) signal is used to initiate the conversion process, while the end-of-conversion (EOC) signal is asserted by the SAR to indicate the conversion completion. Before each conversion cycle, the Reset signal is enabled to initiate the discharge of capacitors through the ground connection and the reset of all the bits in the SAR to the logic low. Note that a clock signal is generally necessary for proper operation of the SAR ADC, even if it does not have to be synchronized with other control signals. Its frequency depends on the conversion resolution and speed. The conversion process consists of a sequence of three operations: the sample phase, the hold phase, and the redistribution phase (or bit testing mode).

In the *sample phase*, all capacitors are connected to the input voltage, V_i, and the comparator feedback switch is closed. The voltage V_C across the capacitor array at the end of the sampling period is actually

$$V_C = V_i - V_{off} \tag{4.1}$$

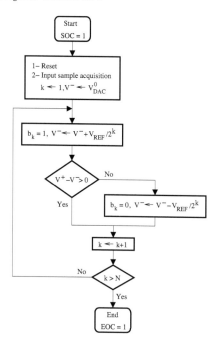

FIGURE 4.7
Binary search algorithm of the SAR ADC.

where V_i is assumed to be positive, and the offset voltage, V_{off}, plays the role of the comparator threshold voltage.

During the *hold phase*, the comparator feedback switch is open and the bottom plates of capacitors are connected to ground. Because the charge on the top plate is conserved, the top plate potential goes to $-V_C$ and the voltage applied to the negative terminal of the comparator can be expressed as

$$V^- = -V_C \tag{4.2}$$

The input node is now connected to the reference voltage, V_{REF}, instead of the input voltage.

The *redistribution phase* begins with the determination of b_1 or MSB. The largest capacitor is then connected to the reference voltage and the equivalent circuit of the capacitor array seen from the input node is a voltage divider consisting of two equal capacitors $2^N C$. Hence, the voltages at the negative and positive terminals of the comparator are given by

$$V^- = V_{REF}/2 - V_C = -V_i + V_{off} + V_{REF}/2 \tag{4.3}$$
$$V^+ = V_{off} \tag{4.4}$$

The logic state of the MSB depends on the sign of the voltage difference,

$V^+ - V^-$, provided by the comparator. It will remain unchanged, that is, $b_1 = 1$, if $V^+ - V^- > 0$, or equivalently $V_i > V_{REF}/2$. But, if $V^+ - V^- < 0$, then $V_i < V_{REF}/2$, and the MSB will be set back to the logic low, $b_1 = 0$, by switching back the largest capacitor to the ground. The determination of the bit from b_2 to b_N also proceeds in the same manner. The bottom plate of the capacitor associated with the bit b_k is connected to the reference voltage, and the previous value of V^- is increased by $V_{REF}/2^k$ as a result of the voltage division carried out by the capacitor array. The final logic state is assigned to the bit under test, and the SAR either maintains the capacitor connected to V_{REF} or connects it to the ground, depending on the polarity of $V^+ - V^-$ detected by the comparator. At the end of the conversion, we have

$$V^- = -V_i + V_{REF} \left(\frac{b_1}{2} + \frac{b_2}{2^2} + \cdots + \frac{b_{N-1}}{2^{N-1}} + \frac{b_N}{2^N} \right) \qquad (4.5)$$

and the value of V^- should be as close to zero as possible. The valid output code can then be stored in the output buffer consisting of latches.

The conversion process of the charge-redistribution unipolar SAR ADC is based on the algorithm shown in Figure 4.7, where $V^0_{DAC} = -V_i + V_{off}$. Furthermore, the code transitions of the transfer characteristic actually occur at k LSB, where k is an integer. However, the quantization error will be reduced if the code transitions can arise at $k/2$ LSB. This can be achieved by switching the lower plate of the capacitor C' to $V_{REF}/2$ rather than the ground after the sample phase.

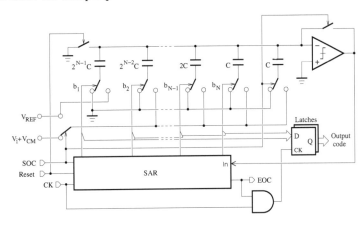

FIGURE 4.8
Block diagram of a charge-redistribution bipolar SAR ADC.

To allow the conversion of positive and negative input voltages, the capacitor switching scheme of the charge redistribution SAR ADC must be somewhat modified. Figure 4.8 shows the block diagram of a charge-redistribution bipolar SAR ADC, whose operation involves the next steps.

In the *sample phase*, the largest capacitor is connected to the reference voltage, V_{REF}, while the remaining capacitors are connected to the input voltage, V_i, and the comparator feedback switch is closed. The charge Q_C stored on capacitors is given by

$$Q_C = 2^{N-1}C(V_i + V_{CM} - V_{off}) \tag{4.6}$$

During the *hold phase*, the comparator feedback switch is open and the bottom plates of capacitors are now switched to the ground. Because the capacitor array is equivalent to a voltage divider consisting of two equal capacitors, the voltage applied to the negative terminal of the comparator now becomes

$$V^- = -(V_i + V_{CM})/2 + V_{off} \tag{4.7}$$

The input node is actually connected to the reference voltage, V_{REF}, instead of the input voltage.

The *redistribution phase* begins with the determination of b_1 or MSB. The SAR is initialized so that the MSB is set to the logic high. With the assumption that the comparator threshold is modeled as an offset voltage, V_{off}, in series with the positive input, we have

$$V^+ = V_{off} \tag{4.8}$$

In the case where $V^+ - V^- > 0$, that is, if $V_i > 0$, the comparator output will be a logic high and the state of the MSB will remain unchanged. However, if $V^+ - V^- < 0$, then $V_i < 0$, and the comparator output will be a logic low. The SAR is then configured to set the first bit of the output code to a logic low and switch the MSB capacitor from the ground to V_{REF}. Here, the MSB is used as a sign bit. For the determination of each of the remaining bits, the corresponding capacitor is switched from the ground to V_{REF} and the comparison trial proceeds as in the unipolar case. At the end of the conversion, the voltage V^- can be written as

$$V^- = -\frac{V_i + V_{CM}}{2} + \frac{V_{REF}}{2}\left(-b_1 + \frac{b_2}{2} + \cdots + \frac{b_{N-1}}{2^{N-2}} + \frac{b_N}{2^{N-1}}\right) \tag{4.9}$$

where V_{CM} denotes the common-mode voltage. An appropriate choice of V_{CM} is useful for handling bipolar input signals. It can be assumed that the converter input signal is biased about

$$V_{CM} = V_{REF} + V_{LSB} = V_{REF}(1 + 1/2^N) \tag{4.10}$$

where $V_{LSB} = V_{REF}/2^N$. Hence, the converter features a two's complement output coding with the zero code, which is actually $1/2$ LSB above the midscale. It should be noted that the converter full-scale still exhibits a dynamic range of V_{REF} as in the case of the charge redistribution unipolar SAR ADC.

An alternative design approach can consist of modifying the switching scheme of the SAR ADC to control the connection between the reference voltage terminal and either V_{REF}^{+} for a positive input or V_{REF}^{-} for a negative input. The *sample phase* and *hold phase* remain similar to the ones of the unipolar structure, except that the sign bit should be detected by the comparator just before the *redistribution phase* and used to select the suitable reference voltage. This latter is required for the determination of the remaining bits of the output code.

The accuracy of the charge redistribution SAR ADC is mainly limited by the achievable capacitor matching. Because the settling time of the comparator increases as the minimum overdrive signal becomes smaller for high resolutions, the time delay of the comparator appears to be a critical limiting factor in the achievable speed of the data conversion. Furthermore, the stability requirements of the comparator should be specified, taking into account the fact that the offset voltage estimation is performed in the unity-gain configuration.

- **Effects of capacitance mismatches on linearity characteristics**

To improve the matching accuracy, large capacitors are generally realized by combining identical unit capacitors. The $j - th$ capacitor in the SAR ADC can be modeled as,

$$C_j = 2^{N-j}C + \delta_{N+1-j} \quad j = 1, 2, \cdots, N \qquad (4.11)$$

where C is the unit capacitor and δ_j is the error term. Assuming that the error distributions of unit capacitors are independent and identically distributed Gaussian random variables, the mean and variance of the error terms can be expressed as,

$$E[\delta_{N+1-j}] = 0 \quad \text{and} \quad E[\delta_{N+1-j}^2] = 2^{N-j}\sigma^2 \qquad (4.12)$$

where σ is the standard deviation of the unit capacitor.

Given a DAC digital input, X, the corresponding analog output provided by the N-bit capacitive-array DAC can be written as,

$$V_{DAC}(X) = \frac{C + \delta_1 + \sum_{j=1}^{N}(2^{N-j}C + \delta_{N+1-j})b_{N+1-j}}{2^N C + \delta_1 + \sum_{j=1}^{N}\delta_{N+1-j}} V_{REF} \qquad (4.13)$$

where the value of b_{N+1-j} is either 0 or 1, the input X corresponds to the code $b_1 b_2 \cdots b_N$, and V_{REF} is the reference voltage.

Capacitance mismatches can affect the SAR ADC accuracy parameters: integral nonlinearity (INL) and differential nonlinearity (DNL) that are especially useful in the high-resolution applications.

The INL specification due to capacitance mismatches can be defined as follows:

$$INL(X) = \frac{V_{DAC,real}(X) - V_{DAC,ideal}(X)}{V_{LSB}} \qquad (4.14)$$

$$= \frac{V_{DAC}(X) - V_{DAC}(X)|_{\delta_j=0}}{V_{LSB}} \qquad (4.15)$$

where $V_{LSB} = V_{REF}/2^N$. The MSB is found out by the first comparison without any capacitor switching. Hence, its determination is not affected by capacitance mismatches.

The maximum INL value is obtained for the code transitions that are performed by switching the largest number of capacitors. Assuming that the mismatch error terms in the denominator of (4.13) can be neglected, it can then be shown that

$$E\{[V_{DAC,real}(X) - V_{DAC,ideal}(X)]^2\} = E\left[\frac{1}{2^{2N}}\left(\sum_{j=1}^{N-1}\delta_j^2\right)\frac{1}{C^2}V_{REF}^2\right]$$

$$= E\left[\frac{1}{2^{N+1}}\frac{(2^N-1)\sigma^2}{2^{N-1}}\frac{1}{C^2}V_{REF}^2\right]$$

$$\simeq \frac{\sigma^2}{2^N C^2}V_{REF}^2 \qquad (4.16)$$

The variance of the maximum INL value can be obtained as,

$$\sigma_{INL,max} \simeq \left(\frac{1}{V_{LSB}}\right)\sqrt{\frac{\sigma^2}{2^N C^2}V_{REF}^2} \simeq \frac{\sigma\sqrt{2^N}}{C} \qquad (4.17)$$

The INL specification provides a measure of how closely the ADC output matches its ideal response.

The definition of the DNL specification can be stated as follows,

$$DNL(X) = INL(X) - INL(X-1) \qquad (4.18)$$

$$= \frac{V_{DAC,real}(X) - V_{DAC,real}(X-1)}{V_{LSB}} \qquad (4.19)$$

The maximum DNL value is expected to occur at the transition between the digital codes $X = [10\cdots0]$ and $X - 1 = [01\cdots1]$. Assuming that the determination of the MSB is not affected by capacitance mismatches, we arrive at

$$\sigma_{DNL,max} = \left(\frac{1}{V_{LSB}}\right)\sqrt{E\left[\frac{1}{2^{2N}}\left(\sum_{j=1}^{N-1}\delta_j^2\right)\frac{1}{C^2}V_{REF}^2\right]}$$

$$\simeq \left(\frac{1}{V_{LSB}}\right)\sqrt{\frac{\sigma^2}{2^N C^2}V_{REF}^2} \simeq \frac{\sigma\sqrt{2^N}}{C} \qquad (4.20)$$

A DNL specification of greater than 1 LSB may indicate non-monotonicity, that is not desirable for control and video applications. An ADC is said to be monotonic when the output code always increases as the input gets larger and always decreases as the input is reduced.

• Calibration

Due to mismatches of passive components, resolutions higher than 12 bits can only be realized by means of self-calibrating converters [4]. This technique is illustrated with the ADC shown in Figure 4.9. The data conversion, which is

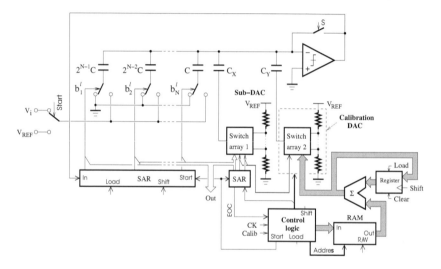

FIGURE 4.9
Block diagram of a self-calibrating SAR ADC. (Adapted from [4], ©1983 IEEE.)

achieved in three steps, is based on the charge redistribution. First, the reset switch is closed and the comparator is configured as a unity gain buffer. The top and bottom plates of capacitors are then connected to the virtual ground and input voltage, respectively. During the next phase, the reset switch is open while the bottom plates are switched to the ground. Due to the charge conservation, the potential at the top plates is changed from V_i to $-V_i$. The last operation consists of finding the digital version of the analog input signal. The conversion process starts with the determination of the MSB. The bottom plate of the largest capacitor is connected to V_{REF}. The connection will be maintained if the comparator output is high. Otherwise, the capacitor is switched to the ground. The following MSB and the remaining bits are determined in the same way. The MSBs are resolved by the binary weighted capacitor DAC. The resistor-string DAC, which is connected to the coupling capacitor C_X, provides the LSBs. Basically, it is not affected by the differential nonlinearity.

Ideally, the voltage contribution of the binary-weighted capacitor DAC at the comparator input is given by

$$V = \frac{V_{REF}}{2^N} \sum_{j=1}^{N} 2^{j-1} b_{N+1-j}^l \qquad (4.21)$$

where V_{REF} is the reference voltage and b_{N+1-j}^l ($l = 0, 1$) is the logic value of the $(N + 1 - j)$-th bit. Due to the fluctuations of the IC fabrication process, the value of a weighted capacitor can be written as

$$C_j = 2^{j-1} C (1 + \epsilon_j), \quad j = 1, 2, \cdots, N, \qquad (4.22)$$

where ϵ_j denotes the matching error and C is the unit capacitor, which is the ratio of the total capacitance C_T to 2^N, that is,

$$C = \frac{C_T}{2^N} \qquad (4.23)$$

The voltage V then becomes

$$\hat{V} = \frac{V_{REF}}{C_T} \sum_{j=1}^{N} C_j b_{N+1-j}^l \qquad (4.24)$$

Substituting Equations (4.22) and (4.23) into Equation (4.24), we obtain

$$\hat{V} = \frac{V_{REF}}{2^N} \sum_{j=1}^{N} 2^{j-1} (1 + \epsilon_j) b_{N+1-j}^l \qquad (4.25)$$

The error voltage can be computed as

$$\triangle V = \hat{V} - V = \sum_{j=1}^{N} V_{\epsilon j} b_{N+1-j}^l \qquad (4.26)$$

where

$$V_{\epsilon j} = \frac{V_{REF}}{2^N} 2^{j-1} \epsilon_j \qquad (4.27)$$

The calibration stage is necessary for the reduction of nonlinearities introduced by the capacitor mismatches. Starting from the largest capacitor to the smallest one, the different error contributions are estimated. For a given capacitor C_k, this is achieved first by connecting V_{REF} to all capacitors except C_k. Then, the charge is redistributed by switching all capacitors except C_k to the ground. It follows that the charge stored at the top plate of the capacitors is

$$\triangle Q = V_{REF} (2C_k - C_T) \qquad (4.28)$$

Substituting Equations (4.22) and (4.23) into Equation (4.28), it can be shown that

$$\triangle Q = CV_{REF}2^k \epsilon_k \qquad (4.29)$$

and the corresponding residual voltage is given by

$$V_{xk} = \frac{\triangle Q}{C_T} = 2V_{\epsilon k} \qquad (4.30)$$

Starting from the MSB capacitor, the residual and error voltages can be computed as

$$V_{\epsilon k} = \begin{cases} \dfrac{V_{xN}}{2} & \text{if } k = N \\ \dfrac{1}{2}\left(V_{xk} - \displaystyle\sum_{j=k+1}^{N} V_{\epsilon j} \right) & \text{otherwise.} \end{cases} \qquad (4.31)$$

The correction signals are obtained from the digital version of the residual voltages. A random access memory (RAM) is used to store the error voltages $V_{\epsilon j}$, which are estimated during the calibration cycle carried out just after the converter power-up. If the k-th bit assumes the high level, the error $V_{\epsilon k}$ will be added to the ones accumulated from the MSB through the $(k-1)$-th bit and the result is stored in the accumulator. Otherwise, $V_{\epsilon k}$ is discarded and the previous error voltage contained in the accumulator remains unchanged. The accumulator output is then converted by the calibration DAC into an analog signal, which is summed with the appropriate sign to the output voltage of the binary-weighted capacitor DAC. The calibration is equivalent to the cancelation of the error voltages due to capacitor mismatches from the corresponding ADC output signal during the normal conversion cycle. It works well provided the linearity error on the coupling capacitor $C_X = C$ is less than $1/2^M$ for an M-bit sub-DAC. In the layout, C_X should preferably be located near the center of the array to counteract the effect of fabrication errors.

- **Differential architecture**

    In a conventional singled-ended SAR ADC, the reference voltage V_{REF} is used to digitize the input signal that is comprised between 0 and V_{REF}. The input signal range can be extended from $-V_{REF}$ to V_{REF}, equivalent to an absolute range of $2V_{REF}$, by using a SAR ADC with differential switched-capacitor (SC) DACs.

    The circuit diagram of a SAR ADC with differential SC DACs is shown in Figure 4.10, where V_i^+ and V_i^- are the differential sampled-and-held input voltages, V_{CM} represents the common-mode voltage and the reference voltages can be defined as, $V_{REF}^+ = V_{CM} + V_{REF}/2$ and $V_{REF}^- = V_{CM} - V_{REF}/2$. It consists of two SC DACs, a latched comparator, and the necessary SAR logic to determine the switching sequence.

    Each conversion starts with a sampling phase during which the input signal and the common-mode voltage are connected to the SC DACs. At the

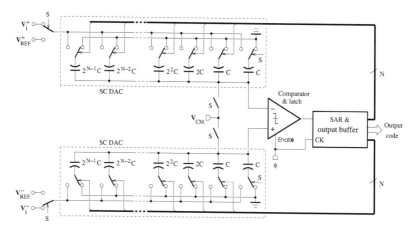

FIGURE 4.10
Circuit diagram of a SAR ADC with differential SC DACs.

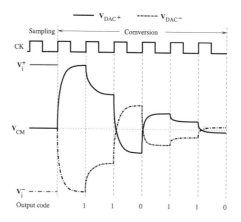

FIGURE 4.11
Waveforms of the SAR ADC with differential SC DACs.

beginning of the conversion phase, the input signal and the common-mode voltage are disconnected. The DAC capacitors are then successively switched between the reference voltages or the ground, starting from the MSB. The voltages applied at the comparator inputs is determined by the difference between the charge stored during the sampling phase and the one related to the reference voltages. Hence,

$$V^+ = -V_i^- + V_{CM} + \frac{V_{REF}^+}{2^N} \sum_{j=1}^{N} 2^{j-1} b_{N+1-j} \qquad (4.32)$$

and

$$V^- = -V_i^+ + V_{CM} + \frac{V_{REF}^-}{2^N} \sum_{j=1}^{N} 2^{j-1} b_{N+1-j} \qquad (4.33)$$

where V^+ and V^- denote the comparator inputs, and the output bits, b_{N+1-j}, are determined after each comparison. Waveforms of the SAR ADC with differential SC DACs are shown in Figure 4.11, where V_{DAC+} and V_{DAC-} represent the reference voltage contributions to the SC DAC outputs, in order to illustrate an example of the conversion process.

- **Split capacitor array**

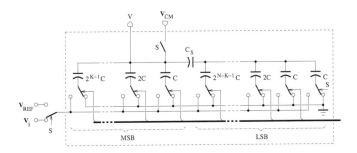

FIGURE 4.12
Circuit diagram of the SC DAC with a split capacitor C_S.

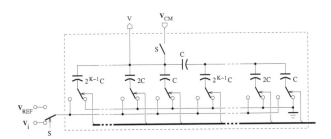

FIGURE 4.13
Circuit diagram of the SC DAC with a unit split capacitor.

The major limitation of the SAR ADC speed is related to the RC time constant of the comparator input network. For a conventional binary-weighted SC DAC, the total capacitance increases exponentially with the resolution in number of bits.

One solution to reduce the total capacitance consists of using split DACs [5], as shown in Figure 4.12, where V denotes the DAC output (or comparator input). For a DAC with an N-bit split capacitor array consisting of a K bit MSB section and an $N - K$ bit LSB section joined together by a split capacitor C_S, we have

$$C_{LSB} = C + \sum_{j=K+1}^{N} 2^{j-K-1}C = 2^{N-K}C \tag{4.34}$$

$$C_{MSB} = \sum_{j=1}^{K} 2^{j-1}C = (2^K - 1)C \tag{4.35}$$

$$\frac{C_S \cdot C_{LSB}}{C_S + C_{LSB}} = C \quad \text{and} \quad C_S = \frac{2^{N-K}}{2^{N-K} - 1}C \tag{4.36}$$

where C_{LSB} and C_{MSB} are the total capacitances of the LSB and MSB capacitor array sections, respectively. The DAC output voltage can be expressed as

$$V = V_{CM} - V_i + \frac{\left(\sum_{j=1}^{K} 2^{j-1} b_{K+1-j}\right)C}{C_{MSB} + \frac{C_S C_{LSB}}{C_S + C_{LSB}}} V_{REF}$$

$$+ \frac{\frac{C_S}{C_S + C_{LSB}}\left(\sum_{j=K+1}^{N} 2^{j-K-1} b_{N+K+1-j}\right)C}{C_{MSB} + \frac{C_S C_{LSB}}{C_S + C_{LSB}}} V_{REF} \tag{4.37}$$

Note that the LSB contribution is scaled by the voltage ratio of the capacitive divider consisting of C_S and C_{LSB}. It then follows that

$$V = V_{CM} - V_i + \frac{V_{REF}}{2^N}\left(2^{N-K}\sum_{j=1}^{K} 2^{j-1} b_{K+1-j} + \sum_{j=K+1}^{N} 2^{j-K-1} b_{N+K+1-j}\right) \tag{4.38}$$

In practice, due to the non-integer value of C_S and IC process limitations, it is difficult to match the capacitors. Furthermore, the split DAC can be affected by the nonlinearity distortion, that is caused by the parasitic capacitances at the terminals of C_S.

Alternatively, the split DAC can be realized with a unit split capacitor, as illustrated in Figure 4.13. However, the suppression of the dummy capacitor C introduces a gain error of 1 LSB, that can be compensated by an adequate digital calibration. The compensated output code is generated by adding the pre-measured digital error codes to the raw output code of the SAR ADC.

4.1.2　Integrating ADC

By using time to quantize a signal, which represents the integral or average of an input voltage over a fixed period of time, an integrating ADC can exhibit good linearity performance and high-frequency noise rejection. In comparison with a successive approximation converter, the integrating ADC is simple, low cost, and slow. Generally, its speed is approximately 500 times slower than the one of a typical successive approximation converter. Integrating ADCs can be implemented using single-slope, dual-slope, or multiple-slope architectures. The primary advantage of a dual-slope ADC over a single-slope architecture is that the final conversion result is insensitive to errors in the component values. A multiple-slope ADC still features the advantages of a dual-slope converter, but its conversion speed can be greatly increased at the cost of extra hardware complexity.

The dual-slope architectures are suitable for digitizing low bandwidth signals in instrumentation devices such as the digital multimeter, which require a high resolution (10 to 20 bits or $3\frac{1}{2}$ to $5\frac{1}{2}$ digits) and a low conversion speed in the range of 100 sps (samples per second).

The dual-slope ADC, as shown in Figure 4.14, uses an analog integrator with switched inputs, a comparator, a control logic, and a counter. Figure 4.15 shows the conversion timing of the converter. The input voltage is assumed to be positive.

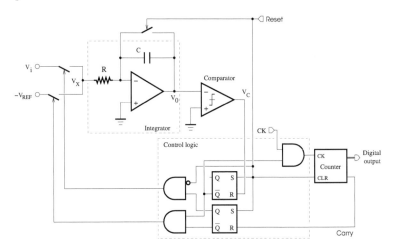

FIGURE 4.14
Circuit diagram of a dual-slope ADC.

The operation of the dual-slope ADC starts with a reset phase, which consists of shorting the capacitor C to drive the integrator output to zero and clearing the counter, followed by the input signal integration phase and the reference signal integration phase.

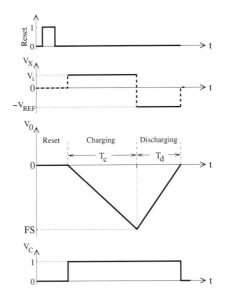

FIGURE 4.15
Conversion timing of the dual-slope ADC.

– **Input signal integration phase**

After the reset phase, the unknown input signal V_i is applied to the integrator, while at the same time the process of counting clock pulses is started. The integrator output voltage is given by

$$V_0(t) = -\frac{1}{\tau} \int_0^t V_i(t) \mathrm{d}t \qquad (4.39)$$

where $\tau = RC$ is the time constant of the integrator. The voltage V_i is then integrated for a duration T_c, which is generally known as the charging (or ramp-up) time. Assuming that V_i is time invariant, the output of the integrator at the end of the integrating phase can be written as

$$V_0(T_c) = -\frac{V_i}{\tau} \cdot T_c \qquad (4.40)$$

Here, the integration of the input signal takes place until the N-bit counter overflows; this corresponds to a fixed time of $T_c = 2^N T$, where T is the clock period. Note that the integrator output voltage is directly proportional to the input signal.

– **Reference signal integration phase**

When the input signal integration phase is completed, the counter is reset. A known reference voltage, V_{REF}, with a polarity opposite to that of V_i is connected to the integrator input and the counting of clock pulses resumes. The integrator output voltage is now ramping down at a constant slope according to the expression

$$V_0(t) = -\int_{T_c}^{t} \frac{-V_{REF}}{\tau} dt + V_0(T_c) = \frac{V_{REF}}{\tau}(t - T_c) - \frac{V_i}{\tau} \cdot T_c \qquad (4.41)$$

The counter is stopped after a discharging (or ramp-down) duration of T_d when the comparator output takes the logic low level because the integrator output reaches zero, that is, $V_0(T_d) = 0$. Hence,

$$\frac{V_{REF}}{\tau} \cdot T_d - \frac{V_i}{\tau} \cdot T_c = 0 \qquad (4.42)$$

The time T_d is then given by

$$T_d = T_c \cdot \frac{V_i}{V_{REF}} \qquad (4.43)$$

On the other hand, T_d can be expressed in terms of the clock period, T, as

$$T_d = N_r \cdot T \qquad (4.44)$$

where N_r is the number of clock periods recorded during the connection of V_{REF} to the integrator input. Thus, it can be found that

$$N_r = 2^N \frac{V_i}{V_{REF}} \qquad (4.45)$$

The digital output of the counter is proportional to the magnitude of the input signal, and is independent of the integrator time constant, which is supposed not to change during an individual conversion cycle. When the input voltage is negative, the dual-slope ADC operates according to the same basic principle as described above for a positive input voltage, except that all the signal polarities are reversed.

The accuracy of the dual-slope ADC seems to be only affected by the fluctuations of the reference voltage and clock timing. But, the behavior of a practical circuit can also be plagued by the nonideal characteristics of the amplifier, MOS switches and capacitors, and the response time or switching delay of the comparator. Figure 4.16 shows the effects of the offset voltage and input over-range on the integrator output waveform. Additional conversion phases are required in order to improve the accuracy of the dual-slope ADC.

The dual-slope ADC can be designed to allow bipolar operation and to incorporate a phase for the converter offset compensation. Figure 4.17 shows the circuit diagram of a dual-slope ADC with an auto-zero (AZ) capacitor [6]. The conversion cycle includes an auto-zero phase, an input signal integration (ISI) phase, a reference signal integration (RSI) phase, and an integrator-output zero (IZ) phase.

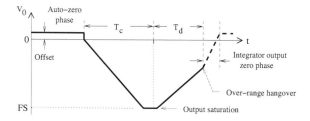

FIGURE 4.16
Effects of the offset voltage and input over-range on the integrator output waveform.

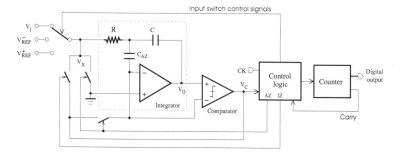

FIGURE 4.17
Circuit diagram of the dual-slope ADC with an auto-zero capacitor.

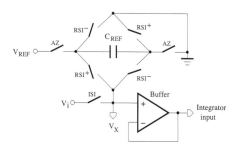

FIGURE 4.18
Circuit diagram of the input section of a bipolar dual-slope ADC with a single reference voltage.

During the *auto-zero phase*, the converter input is connected to the ground and a feedback loop is closed around the system such that the error voltage can appear across the auto-zero capacitor, C_{AZ}, whose charge is used for offset voltage correction during the subsequent phases. By including all active components in the loop, the accuracy of the auto-zero calibration is limited only by the noise of the system. Note that a larger value of C_{AZ} should be used to minimize the noise sensibility for a converter with a small resolution. Typically, the capacitor C_{AZ} is at least two times greater than C.

In the *input signal integration phase*, the auto-zero switch is open and the voltage V_i is connected to the converter input. The input signal polarity is determined at the end of this phase.

During the *reference signal integration phase*, the control circuit connects the reference signal with the polarity such that the integrator output can be driven with a fixed slope to the zero level established in the auto-zero phase.

In the *integrator output zero phase*, the switch controlled by the IZ control signal is closed and the negative feedback from the output of the comparator to the converter input drives the integrator output to zero. This phase is used to completely discharge the feedback capacitor of the integrator following the occurrence of an over-range condition (see Figure 4.16), which is characterized by the integrator output remaining far from the initial level after the maximum time allowed to the reference signal integration phase.

Due to the need to accurately estimate the occurrence time of the zero-crossing at the end of the reference signal integration phase, the operation speed of the dual-slope ADC is generally slow. Although the frequency of the clock signal may be increased, the use of a high clock rate is limited by the delay of the comparator in detecting the zero-crossing. For a typical converter, the comparator delay should not be greater than the duration of one-half clock pulse, yielding a clock frequency on the order of a few hundred kilohertz based on a comparator with a delay of a few microseconds. An improvement in the performance may be achieved by modifying the structure of the conventional dual-slope ADC such that the discharging time of the integrator can be measured with a precision greater than the pulse width of the clock signal.

In the approach relying on the use of two reference voltages to design a dual-slope ADC with a bipolar input range, the symmetry of the converter transfer characteristic may be affected by the independent fluctuation of each reference source. A solution can consist of using the input section of the dual-slope ADC shown in Figure 4.18, where the charge stored by a capacitor can induce balanced reference voltages [7, 8]. Switches controlled by the signal RSI^+ and RSI^- are closed for the positive and negative input signals, respectively.

During the AZ phase or an idle phase, the reference capacitor, C_{REF}, is charged by the reference voltage. For a stable storage of the charge due to the reference voltage, the capacitor C_{REF} must exhibit a low leakage current and the effect of stray capacitances appearing on the reference capacitor nodes must also be minimized. A capacitor in the microfarad range is required to prevent rollover error, which arises as a consequence of the difference in the reference voltage value for positive and negative input signals. The polarity of the input signal determined by the comparator at the end of the signal integration phase is used by the control logic to connect the capacitor C_{REF} to the input buffer such that the integrator output can return to the zero level. Hence, depending on the polarity of the analog input, the reference capacitor is used to generate either a positive or negative voltage during the reference signal integration phase.

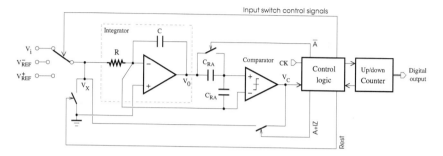

FIGURE 4.19
Circuit diagram of the dual-slope ADC with extra amplification and feedback circuitry.

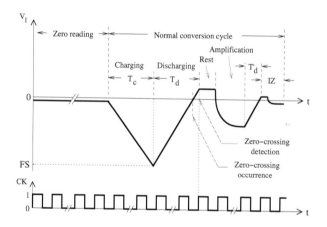

FIGURE 4.20
Integrator output during each conversion phase.

The block diagram of a dual-slope ADC including an amplification and feedback circuitry to improve the conversion accuracy [9] is depicted in Figure 4.19. The conversion phases are illustrated on the integrator output shown in Figure 4.20.

Initially, a conventional dual-slope integration consisting of an *input signal integration phase* and a *reference signal integration phase* is performed. The converter operates by integrating an unknown input analog signal for a predetermined time period, T_c, and then integrating a known reference signal until the integrator output reaches a predetermined zero level. Due to the comparator delay, the zero-crossing is actually detected on the first clock pulse after its occurrence, even if the integrator output reached the zero level within the clock period. The duration, T_d, of the reference signal integration phase measured by counting clock pulses is then larger than its true value. Because the integrator output continues to ramp down below the zero level until the end of the clock period, a residual voltage appears across capacitors C and C'_{RA}.

A *rest phase* is required to maintain constant the residual integrator charge, which is then scaled up by a given negative factor, $-k$, and fed back to the input node of the integrator amplifier in order to account for the measurement error on the phase duration during the *amplification phase*.

In general, the absolute value of the multiplication factor depends on the converter number system and is of the form, $k = 2^p$ for a binary system, where p is the resolution increase in number of bits and $k = 10^q$ for a decimal system, where q represents the resolution increase in number of digits. Typically, the residual voltage is multiplied by -8 in a binary converter or -10 in a decimal converter. In practice, the multiplication of the residual integrator charge is achieved by sizing capacitors C_{RA} and C'_{RA} such that

$$C'_{RA} = kC_{RA} \tag{4.46}$$

At the start of the amplification phase, the comparator output is fed back to the integrator input, the voltage across C_{RA} is zero, and the charge stored on C'_{RA} is

$$Q = C'_{RA}V_{res} = kC_{RA}V_{res} \tag{4.47}$$

where V_{res} is the residual voltage. To establish a virtual ground at the positive input of the comparator, a charge transfer is initiated between C'_{XR} and C_{XR}, ending with the induction of a voltage across the capacitor C_{XR} given by

$$V_{C_{RA}} = -Q/C_{RA} = -kV_{res} \tag{4.48}$$

The voltage across the capacitor C is also $-kV_{res}$ because $V_{C'_{RA}}$ is actually almost equal to zero.

The converter enters a second *reference signal integration phase*, where the capacitor C_{RA} is short-circuited again, and the feedback connection between the comparator output and integrator input is open. To keep the effect of the charge redistribution between the capacitors negligible, the capacitor C is designed to be much larger than the capacitor C'_{RA}. Thus, for this phase, the initial value of the voltage at the positive input of the comparator remains close to $-kV_{res}$.

A second *reference signal integration phase* is initiated. The time, T'_d, which is measured as the number of clock pulses required by the integrator output to cross the zero level again, is proportional to the residual error. The net count resulting from the subtraction in the same scale of T'_d from T_d represents the value of the time effectively needed by the integrator output to cross the zero level. This calibration scheme can be implemented using up-down counters incremented and decremented by the control logic.

The accuracy of the actual time measurement can be further improved by resorting to subsequent *rest, amplification,* and *reference signal integration phases.*

After the occurrence of an over-range condition, the comparator output and the input analog signal are of opposite polarity. The feedback connection realized between the comparator output and the integrator input during the *integrator output zero phase* forces the integrator output voltage to change in a direction such that its magnitude is decreased toward zero, thereby causing the discharge of the capacitor C.

In this approach, the zero reading, that is, the result of the conversion with the input voltage shorted to the ground, should be subtracted from each measurement. This helps minimize the effect of offset voltages on the conversion process.

For proper operation of a dual-slope ADC, the period of each clock pulse should be greater than the comparator delay. Hence, the conversion speed is primarily limited by the comparator delay, which is dependent on the converter overdrive, defined as the maximum integrator output voltage swing divided by the maximum number of clock pulses during the reference signal integration phase. By augmenting the number of times the rest, amplification, and reference signal integration phases are repeated, a decrease in the duration of each phase becomes feasible, thereby yielding an overall conversion with a faster speed.

4.1.3 Flash ADC

The most straightforward way to perform the N-bit analog-to-digital conversion is to compare the sampled-and-held version of an analog signal with 2^N reference voltages. The flash ADC, which generally consists of a resistive divider, comparators, and a binary encoder, is based on this principle. The foregoing track-and-hold (T/H) circuit samples the analog input and holds it for half a clock cycle. This operation can be performed by exploiting the inherent sampling properties of latched comparators. However, the use of a T/H circuit is preferred because the operation of the comparator array is often affected by clock skews, which can degrade converter linearity.

The block diagram of a 3-bit flash ADC is shown in Figure 4.21. In this case, $N = 3$ and $2^3 - 1 = 8$ comparators are used. Furthermore, the input signal is supposed to be positive. The reference voltage at the k-th node can be expressed as

$$V_k^- = V_{REF} \frac{R/2 + (k-1)R}{R/2 + (2^N - 2)R + 3R/2} = k\frac{V_{REF}}{2^N} - \frac{V_{REF}}{2^{N+1}} \qquad (4.49)$$

The difference between the reference voltages of two adjacent comparators is $V_{REF}/2^N$, that is,

$$\triangle = V_{k+1}^- - V_k^- = \frac{V_{REF}}{2^N} \qquad (4.50)$$

where $\triangle$ denotes the voltage level of an LSB. The output of a comparator will

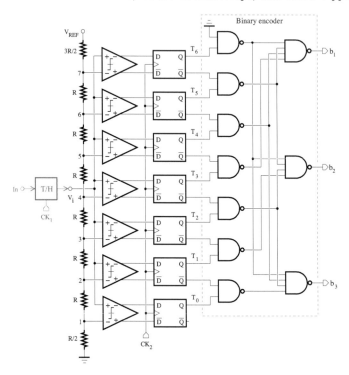

FIGURE 4.21
Block diagram of a 3-bit unipolar flash ADC.

TABLE 4.2
Thermometer Code for a 3-Bit Unipolar Flash ADC (with $\triangle = V_{REF}/8$)

	Input Range	Thermometer Code							Binary Code		
		T_6	T_5	T_4	T_3	T_2	T_1	T_0	b_1	b_2	b_3
7	$13\triangle/2 < V_i < 8\triangle$	1	1	1	1	1	1	1	1	1	1
6	$11\triangle/2 < V_i < 13\triangle/2$	0	1	1	1	1	1	1	1	1	0
5	$9\triangle/2 < V_i < 11\triangle/2$	0	0	1	1	1	1	1	1	0	1
4	$7\triangle/2 < V_i < 9\triangle/2$	0	0	0	1	1	1	1	1	0	0
3	$5\triangle/2 < V_i < 7\triangle/2$	0	0	0	0	1	1	1	0	1	1
2	$3\triangle/2 < V_i < 5\triangle/2$	0	0	0	0	0	1	1	0	1	0
1	$\triangle/2 < V_i < 3\triangle/2$	0	0	0	0	0	0	1	0	0	1
0	$0 < V_i < \triangle/2$	0	0	0	0	0	0	0	0	0	0

be at the high state if its input voltage is higher than the reference voltage. Otherwise, the comparator output is at the low state. The outputs of all comparators form a thermometer code that is then converted into a binary code by an encoder. Table 4.2 lists the thermometer and binary coding schemes for a 3-bit word. Note that the number of consecutive 1's in the thermometer code corresponds to the count of reference voltages less than the input signal. By choosing the bottom resistor in the comparator resistor string to be $R/2$, the transitions of the ADC transfer characteristic are at multiples of $\triangle/2$.

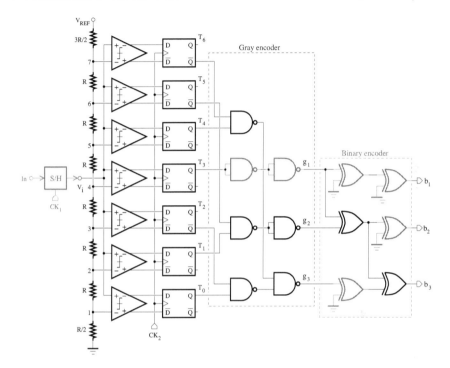

FIGURE 4.22
Block diagram of a flash ADC with improved encoder.

In practice, due to mismatches and imperfections in the reference resistor string and in the latched comparators, as well as high speed limitations, errors can be introduced in the thermometer code produced at the comparator outputs. They show up as bubbles,[1] which are zeros surrounded by ones, or vice versa. This may happen in the specific case where a comparator switches more slowly than expected, so that the output is latched before reaching the final state. When the bubble error cannot be detected by the digital encoder, the ADC will produce an output code not representative of the input signal value.

The effect of bubble errors can be reduced by first converting the thermometer code into the Gray code, which is then used for the computation of

---

[1]The use of the term "bubble" is justified by the fact that such a code error is analogous to a bubble occurring in the mercury of a thermometer.

TABLE 4.3
Encoder Truth Table ($\triangle = V_{REF}/8$)

	Thermometer Code							Gray Code			Binary Code		
	T_6	T_5	T_4	T_3	T_2	T_1	T_0	g_1	g_2	g_3	b_1	b_2	b_3
7	1	1	1	1	1	1	1	1	0	0	1	1	1
6	0	1	1	1	1	1	1	1	0	1	1	1	0
5	0	0	1	1	1	1	1	1	1	1	1	0	1
4	0	0	0	1	1	1	1	1	1	0	1	0	0
3	0	0	0	0	1	1	1	0	1	0	0	1	1
2	0	0	0	0	0	1	1	0	1	1	0	1	0
1	0	0	0	0	0	0	1	0	0	1	0	0	1
0	0	0	0	0	0	0	0	0	0	0	0	0	0

TABLE 4.4
Thermometer-to-Gray and Gray-to-Binary Encodings

$N = 3$	$g_1 = T_3$	$b_1 = g_1$
	$g_2 = T_1 \cdot \overline{T_5}$	$b_2 = g_2 \oplus b_1$
	$g_3 = T_0 \cdot \overline{T_2} + T_4 \cdot \overline{T_6}$	$b_3 = g_3 \oplus b_2$
$N = 4$	$g_1 = T_7$	$b_1 = g_1$
	$g_2 = T_3 \cdot \overline{T_{11}}$	$b_2 = g_2 \oplus b_1$
	$g_3 = T_1 \cdot \overline{T_5} + T_9 \cdot \overline{T_{13}}$	$b_3 = g_3 \oplus b_2$
	$g_4 = T_0 \cdot \overline{T_2} + T_4 \cdot \overline{T_6} + T_8 \cdot \overline{T_{10}} + T_{12} \cdot \overline{T_{14}}$	$b_4 = g_4 \oplus b_3$

the binary code. Gray encoding itself has no correction ability. But its tolerance to bubbles is due to the fact that only one bit changes between adjacent codes, leading to a small difference between the ideal and incorrect codes. The logic equations for the thermometer-to-Gray and Gray-to-binary encodings are summarized in Table 4.4 for the 3-bit and 4-bit flash ADCs. The circuit diagram of a 3-bit flash ADC including a Gray-code-based encoder is shown in Figure 4.22. In order to equalize the propagation delay of the different signal paths, additional NAND and XOR gates were inserted in the encoder according to the following Boolean algebraic identities:

$$A = \overline{\overline{A}} \qquad (4.51)$$

$$A = A \oplus 0 \qquad (4.52)$$

where A is a logic variable. Note that long wiring structures can be required to logically combined signals as the number of bits is increased, resulting in an irregular circuit layout. Furthermore, latches can be introduced between the logic stages to improve the operation speed and to reduce the metastability error probability.

Due to their regular structure, ROM-based encoders are preferred over gate-based encoders for converter implementations with a resolution greater than 5 bits. A ROM can consist of bit lines, word lines, and MOS-type memory cells. The storage of a logic 0 is carried out using an n-channel pull-down transistor, while no connection is required for the storage of a logic 1. A precharged logic is generally used to eliminate the static power dissipation and to help keep the pull-up and pull-down transistors as close as possible to the minimum size. However, the converter speed can be limited by the ROM precharge time. Ideally, the 1-out-of-2^N code obtained by detecting the location of the 1-to-0 transition in the thermometer code is commonly used to enable a single word line of the ROM. Hence, only the bit representing the location of the detected transition can take the logic 1 and all the remaining bits are at the logic 0. However, if bubbles exist in the thermometer code, there will be multiple 1-to-0 transition points and the 1-out-of-2^N encoder will select more than one ROM line. As a result, errors are introduced in the binary output, which is actually the representation of the bitwise logical OR between the ROM lines.

An approach for the bubble correction consists of using a majority logic function [10] whose output takes the same logic state as the greater number of inputs. The block diagram of the 3-bit flash ADC with the bubble error correction based on a majority logic (ML) is illustrated in Figure 4.23. With the assumption that $T_{-1} = 1$ and T_7 represents the over-range signal, the corrected thermometer code, T_i^*, is given by the Boolean equation,

$$T_i^* = T_{i-1} \cdot T_i + T_i \cdot T_{i+1} + T_{i-1} \cdot T_{i+1} \tag{4.53}$$

where $i = 0, 1, \cdots, 6$. The state of a given bit in the thermometer code is then determined by one of its two neighbors. The ML encoder can efficiently suppress the bubbles, which affect the 1-0 transition in the thermometer code. However, the power consumption and chip area can be increased because the required number of gates tends to be high. The ROM line selection can be carried out using a thermometer to 1-out-of-2^N encoder based on two-input AND gates. This simple encoder will operate correctly only if there is no bubble error in the thermometer code.

In general, the occurrence likelihood of bubbles becomes less important as their number becomes greater. Because the bubbles are mainly introduced near the one-to-zero transition point in the thermometer code, the encoder can be designed to correct only certain single bubbles. Figure 4.24 shows the block diagram of the 3-bit flash ADC with the bubble error correction based on three-input AND gates [11]. The correction is limited to bubble errors, which do not affect the 1-to-00 transition in the thermometer code. The possible states of the different signals in response to a given input are shown in Table 4.5 for the 3-bit flash ADC. The appropriate word line in the Gray ROM encoder is activated by the signals t_i, which are given by the

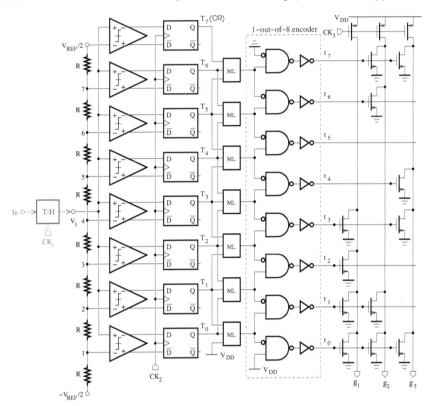

FIGURE 4.23
Block diagram of a 3-bit flash ADC with the bubble error correction based on a majority logic (OR: over-range).

following Boolean equation,

$$
t_i = \begin{cases}
T_6 \cdot \overline{T_7} & \text{for} \quad i = 7 \\
T_{i-1} \cdot \overline{T_i} \cdot \overline{T_{i+1}} & \text{for} \quad i = 1, 2, \cdots, 6 \\
\overline{T_0} \cdot \overline{T_1} & \text{for} \quad i = 0
\end{cases} \tag{4.54}
$$

where T_i denotes a bit of the thermometer code. Note that T_7 represents the over-range signal generated as a result of the comparison between the input signal and $V_{REF}/2$.

The fundamental difference between the ML and three-input AND encoders can be illustrated by comparing the ideal thermometer code and error-correction results of the examples included in Table 4.6, where the first T_i column represents the ideal case and bold characters are used to show the bits affected by bubble errors. The ML encoder relies on the best-expectation principle and the detection of the 1-to-0 transition, while the three-input AND

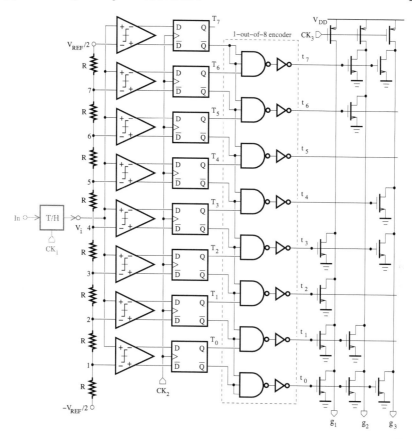

FIGURE 4.24
Block diagram of a 3-bit flash ADC with the bubble error correction based on
three-input AND gates.

encoder is based on the detection of the 1-to-00 transition. In both cases, the
error correction cannot be ensured for all bubble types.

An N-bit flash ADC generally requires at least $2^N - 1$ comparators, whose
reference voltages are set by a resistor string. The value of each comparator
output depends on whether or not the input voltage exceeds the corresponding
reference voltage. A set of all comparator outputs, which form the so-called
thermometer code, is first scaled to the appropriate logic levels by latches
in order to mitigate the metastability problem and then converted into a
form of binary data. The resulting data with N-bit resolution are applied to
the output buffer, which can be implemented using D latches. Note that an
additional comparator is often used to indicate the presence of a signal over-
range. Because the metastability, which is characterized by an output state
between the logic level high and logic level low, is due to the violation of
setup and hold time, it can be reduced by allowing more time for comparator

TABLE 4.5

Truth Table of the 3-Bit Bipolar Flash ADC with a Three-Input AND Encoder

Input Range	Thermometer Code T_6 T_5 T_4 T_3 T_2 T_1 T_0							1-out-of-8 Code t_7 t_6 t_5 t_4 t_3 t_2 t_1 t_0								Gray Code g_1 g_2 g_3		
$3\Delta < V_i < 4\Delta$	1	1	1	1	1	1	1	1	0	0	0	0	0	0	0	1	0	0
$2\Delta < V_i < 3\Delta$	0	1	1	1	1	1	1	0	1	0	0	0	0	0	0	1	0	1
$\Delta < V_i < 2\Delta$	0	0	1	1	1	1	1	0	0	1	0	0	0	0	0	1	1	1
$0 < V_i < \Delta$	0	0	0	1	1	1	1	0	0	0	1	0	0	0	0	1	1	0
$-\Delta < V_i < 0$	0	0	0	0	1	1	1	0	0	0	0	1	0	0	0	0	1	0
$-2\Delta < V_i < -\Delta$	0	0	0	0	0	1	1	0	0	0	0	0	1	0	0	0	1	1
$-3\Delta < V_i < -2\Delta$	0	0	0	0	0	0	1	0	0	0	0	0	0	1	0	0	0	1
$-4\Delta < V_i < -3\Delta$	0	0	0	0	0	0	0	0	0	0	0	0	0	0	1	0	0	0

TABLE 4.6

Illustration of Three Correction Examples of Bubbles in the ML and Three-Input AND Encoders

		ML Encoder									Three-input AND Encoder					
		Case 1			Case 2			Case 3			Case 1		Case 2		Case 3	
i	T_i	T_i	T_i^*	t_i	T_i	T_i^*	t_i	T_i	T_i^*	t_i	T_i	t_i	T_i	t_i	T_i	t_i
7	0	0	0	0	0	0	0	0	0	0	0	0	0	0	0	0
6	0	0	0	0	0	0	0	0	0	0	0	0	0	0	0	0
5	0	0	0	0	0	0	0	0	0	0	0	1	0	0	0	0
4	0	**1**	0	1	0	0	0	0	0	0	**1**	0	0	1	0	**1**
3	1	**0**	1	0	1	0	1	1	0	0	0	0	1	0	**1**	0
2	1	1	1	0	**0**	1	0	**0**	0	0	1	0	**0**	0	**0**	0
1	1	1	1	0	1	1	0	**0**	0	1	1	0	1	0	**0**	1
0	1	1	1	0	1	1	0	1	1	0	1	0	1	0	1	0

regeneration. The speed of the overall structure depends on the comparator speed and the propagation delay of the digital section. But with the use of pipeline latches, the comparison and code conversion can ideally be achieved in a single clock phase allowing the flash ADC to operate at higher frequencies.

The block diagram of a flash ADC based on the above principle is shown in Figure 4.25. For a given analog input voltage, the thermometer code should ideally exhibit only a single 1-to-0 transition. But in practice, the converter components may be subject to nonidealities such as finite bandwidth, noise, and mismatches, leading to the introduction of bubbles in the thermometer code. This problem can be alleviated by using an intermediate Gray encoding stage instead of converting directly from the thermometer code to the binary

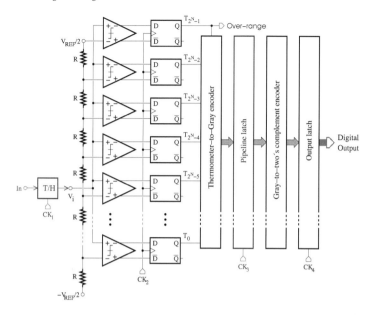

FIGURE 4.25
Block diagram of a flash ADC.

code. Furthermore, the robustness to bubbles can be enhanced by inserting an adequate error correction stage before the digital encoder.

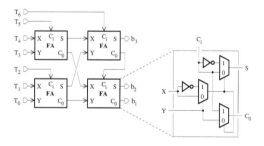

FIGURE 4.26
Tree encoder based on ones-counter for a 3-bit flash ADC.

Note that the output generated by a thermometer-to-binary encoder is a binary representation of the number of 1s in the thermometer code. Hence, a ones-counter based on full adders can also be used as a digital encoder. Figure 4.26 shows the block diagram of a tree encoder for a 3-bit flash ADC. This approach is insensitive to bubbles, which maintain constant the overall number of 1s in the thermometer code. However, the propagation delay of the tree encoder increases linearly with the converter resolution. Bit-level pipelining, which consists of inserting a register between logic blocks, can then be used to reduce the critical path. This will incur an increase in the hardware overhead and power consumption.

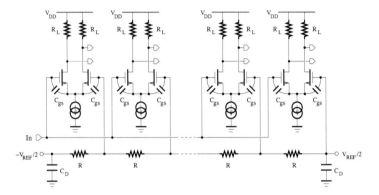

FIGURE 4.27
Equivalent circuit model of the flash ADC input stage.

High-resolution flash ADCs generally require a large number of comparators. This can result in a large input capacitor, considerable chip area and very high power consumption. Figure 4.27 shows the equivalent circuit model of the flash ADC input stage, where C_D is used to decouple the reference voltages. The gate-source capacitors, C_{gs}, of the transistors in the comparator input stage form a parasitic capacitor between the comparator inputs. The input signal is then capacitively coupled to the resistive reference network, thereby leading to a variation in the reference voltages. The total resistance of the resistive reference network should be kept low enough to maintain the maximum feedthrough error lower than 1 LSB. This requirement can be relaxed by dividing the resistive reference network into subsections with an adequate size using decoupling capacitors, especially in cases where there is an important increase in the power consumption.

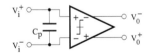

FIGURE 4.28
Comparator with a parasitic capacitor between the inputs.

At high frequencies, the input parasitic capacitor of the comparator, as illustrated in Figure 4.28, couples the input signal with the resistive network used for the generation of the reference voltages. To avoid a variation in reference voltages, which can affect the comparator threshold levels, the maximum resistance of the resistive network must be estimated using the equivalent circuit model shown in Figure 4.29, where it is assumed that

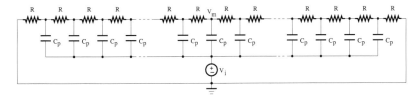

FIGURE 4.29
Equivalent circuit model for the maximum resistance derivation in the flash ADC.

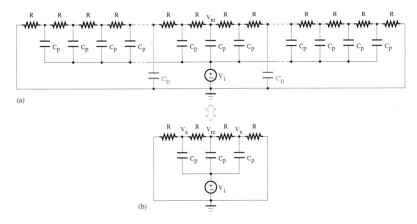

FIGURE 4.30
(a) Equivalent circuit model with decoupling capacitors; (b) equivalent circuit model of a subsection of the resistive reference network.

the high and low reference voltages are fully decoupled. For a flash ADC with a resolution of N bits, we have

$$R = R_T/2^N \tag{4.55}$$

and

$$C_p = C_T/2^N \tag{4.56}$$

where R_T denotes the total resistance of the resistive network and C_T is the total parasitic capacitance. Generally, the resulting resistance is low for high-speed ADCs, and the power consumption of the reference network, which is of the form V_{REF}^2/R_T, is increased. With the use of decoupling capacitors as shown in Figure 4.30(a), where C_D' is the decoupling capacitor, the total resistance of the resistive network can be increased without deteriorating the level of the feedthrough error. The effect of the decoupling is to split the equivalent circuit model required for the worst-case estimation of R_T into q almost identical subcircuits. Simulations of the equivalent subcircuit depicted in Figure 4.30(b) show that the voltage at the middle node, V_m, is most affected by the input signal feedthrough error. Using Kirchhoff's current law, the next node equations can be obtained,

$$(V_i - V_m)sC_p = 2\frac{V_m - V_x}{R} \tag{4.57}$$

$$(V_i - V_x)sC_p = \frac{V_x}{R} + \frac{V_x - V_m}{R} \tag{4.58}$$

Combining Equations (4.57) and (4.58), we get

$$\frac{V_m}{V_i} = \frac{(sRC_p)^2 + 4sRC_p}{(sRC_p)^2 + 4sRC_p + 2} = \frac{(sR_TC_T)^2 + 2^{2N+2}sR_TC_T}{(sR_TC_T)^2 + 2^{2N+2}sR_TC_T + 2^{4N+1}} \tag{4.59}$$

Let f be the frequency of the input voltage V_i. With the assumption that $s = j2\pi f$ and $\pi f R_T C_T \ll 1$, we can find the following expression:

$$\left|\frac{V_m}{V_i}\right| \simeq \frac{\pi}{4}f R_T C_T \tag{4.60}$$

In the worst case, the voltages V_m and V_i can be expressed in LSB units, leading to

$$\left|\frac{V_m}{V_i}\right| = q\frac{k}{2^N} \tag{4.61}$$

where k represents the feedthrough level and q is the decoupling period. The maximum resistance is then defined by

$$R_T \leq q\frac{4k}{\pi f C_T 2^N} \tag{4.62}$$

Because q is greater than 1, the decoupling capacitors will increase the total resistance by a factor of q.

Furthermore, the attainable resolution is limited by the accuracy of the reference voltages and comparator offset voltage. Typically, for a 10-bit converter with 2 V_{p-p} input signal, the comparator should resolve less than about 2 mV. This requirement is difficult to meet in CMOS technology, and various techniques for reducing the offset-voltage effects such as chopper stabilization and auto-zeroing result in an increase in the power consumption and a reduction in the conversion speed. Therefore, a good compromise may be to design a flash ADC for applications requiring less than 8 bits for the data representation.

Note that flash ADCs can also be implemented without the input T/H circuit, which is generally based on open-loop structures in high-speed applications. In this case, a higher dynamic performance is required for comparators.

4.1.4 Averaging ADC

Device mismatches often limit the differential and integral nonlinearity characteristics and the resolution of classical flash ADC architectures. Averaging techniques can be used to minimize the mismatch effect.

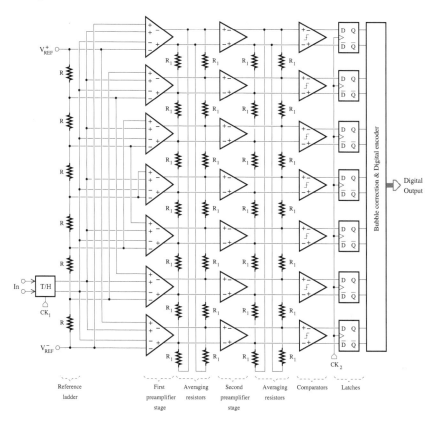

FIGURE 4.31
Circuit diagram of a flash ADC with averaging stages.

Figure 4.31 shows the circuit diagram of a flash ADC with two averaging stages. The inherent symmetry of differential circuits is exploited to use the same number of reference nodes as in single-ended structures. The mirror-image preamplifiers with respect to the reference midpoint can then be connected oppositely to reference voltage nodes. The requirement of maintaining the gain high enough in order to reduce the effect of the input-referred offset on the subsequent stages can be met by cascading two stages of preamplification and averaging. By inserting lateral resistors between the outputs of neighboring preamplifier stages, the random component of currents is reduced and the effect of offsets on the ADC performance is attenuated [12]. The offset reduction is dependent on the value of the preamplifier output resistance, R_0, and the averaging resistors, R_1. Using the superposition principle with the consideration that the signal and reference voltage sources are disconnected, the inputs of the first stage of preamplifiers are reduced to the offset voltages.

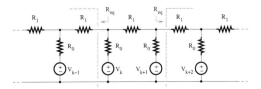

FIGURE 4.32
Equivalent circuit of the preamplifier array with an averaging resistor network.

Figure 4.32 shows the equivalent circuit of a preamplifier array with an averaging resistor network supposed to be infinite. The equivalent resistance, R_{eq}, seen to the right and to the left of each output node is given by

$$R_{eq} = R_1 + R_0 \parallel R_{eq} \Rightarrow R_{eq} = \frac{1}{2}R_1 + \sqrt{\frac{1}{4}R_1^2 + R_1 R_0} \qquad (4.63)$$

In practice, the length of the averaging resistive network is limited and the equivalent resistance, especially at the termination nodes, may differ from the above value.

The preamplifier transfer characteristics are illustrated in Figure 4.33(a). The zero-crossing points of the preamplifiers at the array edges are shifted inward. This leads to undesirable variations in the INL mean value near the two edges of the preamplifier array, as illustrated in Figure 4.33(b). Furthermore, the standard deviation of the offset voltage is larger for preamplifiers at the boundaries than the one at the center.

To avoid any discontinuity at the edges of the averaging network, over-range dummy preamplifiers, whose outputs remain unused, are often added at each edge of the array. The complementary outputs of the first and last over-range preamplifiers are cross-connected through a pair of resistors to ensure that every preamplifier in the array sees the same effective load resistance and has a balanced number of preamplifiers contributing to its output [13]. A 6-bit ADC, for instance, includes an array of 63 preamplifiers, and about 9 dummy preamplifiers are required at each array end to create a linear behavior at the edges of the input full scale. However, the addition of dummy preamplifiers results in a reduction in the available headroom for the input voltage, and an increase in the input capacitance and power dissipation.

In the specific case of resistor-loaded preamplifiers, the number of dummy preamplifiers can be reduced by resizing the resistors at either edge of the averaging network, as shown in Figure 4.34 [14], where the index k is used for the lowermost or uppermost in-range preamplifier. Provided $R_1 > R_0$, the electrical load symmetry at the averaging network edges can be restored without resorting to the use of dummy preamplifiers by changing the first and

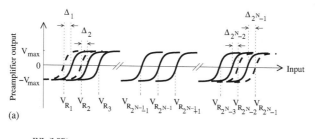

(a)

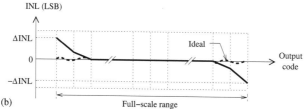

(b)

FIGURE 4.33
(a) Preamplifier transfer characteristics; (b) effect of the asymmetrical preamplifier boundaries on the converter linearity.

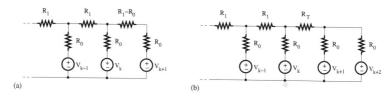

(a)   (b)

FIGURE 4.34
Equivalent circuits of one edge of the averaging network with a resized termination resistor: (a) Case $R_1 > R_0$, (b) case $R_1 < R_0$.

last resistances to $R_1 - R_0$. In the cases where $R_1 < R_0$, the first and last averaging resistors will be short-circuited and the value of the next resistors will be modified as follows:

$$
R_T = \begin{cases}
\dfrac{3R_1 - R_0}{2} & \text{if} \quad R_1 < 3R_0 \\[2mm]
2R_1 - \dfrac{R_0}{3} & \text{if} \quad R_1 < 6R_0
\end{cases}
\tag{4.64}
$$

Otherwise, additional preamplifiers with short-circuited averaging resistors will be connected at the network edges to reduce the equivalent load resistance so that a positive value can be found for R_T.

For a given preamplifier, the magnitude of offset reduction provided by the averaging network depends on the ratio R_0/R_1 and the number of neighboring preamplifiers operating in the nonsaturated region. By increasing the ratio R_0/R_1 to improve the offset reduction, the overall output resistance seen by the preamplifiers can be reduced, leading to a decrease in the gain and an increase in the input-referred offset for the subsequent stages (amplifiers, or comparators). On the other hand, an augmentation in the number of nonsaturated preamplifiers can also improve the offset reduction and can contribute to a substantial increase in the preamplifier gain. Because this can be achieved by stepping up the overdrive voltage, there is a great repercussion on the power consumption. In practice, it is necessary to use a high number of nonsaturated preamplifiers and a low value of the ratio R_0/R_1 to maintain the effects of resistor mismatches to an acceptable level. At least a two times reduction in the offset voltage of a flash converter can be achieved by choosing the ratio R_0/R_1 between 0.5 and 1. The use of edge termination resistors does not significantly increase the power consumption and area of the converter. However, it can effectively mitigate the effect of preamplifier offset voltages only when the averaging window is narrow and the specification of matching termination resistors is less stringent.

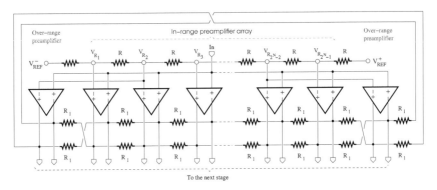

FIGURE 4.35
Preamplifier array with the averaging resistor network based on the triple cross-connection.

An alternative low-power and low-area design solution is to use an averaging resistor network based on the triple cross-connection [15], as shown in Figure 4.35. The effect of zero-crossing shifts can be compensated by introducing a cross-connection and an over-range preamplifier at each boundary. A proper termination can be realized due to the symmetry provided by the cross-connection at the center. To counteract the reduction in the effective transconductance at the boundary due to the negative transconductance of the over-range preamplifier, the over-range preamplifier should be designed such that its input linear region extends along that of the adjacent in-range preamplifier. Furthermore, it is desirable that the over-range preamplifier operate with the same reference voltage as the penultimate preamplifier.

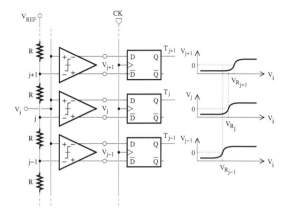

FIGURE 4.36
A section of the processing path in a flash ADC.

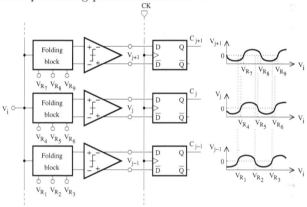

FIGURE 4.37
A section of the processing path in a folding ADC.

4.1.5 Folding and interpolating ADC

An analog preprocessing structure, which performs the folding and interpolation operations, can be used to reduce, respectively, the number of comparators and preamplifiers needed in the flash ADC [16–19]. Figure 4.36 shows a section of the processing path in a flash ADC. For N-bit conversion, the input signal should be compared with at least $2^N - 1$ voltage reference levels and a different comparator is used for each possible digital code. By introducing folding stages, as shown in Figure 4.37 for three reference voltages, the locations of the zero-crossings in the signals now determine the cyclic code transitions to be identified by comparators. The folding factor F (here $F = 3$) indicates the number of zero-crossings included in the transfer characteristic.

One comparator detects F reference voltages rather than one, as is the case in the full flash architecture, and the reference voltages of the folding amplifier should be sufficiently far apart to ensure a proper operation of comparators.

TABLE 4.7
Thermometer and Circular Codes of Numbers from 0 to 3

Circular Code				Thermometer Code			
C_3	C_2	C_1	C_0	T_2	T_1	T_0	
0	0	0	0	0	0	0	0
0	0	0	1	0	0	1	1
0	0	1	1	0	1	1	2
0	1	1	1	1	1	1	3
1	1	1	1	0	0	0	0
1	1	1	0	0	0	1	1
1	1	0	0	0	1	1	2
1	0	0	0	1	1	1	3

Let us consider a folding ADC including four folding amplifiers, whose reference voltages are chosen as follows:

$$F_1 : \quad (1/16)V_{REF} \quad (5/16)V_{REF} \quad (9/16)V_{REF} \quad (13/16)V_{REF}$$
$$F_2 : \quad (2/16)V_{REF} \quad (6/16)V_{REF} \quad (10/16)V_{REF} \quad (14/16)V_{REF}$$
$$F_3 : \quad (3/16)V_{REF} \quad (7/16)V_{REF} \quad (11/16)V_{REF} \quad (15/16)V_{REF}$$
$$F_4 : \quad (4/16)V_{REF} \quad (8/16)V_{REF} \quad (12/16)V_{REF} \quad (16/16)V_{REF}$$

The first cycle of the circular code, which is produced at the comparator outputs, is shown in Table 4.7. The equivalent thermometer code can be derived using a circular-to-thermometer encoder based on the following Boolean equations:

$$T_2 = C_2 \oplus C_3 \quad T_1 = C_1 \oplus C_3 \quad \text{and} \quad T_0 = C_0 \oplus C_3 \qquad (4.65)$$

and implemented using XOR gates. There are two folds in one cycle. With a folding factor of 4 as in this example, the input full range is divided into two cycles or four folds. Because the folding characteristic is redundant, a cycle pointer is generally necessary to resolve the ambiguity in the output code.

A high degree of folding can result in a decrease in the signal bandwidth, which is caused by the presence of parasitic capacitors at the folder output nodes. In practice, F is then limited to eight and more quantization levels can be obtained using a parallel configuration of folding blocks. The necessary number of folding amplifiers or folders can still be high. The amplifier number is reduced using an interpolator with the factor I between two consecutive folding blocks. The interpolation is based on the signal division and can be realized by a resistor ladder, as illustrated in Figure 4.38, where we have $V_{R4} = (V_{R1} + V_{R7})/2$, $V_{R4} = (V_{R2} + V_{R8})/2$, and $V_{R6} = (V_{R3} + V_{R9})/2$.

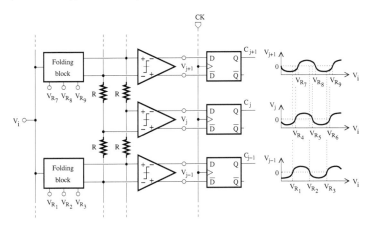

FIGURE 4.38

A section of the processing path in a folding and interpolating ADC.

The folding operation relies on mapping the input waveform into a repetitive signal, whose frequency is multiplied by the folding factor. Here, the amplitude quantization is transformed into the detection of zero-crossings of the folding signal. Figure 4.39 shows the circuit diagram of the j-th folding stage, where F is the folding factor. The output of the odd- and even-numbered differential transistor pairs are cross-coupled. The input reference voltages are defined by a resistive network. The polarity of the signal changes each time the input voltage attains a reference level.

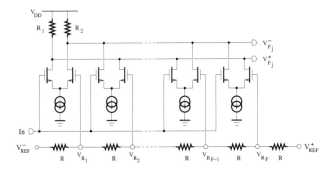

FIGURE 4.39

Implementation of a folding amplifier.

The interpolation can be realized using passive elements, and active elements and components. The circuit diagram of an interpolation circuit with a differential structure is shown in Figure 4.40(a). The voltage division is

realized by a resistor ladder driven by nMOS source followers [19]. The interpolated signals are denoted by V_{i_k}, where $k = 1, 2, \cdots, I + 1$, and I is the
interpolation factor. The output signals of two neighboring folding stages can
also be interpolated as shown in Figure 4.40(b). Due to the fact that only the
zero-crossings contain useful information, an accurate gain is not required for
any amplifier of the interpolation circuit. The interpolation amplifiers can simply consist of differential transistor pairs, which can operate with a low supply
voltage. Here, the interpolation by a factor of the form $I = 2^p$ is achieved by
cascading p amplifier sections and $2^q + 1$ amplifiers are required in the q-th
section.

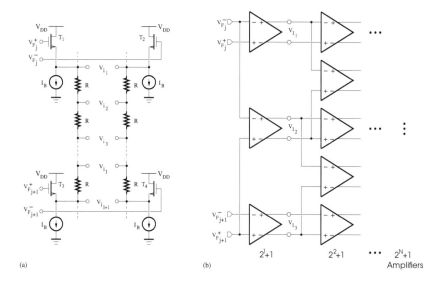

FIGURE 4.40
Implementations of interpolating circuits using (a) resistors and (b) differential
amplifiers.

Note that current-division techniques can also be exploited in order to
implement the interpolation stage. This approach relies on the use of current
comparators or I-to-V converters and can have the drawback of reducing the
bandwidth of the folder circuit due to the extra node, which can be introduced
in the signal path.

In the case of interpolations by a factor higher than 2, a systematic error
is introduced in the position of zero-crossings due to the effect of the nonlinear transfer characteristic of folding amplifiers on the signal values near
the boundaries of the input range. Figure 4.41 shows signals obtained as a
result of the interpolation by a factor of 4. Because of the symmetry of the
transfer characteristic, the zero-crossing location of the interpolated signal
in the middle remains unaffected. In comparison with the ideal case, where
the zero-crossings should be uniformly distributed, the other two interpolated

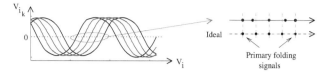

FIGURE 4.41
Illustration of the input stage nonlinearity effects on interpolated signals.

zero-crossings tend to move outward. This can be attributed to the amplitude mismatch between the folding signals and the interpolated signals. One way to restore the ideal zero-crossing points can be to extend the operation range of the folding amplifier.

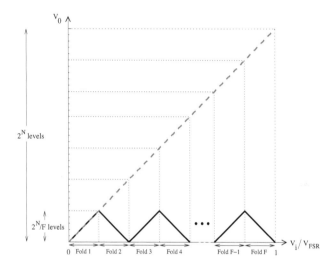

FIGURE 4.42
Transfer characteristic of an ideal folding ADC.

Ideally, the input-output characteristic of the folding amplifier is periodic and consists of piece-wise linear segments, or folds, whose number is related to the folding factor. As this leads to a repetitive code at the comparator output, a cycle pointer or coarse converter is needed in addition to a fine converter. The coarse converter ascertains in which period of the folding amplifier transfer characteristic the input signal lies. The transfer characteristic of a converter based on the folding principle is illustrated in Figure 4.42, where the input range is assumed to be divided into F regions or folds. Note that two folds constitute a folding cycle. An N-bit full flash ADC requires 2^N quantization levels, while a folding ADC needs $2^N/F$ levels, allowing the use of the folding signal signs for the determination of the LSBs and F levels for the generation of the MSBs of the output code.

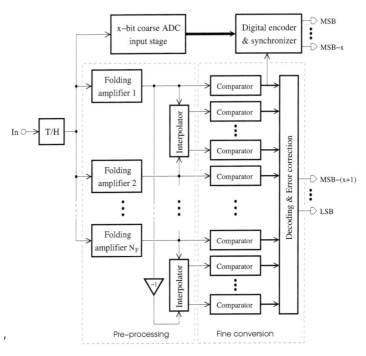

FIGURE 4.43
Block diagram of a folding and interpolating ADC.

The folding and interpolating converter, as shown in Figure 4.43, consists of two ADCs operating in parallel. The coarse ADC quantizes the input signal and provides the MSBs, while the fine ADC, which includes an analog preprocessing stage, is used for the generation of the LSBs. The outputs at the boundaries of the resistor interpolation structure are generally cross-connected to alleviate the effects caused by the asymmetrical nature of the network edges. As a result, the translational symmetry of the impulse response of the resistor network is preserved. The mitigation of the delay effect introduced by the analog preprocessing between the MSBs and the other bits is achieved by a bit-synchronization block. Otherwise, a misalignment of the different bits can be observed for high-frequency inputs. The decoder transforms the comparator output signals into a binary code.

The total resolution of the folding ADC is given by

$$N = N_{MSB} + N_{LSB} \qquad (4.66)$$

where N_{MSB} and N_{LSB} are the numbers of bits resolved in the coarse and fine converters, respectively. Let F be the folding factor, I denote the interpolation factor, N_F represent the number of primary folding signals, and N_I be the

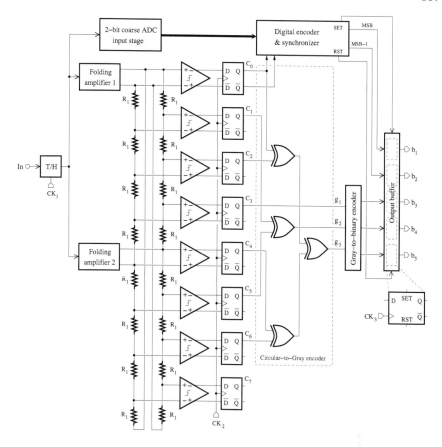

FIGURE 4.44
Block diagram of a 5-bit folding and interpolating ADC.

total number of interpolated signals [20]. We have

$$F = 2^{N_{MSB}} \tag{4.67}$$

and

$$N_I = 2^{N_{LSB}} \tag{4.68}$$

while the total number of folding amplifier is computed as the ratio between the total number of primary folds and the interpolation factor,

$$N_F = N_I/I \tag{4.69}$$

It can be shown that $N_F F = 2^N/I$. In practice, F and I are assumed to be a power of two. For a given resolution, the choice of F, I, and N_F is determined by the trade-off to be made between the bandwidth, speed, and power consumption of the converter.

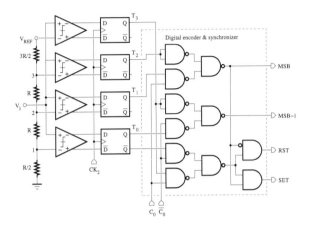

FIGURE 4.45
Block diagram of the coarse ADC.

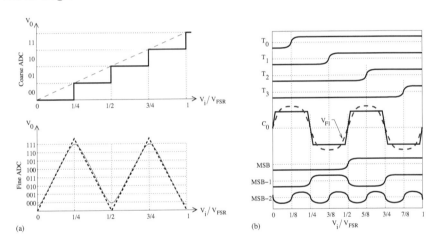

FIGURE 4.46
(a) Transfer characteristic of a 5-bit folding and interpolating ADC; (b) illustration of the synchronization between the coarse ADC bits and MSB-2.

The numbers of comparators in flash and folding ADCs are summarized in Table 4.8. For the same resolution, the flash ADC requires more comparators than the folding ADC. The use of the folding technique then results in an important reduction in the comparator number as the converter resolution is increased.

A 5-bit full flash ADC includes at least thirty-one comparators. Using a folding and interpolating architecture based on a 2-bit coarse ADC and a 3-bit fine ADC, as shown in Figure 4.44, the number of comparators can be reduced, leading to a decrease in chip area and power consumption [21]. In this case,

TABLE 4.8

Number of Comparators in Flash ADC and Folding ADC versus the Converter Resolution

	Number of Comparators			
	5 bits	6 bits	7 bits	8 bits
Flash ADC	31	63	127	255
Folding ADC with a 2-bit coarse ADC	11	19	35	67
Folding ADC with a 3-bit coarse ADC	11	15	23	39

the reference voltages for the folding amplifier F_1 and F_2 are, respectively, as follows:

$$F_1: \quad (1/32)V_{REF} \quad (9/32)V_{REF} \quad (17/32)V_{REF} \quad (25/32)V_{REF}$$
$$F_2: \quad (5/32)V_{REF} \quad (13/32)V_{REF} \quad (21/32)V_{REF} \quad (29/32)V_{REF}$$

Only two primary folded signals are generated and the remaining ones are recovered by interpolation. By using two interpolators with a factor of 4, the fine ADC required eight comparators. The output code provided by these comparators is repeated for every sixteen quantization levels and the overall input range of the converter involves two folding cycles. Taking into account the four comparators required in the coarse ADC, which is based on the flash architecture, as shown in Figure 4.45, the total comparator count for the folding and interpolating converter is twelve.

The transfer characteristic of the 5-bit folding and interpolating ADC is shown in Figure 4.46(a). Due to the finite slew rate of the folding amplifier, the triangular characteristic is approximated by a sinusoidal-like waveform. The comparator outputs of the fine ADC form a circular code, which can be converted into Gray representation using XOR gates. The circular-to-Gray encoder is based on the next Boolean expressions:

$$g_1 = C_3 \tag{4.70}$$
$$g_2 = C_5 \oplus C_1 \tag{4.71}$$
$$g_3 = (C_6 \oplus C_4) \oplus (C_2 \oplus C_0) \tag{4.72}$$

Note that the bit C_7, which is not actually connected to the encoder, can be useful in the case where the circular code must be represented as a thermometer code.

Note that the transition between a group of 1s and a group of 0s in the circular code can also be detected using two-input XOR gates, whose Boolean equations are expressed as follows:

$$c_j = \begin{cases} \overline{C_0 \oplus C_7}, & \text{if} \quad j = 0 \\ C_j \oplus C_{j-1}, & \text{if} \quad j = 1, 2, \cdots, 7. \end{cases} \tag{4.73}$$

The XOR gate outputs, c_j, are then to be applied to a ROM structure for the conversion to Gray or binary representation.

For the coarse ADC, which is based on the flash architecture, the comparator outputs constitute a thermometer code to be converted into the binary representation. In addition to the minimum of three comparators required by a 3-bit flash converter, the coarse ADC also includes a comparator for the generation of the over-range signal. Ideally, the bit transitions of the coarse and fine ADCs should be exactly synchronized, as illustrated in Figure 4.46(b) for the 3 MSBs. However, this is not the case in practice because the coarse and fine ADCs operate independently and exhibit different delays. A bit-synchronization section is associated with the digital encoder of the coarse ADC to prevent errors from occurring in the output code, as shown in Figure 4.47. It should be noted that the critical regions are located near the MSB transitions. The coarse ADC bits can then be expressed as

$$MSB = C_0 \cdot T_1 + T_2 \tag{4.74}$$

$$MSB - 1 = \overline{C_0} \cdot T_0 + T_3 \tag{4.75}$$

where C_0 is obtained from the fine ADC and is used as a bit-synchronization signal.

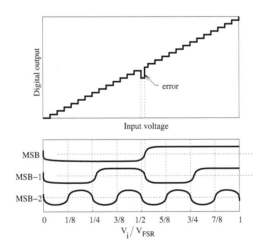

FIGURE 4.47
Effect of the misalignment between the coarse ADC bits and MSB-2.

For proper operation of the folding and interpolating ADC, the output code should be set to the maximum or minimum value, respectively, provided the input voltage is greater than the highest reference voltage or lower than the lowest reference voltage. However, the MSBs saturate at the end of the input range while the LSBs wrap around due to the circular nature of the folding signals. For instance, the code 00000 is changed to 00111 as the input voltage decreases; and with the input voltage increasing, the code 00111 becomes

TABLE 4.9
Circular and Binary Codes for the 5-Bit Folding and Interpolating ADC

Circular Code								Binary Code				
C_7	C_6	C_5	C_4	C_3	C_2	C_1	C_0	MSB	MSB-1	MSB-2	MSB-3	LSB
0	0	0	0	0	0	0	0	0	0	0	0	0
0	0	0	0	0	0	0	1	0	0	0	0	1
0	0	0	0	0	0	1	1	0	0	0	1	0
0	0	0	0	0	1	1	1	0	0	0	1	1
0	0	0	0	1	1	1	1	0	0	1	0	0
0	0	0	1	1	1	1	1	0	0	1	0	1
0	0	1	1	1	1	1	1	0	0	1	1	0
0	1	1	1	1	1	1	1	0	0	1	1	1
1	1	1	1	1	1	1	1	0	1	0	0	0
1	1	1	1	1	1	1	0	0	1	0	0	1
1	1	1	1	1	1	0	0	0	1	0	1	0
1	1	1	1	1	0	0	0	0	1	0	1	1
1	1	1	1	0	0	0	0	0	1	1	0	0
1	1	1	0	0	0	0	0	0	1	1	0	1
1	1	0	0	0	0	0	0	0	1	1	1	0
1	0	0	0	0	0	0	0	0	1	1	1	1
0	0	0	0	0	0	0	0	1	0	0	0	0
0	0	0	0	0	0	0	1	1	0	0	0	1
0	0	0	0	0	0	1	1	1	0	0	1	0
0	0	0	0	0	1	1	1	1	0	0	1	1
0	0	0	0	1	1	1	1	1	0	1	0	0
0	0	0	1	1	1	1	1	1	0	1	0	1
0	0	1	1	1	1	1	1	1	0	1	1	0
0	1	1	1	1	1	1	1	1	0	1	1	1
1	1	1	1	1	1	1	1	1	1	0	0	0
1	1	1	1	1	1	1	0	1	1	0	0	1
1	1	1	1	1	1	0	0	1	1	0	1	0
1	1	1	1	1	0	0	0	1	1	0	1	1
1	1	1	1	0	0	0	0	1	1	1	0	0
1	1	1	0	0	0	0	0	1	1	1	0	1
1	1	0	0	0	0	0	0	1	1	1	1	0
1	0	0	0	0	0	0	0	1	1	1	1	1

00000. An over-range and under-range detection mechanism is required to set or reset the output latches. The signals to detect the over-range and under-range conditions are given by

$$RST = \overline{MSB} \cdot \overline{IR} \tag{4.76}$$

and

$$SET = MSB \cdot \overline{IR} \qquad (4.77)$$

where

$$IR = (C_0 + T_0) \cdot (\overline{C_0} + \overline{T_3}) \qquad (4.78)$$

The logic state of IR can be used to flag out-of-range signals. Table 4.9 summarizes the output codes of the 5-bit folding and interpolating ADC. The MSB and MSB-1 delivered by the coarse ADC are necessary for the identification of each fold, and the MSB-2, MSB-3, and LSB are derived from the circular code available at the comparator outputs of the fine ADC.

CMOS folding and interpolating converters can achieve resolutions of 8 to 10 bits at sampling frequencies comparable to that of a flash ADC and are suitable for low-power applications. However, the converter operation can be affected by nonideal effects, such as offsets in the comparators required for the zero-crossing detection. Furthermore, without a front-end track-and-hold circuit, the converter performance can be limited by distortions due to the nonlinear transfer characteristic of folders.

4.1.6   Sub-ranging ADC

A solution for the reduction of the hardware growth with the number of bits resolved can consist of achieving the analog-to-digital conversion in two steps as shown in Figure 4.48 [22]. The input is first tracked and held and then digitized by the L-bit ADC to produce the MSBs, which are applied to the DAC and the result is subtracted from the T/H output signal. This difference denotes the residue of the first step and is extended to the full scale by a 2^L amplification before being processed by the (N-L)-bit ADC, which determines the LSBs. A summer can then combine a delayed version of the MSBs and the LSBs to yield the final N-bit word.

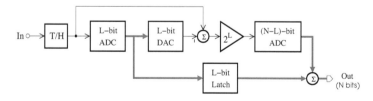

FIGURE 4.48
Block diagram of a sub-ranging ADC.

Sub-ranging ADCs can reach an accuracy of 10 to 12 bits and possess the latency of two clock cycles. This latter is the delay between the instant of the conversion initiation of an input sample and the instant at which the corresponding digital data are being made available. With a speed comparable to the one of a flash ADC, a sub-ranging ADC uses far less hardware for the same resolution. For example, an N-bit flash converter requires $2^N - 1$ comparators, while a two-stage sub-ranging structure needs only $2(2^{N/2} - 1)$ comparators. However, the design of high-resolution sub-ranging ADCs remains limited by practical circuit nonidealities such as component mismatches, charge injection, offset, noise, and finite amplifier gain and bandwidth.

4.1.7 Pipelined ADC

A pipelined ADC is another type of sub-ranging ADC, derived by breaking a high-resolution conversion into multiple steps. Pipelined converters are attractive for applications, such as image and video processing, digital communication, and instrumentation, that require a resolution from 10 to 16 bits and data throughput greater than 5 Ms/s.

The pipelined ADC architecture shown in Figure 4.49 includes a T/H circuit, a cascade of $M - 1$ coarse conversion stages, followed by a fine converter. The cascaded stages are structurally similar, consisting of a sub-ADC (SADC), a DAC, and an amplifier with a gain factor of 2^{n_k}. Each of these stages, for instance, performs a coarse conversion of the incoming full-scale ramp signal and generates a residue signal that corresponds to an amplified version of the quantization error. The parameter n_k $(k = 1, 2, \cdots, M - 1)$ represents the number of bits resolved by the k-th stage of the converter, and the fine converter exhibits a resolution of n_M bits.

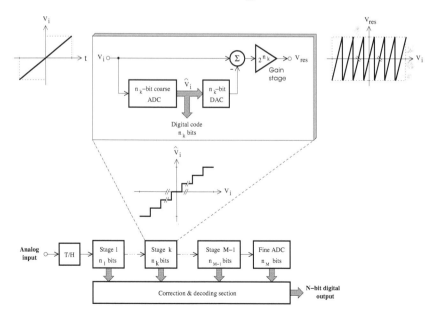

FIGURE 4.49
Block diagram of a pipelined ADC.

The first processing step is performed by the T/H circuit and amounts to acquiring a sample of the input signal. The resulting output is digitized by the SADC to provide the first n_1 MSBs, which are also transformed into the corresponding analog voltage by the n_1-bit DAC. The residue obtained by subtracting the n_1-bit DAC output from the sampled-and-held input is amplified by a factor of 2^{n_1} so that its maximum swing is readjusted to the ADC full scale. This amplified residue is passed to the subsequent stages, where identical operations are performed. Finally, the different digital codes are appropriately synchronized to eliminate the clock delay and combined to produce the required N-bit full ADC resolution. Thus,

$$N = \sum_{k=1}^{M-1} n_k + n_M \tag{4.79}$$

Note that the amplification of the residue signal helps keep the signal level constant, allowing the use of the same reference source for all sub-ADCs and yielding a reduced sensibility of the last stages to circuit imperfections such as noise, offset voltages, switch charge injection, and parasitic-loading capacitors.

Furthermore, the insertion of a track-and-hold circuit between the pipelined stages to allow a concurrent operation of all stages can result in a high conversion throughput. However, this is achieved at the cost of increased latency and power consumption.

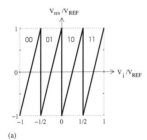 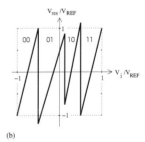

(a)   (b)

FIGURE 4.50
(a) Ideal and (b) practical residue-input signal transfer characteristics.

Let us consider a 2-bit pipelined stage for the analysis of the circuit imperfection effects on the residue waveform for a full-scale ramp input signal [23, 24]. The ideal relationship between the amplified residue signal, V_{res}, and the input signal is shown in Figure 4.50(a). The reference levels $(0, \pm V_{REF}/2)$ of the flash SADC set the position of transitions, whose magnitude at the code boundary is determined by the DAC and the gain of the interstage amplifier. The different output voltage contributions of the DAC are $\pm V_{REF}/4$ and $\pm 3V_{REF}/4$. The signal V_{res} is a linearly increasing function of the input, the magnitude of which is between two adjacent thresholds of the SADC comparator. Once the input signal reaches a threshold level, the signal V_{res} abruptly changes in the opposite direction while remaining within the full scale. Note that the residue is multiplied by the interstage gain of 4 to exactly fit the full-scale of the next stage.

Practical converters can exhibit some linearity errors (see Figure 4.50(b)) such as missing codes. This can be the result of a missing decision level or the loss of a digital code at a DAC input caused by a residual voltage, which exceeds the actual conversion range due to the gain error of the interstage amplifier and offset voltages.

The over-range errors of the residual voltage can be corrected by reducing the interstage gain to 2 (see Figure 4.51(a)) [23]. For a full-scale input, the signal V_{res} of the pipelined stage k should remain between $-V_{REF}/2$ and $V_{REF}/2$. Any excursion of V_{res} outside this range is considered an error, which can be detected by adding extra quantization levels to the stage $k + 1$. The SADC outputs representing these quantization levels are used by the digital correction to either increment or decrement the output code of the stage k for a residue signal greater than $V_{REF}/2$ or less than $-V_{REF}/2$, respectively. This correction approach works successfully for decision errors as large as $\pm V_{REF}/4$ or $\pm 1/2$ LSB.

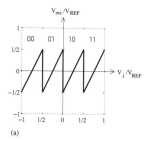 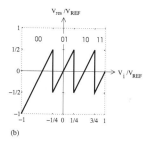

FIGURE 4.51
Residue-input signal transfer characteristics (a) with a gain of 2 and (b) with $V_{REF}/4$ offset voltage.

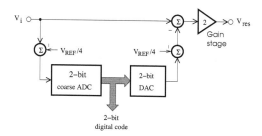

FIGURE 4.52
An equivalent model of the 2-bit stage with offset adjustments.

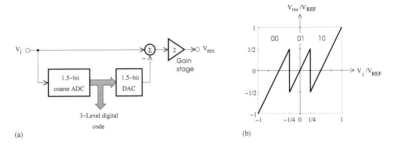

(a)

(b)

FIGURE 4.53

(a) A 1.5-bit stage of the pipelined ADC; (b) residue-input signal transfer characteristic.

The arithmetic of the correction section can be reduced to an addition operation using the stage architecture of Figure 4.52. By appending a $-V_{REF}/4$ offset voltage to the SADC and DAC, the signal V_{res} now varies from $-V_{REF}$ to $V_{REF}/2$. Because the locations of the SADC decision levels and the DAC reference voltages are uniformly shifted by the offset value, a negative over-range condition is prevented for errors up to $\pm V_{REF}/4$. This eliminates the requirement for the subtraction function in the correction logic.

As shown in Figure 4.51(b), the top decision level can be fixed at $V_{REF}/4$ for each stage except the last one. The comparator with the threshold at $3V_{REF}/4$ is not required because the amplified residue signal falling above the expected range can be detected by the next stage. The MSB of each pipelined stage can then be resolved by an SADC with only 1.5-bit resolution. In this case, the comparator reference voltages of the SADC are at $\pm V_{REF}/4$ and the output levels of the DAC include $\pm V_{REF}/2$ and 0. Table 4.10 summarizes the characteristics of the 2-bit and 1.5-bit pipelined stages.

TABLE 4.10

Characteristics of 2-Bit and 1.5-Bit Pipelined Stages

	2-Bit/stage	1.5-Bit/stage
Input range	From $-V_{REF}$ to V_{REF}	From $-V_{REF}$ to V_{REF}
SADC threshold levels	$-V_{REF}/2$, 0, $V_{REF}/2$	$-V_{REF}/4$, $V_{REF}/4$
Number of comparators	3	2
DAC reference levels	$-3V_{REF}/4$, $-V_{REF}/4$, $V_{REF}/4$, $3V_{REF}/4$	$-V_{REF}/2$, 0, $V_{REF}/2$
Output digital code	00, 01, 10, 11	00, 01, 10
Inter-stage gain	4	2

Here, the nonideal effect of the component is reduced by introducing a redundancy in the pipelined ADC. That is, the resolution of the converter should be less than the sum of the ones provided by single stages. The extra bits are eliminated at the output by the digital correction.

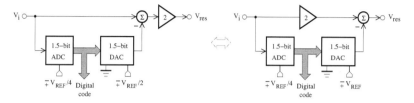

FIGURE 4.54
Scaling of the 1.5-bit pipelined stage for the switched-capacitor implementation.

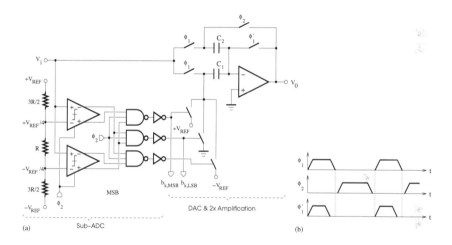

FIGURE 4.55
(a) A switched-capacitor implementation of the 1.5-bit stage circuit; (b) timing diagram.

For the switched-capacitor (SC) implementation, the block diagram of the 1.5-bit pipelined stage can be scaled as shown in Figure 4.54. This allows the use of an adequate clocking scheme to maintain a synchronization between the signal paths. Let V_i be the input voltage and V_{REF} denote the reference voltage. Ideally, the amplified residue signal generated by the k-th stage can then be written as

$$V_{res(k+1)} = 2V_{res(k)} - D_k V_{REF} \qquad k = 1, 2, \cdots, M - 1 \qquad (4.80)$$

where

$$D_k = \begin{cases} -1 & \text{if} \quad V_{res(k)} < -V_{REF}/4 \\ 0 & \text{if} \quad -V_{REF}/4 < V_{res(k)} < V_{REF}/4 \\ 1 & \text{if} \quad V_{res(k)} > V_{REF}/4 \end{cases} \qquad (4.81)$$

and $V_{res(1)} = V_i$. A single-ended SC implementation of the 1.5-bit pipelined stage is depicted in Figure 4.55. This circuit includes a flash SADC and a mul-

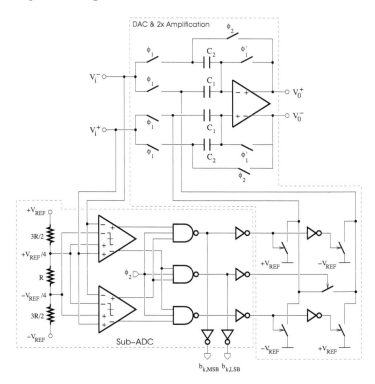

FIGURE 4.56
Differential version of the switched-capacitor implementation of the 1.5-bit stage circuit.

tiplying DAC (MDAC), which performs the multiplication by two followed by the digital-to-analog conversion of the SADC output, the track-and-hold function and the two times amplification of the input signal, and the subtraction. It operates with two-phase nonoverlapping clock signals. To reduce the effect of the difference in propagation delay between the signal paths, the charges due to the held and reconverted versions of the input signal are transferred toward the output during the same clock phase, ϕ_2. The comparators used in the SADC can be designed using an input-sensing preamplifier followed by a regenerative latch, and the control signals of the MDAC are derived from the outputs of a thermometer-to-binary encoder. The amplifier should be designed with a dc gain and bandwidth such that it can settle within $\pm 1/2$ LSB accuracy in one-half of the clock signal period. The output signal, V_0, is given by

$$
V_0 = \begin{cases}
\left(1 + \dfrac{C_1}{C_2}\right) V_i - V_{REF} & \text{if} \quad V_i > V_{REF}/4 \\[2ex]
\left(1 + \dfrac{C_1}{C_2}\right) V_i & \text{if} \quad -V_{REF}/4 \le V_i \le V_{REF}/4 \\[2ex]
\left(1 + \dfrac{C_1}{C_2}\right) V_i + V_{REF} & \text{if} \quad V_i < -V_{REF}/4
\end{cases} \qquad (4.82)
$$

The stage gain of 2 is realized for $C_1 = C_2$. By requiring only two capacitors, the amplifier loading requirement of the resulting 1.5-bit pipelined stage is relaxed, yielding a reduced sensibility to noise and an improved operation speed.

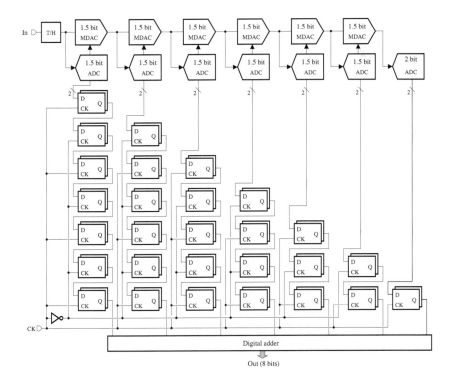

FIGURE 4.57
Block diagram of an 8-bit pipelined ADC including a digital correction stage.

A differential version of the 1.5-bit pipelined stage is shown in Figure 4.56 [25]. The zero reference is implemented by shorting the DAC output together. It should be noted that differential circuits have the advantage of increasing the signal dynamic range. In a pipelined ADC, the first stage generates the MSB of the resulting digital code, while subsequent stages increase

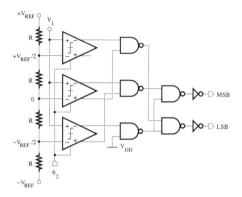

FIGURE 4.58
Circuit diagram of a 2-bit flash ADC.

the resolution of the input signal conversion by delivering additional bits in less significant positions of the output code. Figure 4.57 shows the block diagram of an 8-bit pipelined ADC, which is based on 1.5-bit stage with an inter-stage amplifier gain of 2. The amplified residue of each pipelined stage, except the last one, is quantized by the following stage. Because the last stage does not need to generate a residue, it can be implemented as the 2-bit flash ADC shown in Figure 4.58. A resistor network is used to set the reference levels of the three latched comparators to $-V_{REF}/2$, 0, and $V_{REF}/2$, respectively, and the thermometer-to-binary encoder is based on few logic gates.

Adjacent stages of the pipelined ADC should be driven by the clock signals with opposite phases to ensure a concurrent operation. There is a delay of a half clock cycle between the instants at which the outputs of two consecutive stages are available. One input signal sample then requires several clock cycles to be processed by the overall pipelined ADC. The output codes ($b_{k,MSB}$ and $b_{k,LSB}$, $k = 1, 2, \cdots, M$) of the different pipelined stages are synchronized using an array of D latches, and then summed to produce the converter output code with B_1 being the MSB.

By exploiting the 0.5-bit redundancy on each stage to correct any decision error in the previous adjacent stage, the concept of the error correction circuit can be illustrated as follows:

$$
\begin{array}{cccccccc}
b_{1,MSB} & b_{1,LSB} & & & & & & \\
& b_{2,MSB} & b_{2,LSB} & & & & & \\
& & \cdots & \cdots & & & & \\
& & & \cdots & \cdots & & & \\
& & & & b_{M-2,MSB} & b_{M-2,LSB} & & \\
& & & & & b_{M-1,MSB} & b_{M-1,LSB} & \\
& & & & & & b_{M,MSB} & b_{M,LSB} \\
\hline
+ & & & & & & & \\
\hline
B_1 & B_2 & \cdots & \cdots \; \cdots & & B_{N-2} & B_{N-1} & B_N
\end{array}
$$

This correction scheme works well as long as the comparator offset magnitudes are not so high (i.e., less than $V_{REF}/4$) as to cause missing codes. In this example, the digital correction should remove the redundancy and deliver an 8-bit output data. It can simply be based on addition. Thus, the correction consists of summing the MSB of a given stage with the LSB of the previous stage. Here, the digital correction cannot be extended to the LSB of the last stage. This implies that either the last stage should be implemented as a 2-bit full flash ADC or using two 1.5-bit stages with the LSB of the final stage being excluded from the converter output code. Figure 4.59 shows the block diagram of a digital adder that may be used in the 1.5-bit/stage pipelined ADC. The propagation of the carry bit is achieved in the direction of the output code MSB. Each 1.5-bit pipelined stage effectively provides 1 bit to the overall resolution. The resulting output data is coded in offset binary format, which is identical with the two's complement representation except that the MSB must be inverted.

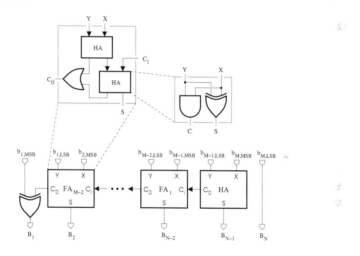

FIGURE 4.59
Block diagram for the digital correction stage (HA: half adder; FA: full adder).

For the 1.5-bit/stage pipelined ADC, the sequence of output digits does not form a conventional binary number representation because the bits are signed. The 1.5-bit stage generates three digits and exhibits a radix of 2. Because the number of digits is greater than the radix, there is a redundancy in the sequence of output digits. Hence, more than one combination of digits can be used to represent the same magnitude of a signal sample. The 1.5-bit/stage pipelined ADC is then equivalent to a redundant signed digit system.

It should be noted that, from the input to the output of the converter, the input-referred noise contribution of the pipelined stages is reduced after each stage due to the cumulative scaling effect of the interstage gain on the signal level. Hence, capacitors can be scaled down toward the later stages to reduce the power consumption [26].

Consider both circuit diagrams of Figure 4.60, which represent the different versions of the MDAC generally used in the implementation of the 1.5-bit-per-stage pipelined ADC architecture.

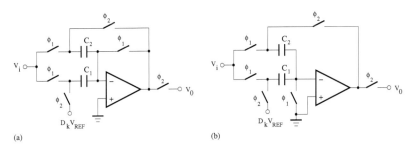

FIGURE 4.60
Two configurations of the SC MDAC for the 1.5-bit stage.

During the clock phase ϕ_1, $(n-1 < t \leq n-1/2)$, according to the MDAC circuit depicted in Figure 4.60(a), the input voltage is sampled onto capacitors C_1 and C_2, and the amplifier is configured as a unity-gain follower. Assuming that A_0 is the amplifier dc gain, it can be shown that

$$V_0(n-1/2) = A_0[V^+(n-1/2) - V^-(n-1/2)] \tag{4.83}$$

where $V^+(n-1/2) = V_{off}$ and $V^-(n-1/2) = V_0(n-1/2)$. During the clock phase ϕ_2, $(n-1/2 < t \leq n)$, the capacitor C_1 is connected to $D_k V_{REF}$, while C_2 acts as a feedback capacitor. The application of the charge conservation law at the negative input terminal of the amplifier yields

$$C_1 \{[D_k V_{REF} - V^-(n)] - [V_i(n-1/2) - V^-(n-1/2)]\}$$
$$= C_2 \{[V^-(n) - V_0(n)] - [V^-(n-1/2)) - V_i(n-1/2)]\} \tag{4.84}$$

and

$$V_0(n) = A_0[V^+(n) - V^-(n)] \tag{4.85}$$

where $V^+(n) = V_{off}$. Because

$$V^-(n-1/2) = V_0(n-1/2) = \frac{\mu}{1+\mu}V_{off} \tag{4.86}$$

and

$$V^-(n) = V_{off} - \mu V_0(n) \tag{4.87}$$

where $\mu = 1/A_0$, it can be shown that

$$
\left[1 + \mu\left(1 + \frac{C_1}{C_2}\right)\right]V_0(n)
$$
$$
= \left(1 + \frac{C_1}{C_2}\right)V_i(n - 1/2) - \frac{C_1}{C_2}D_k V_{REF} + \frac{\mu}{1+\mu}\left(1 + \frac{C_1}{C_2}\right)V_{off}
\tag{4.88}
$$

Finally, the output-input relationship in the z-domain can be derived as

$$
V_0(z) = \frac{\left(1 + \dfrac{C_1}{C_2}\right)z^{-1/2}V_i(z) - \dfrac{C_1}{C_2}D_k V_{REF} + \dfrac{\mu}{1+\mu}\left(1 + \dfrac{C_1}{C_2}\right)V_{off}}{1 + \mu\left(1 + \dfrac{C_1}{C_2}\right)}
\tag{4.89}
$$

An alternative MDAC structure is shown in Figure 4.60(b). Its operation is similar to the first one, except during the phase ϕ_2, where the amplifier is now used in the open-loop configuration and $V^-(n - 1/2) = 0$. The output-input relationship in the z-domain is then of the form

$$
V_0(z) = \frac{\left(1 + \dfrac{C_1}{C_2}\right)z^{-1/2}V_i(z) - \dfrac{C_1}{C_2}D_k V_{REF} + \left(1 + \dfrac{C_1}{C_2}\right)V_{off}}{1 + \mu\left(1 + \dfrac{C_1}{C_2}\right)}
\tag{4.90}
$$

For a sufficiently high dc gain, μ tends to zero and it appears that the second MDAC exhibits an increased sensibility to the effect of the offset voltage. Ideally, the input signal is multiplied by a factor of 2 for $C_1 = C_2$.

Assuming that the effects of the amplifier imperfections are negligible, the linearity of the converter can still be limited by the capacitor mismatch. With α_k being the ratio mismatch between C_1 and C_2, that is, $C_1 = (1 + \alpha_k)C_2$, the output voltage of the MDAC can be expressed as

$$
V_0(z) = 2(1 + \alpha_k/2)z^{-1/2}V_i(z) - D_k(1 + \alpha_k)V_{REF}
\tag{4.91}
$$

In order to fulfill the N-bit resolution requirement, any deviation error due to a component imperfection should be no larger than LSB/2, or $1/2^{N-k-1}$ for the k-th pipelined stage.

The correction logic in the 1.5-bit/stage pipelined ADC is implemented as an addition with the carry propagation of the overlapping correction bits. Although this technique is very simple, it can only correct the offset voltage effect of SADC comparators. The resulting resolution is then limited to 10 bits due to the effect of errors introduced by the finite gain and bandwidth of the amplifier, capacitor mismatches, and noise on the accuracy of the digital-to-analog conversion and interstage amplifier gain.

The use of laser wafer trimming to adjust the values of circuit components during the IC test results in the enhancement of the converter resolution. But this approach can be limited by aging and temperature variations, which can affect the matching accuracy, and by the extra production cost.

To enable a high resolution, various other circuit techniques (digital calibration, capacitor averaging, dithering, gain-boosting method) can be used [27–30], generally at the price of an increased circuit complexity and power consumption, and a reduced conversion speed.

With a uniform per-stage resolution, the design is modular, but the use of a multi-bit stage at the converter input can greatly relax the matching and noise requirements for the following stages. In [31], a converter resolution of 14 bits is achieved using a 4-bit pipelined stage followed by eight 1.5-bit stages and a 3-bit flash ADC. Note that the stage resolution is generally not greater than 4 bits to maintain the converter power consumption at an acceptable level.

The principle of the redundant sign digit coding can also be exploited in the design of pipelined stages with a resolution greater than 1.5 bits.

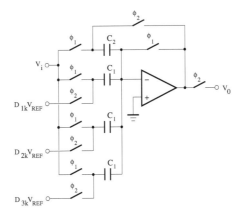

FIGURE 4.61
Circuit diagram of the 2.5-bit SC MDAC.

In the specific case of a 2.5-bit stage, the SADC has six reference levels of the forms, $\pm V_{REF}/8$, $\pm 3V_{REF}/8$, and $\pm 5V_{REF}/8$. In the ideal case, the amplified residue signal generated by the k-th stage can be obtained as

$$V_{res(k+1)} = 4V_{res(k)} - D_k V_{REF} \qquad k = 1, 2, \cdots, M - 1 \qquad (4.92)$$

where

$$D_k = \begin{cases} -3 & \text{if} & V_{res(k)} < -5V_{REF}/8 \\ -2 & \text{if} & -5V_{REF}/8 < V_{res(k)} < -3V_{REF}/8 \\ -1 & \text{if} & -3V_{REF}/8 < V_{res(k)} < -V_{REF}/8 \\ 0 & \text{if} & -V_{REF}/8 < V_{res(k)} < V_{REF}/8 \\ 1 & \text{if} & V_{REF}/8 < V_{res(k)} < 3V_{REF}/8 \\ 2 & \text{if} & 3V_{REF}/8 < V_{res(k)} < 5V_{REF}/8 \\ 3 & \text{if} & V_{res(k)} > 5V_{REF}/8 \end{cases} \qquad (4.93)$$

and $V_{res(1)} = V_i$. To maintain the residue output voltage within $\pm V_{REF}$, the value of the comparator offset voltage should not exceed $\pm V_{REF}/8$. The representation in the binary format of the SADC comparator outputs can result in seven codes, which are 000, 001, 010, 011, 100, 101, and 110.

The circuit diagram of a 2.5-bit MDAC is shown in Figure 4.61, where C_1 and C_2 should have the same value. During the first phase of the clock signal, the capacitors C_1 are connected to the input voltage. In the second clock phase, the capacitor C_2 is included in the amplifier feedback loop, while the capacitors C_1 are set to the appropriate values of the reference voltage. Ideally, the output voltage is given by

$$V_0(z) = \left(1 + \frac{3C_1}{C_2}\right) z^{-1/2} V_i(z) - \frac{C_1}{C_2}(D_{1k} + D_{2k} + D_{3k}) V_{REF} \qquad (4.94)$$

where D_{1k}, D_{2k}, and D_{3k} are equal to -1, 0, or 1, depending on the output of the SADC. The MDAC output signal corresponding to a given SADC digital code is generated by appropriately switching each of the three capacitors C_1 to one of the reference levels, $-V_{REF}$, 0, or V_{REF}.

4.1.8 Algorithmic ADC

Algorithmic or cyclic analog-to-digital converters (ADCs) [32–35], which are based on the binary division principle, are useful in applications requiring low power consumption and chip area. The core of an algorithmic ADC, as shown

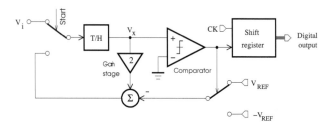

FIGURE 4.62
Block diagram of an algorithmic ADC.

in Figure 4.62, consists of a T/H circuit, precision multiply-by-two amplifier, comparator, and summer.

Let k be the number of conversion cycles and b_k the bit to be determined. The residue signal, V_x, which is considered the partial remainder of a division, can be expressed as

$$V_{x(k)} = \begin{cases} 2V_i & \text{for} \quad k = 1 \\ 2V_{x(k-1)} + (-1)^{b_k-1}V_{REF} & \text{otherwise,} \end{cases} \qquad (4.95)$$

where

$$b_k = \begin{cases} 1 & \text{if} \quad V_{x(k-1)} \geq V_{REF} \\ 0 & \text{if} \quad V_{x(k-1)} < V_{REF} \end{cases} \qquad (4.96)$$

and $k = 1, 2, 3, \cdots, N$. The iterative execution of this algorithm produces a digital output code in the offset binary representation. At each conversion step, the operation of the algorithmic ADC involves a magnitude comparison, a bit selection, and a decision-dependent summation.

The input signal is first selected using the input switch before being sampled by the T/H circuit. The resulting signal, V_x, is applied at the input of the comparator. If V_x is greater than zero, the selected bit will be set to the high logic state and V_{REF} will be subtracted from $2V_x$; otherwise, this bit will take the low logic state and V_{REF} will be added to $2V_x$. By closing the feedback loop through the operation of the input switch, the residue is sent back to the input of the T/H circuit and the determination of the remaining LSB bits proceeds in the same manner. It should be emphasized that each bit of the digital output code is kept in the shift register after its determination.

Starting with the MSB, each bit b_k is determined sequentially, depending on the polarity of $V_{x(k)}$. By iterating up to $k = N$, we can obtain

$$V_{x(N)} = 2^{N-1}\left[V_i + \left(\sum_{k=1}^{N-1}(-1)^{b_k}2^{-k}\right)V_{REF}\right] \qquad (4.97)$$

The signal V_x can be considered the residue generated after the determination of each bit. Hence, the set of bits b_k $(k = 1, 2, \cdots N)$ is a binary representation

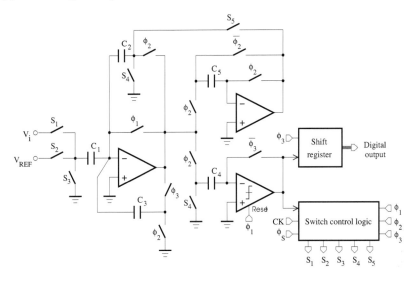

FIGURE 4.63
Circuit diagram of an algorithmic ADC.

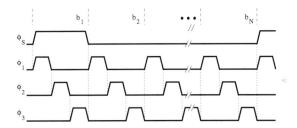

FIGURE 4.64
Clock signals for the algorithmic ADC.

of a fractional number equal to V_i/V_{REF}, where V_i is a bipolar signal with a range from $-V_{REF}$ to V_{REF}.

To achieve a high resolution, the algorithmic ADC circuit should be designed such that its sensitivity to component imperfections is minimized [36, 37]. For a given resolution, the deviation of the converter characteristic due to these nonidealities should be maintained well below 1/2 LSB. Figure 4.63 shows the circuit diagram of a ratio-independent algorithmic ADC, which requires a three-phase clock signal to overcome the effects of capacitor mismatches and offset voltages. To perform the voltage multiplication by a factor of 2 independently of the capacitor ratio, the charge of the input capacitor C_3 due to the residue voltage, V_x, is transferred onto the feedback capacitor, C_2, initially charged to V_x. The operation of the ratio-independent algorithmic

ADC is better understood by analyzing the charge transfer between capacitors during the three phases of the clock signal.

During the first phase, a sample of the input signal is connected to the converter and the charge due to the input voltage is stored on the capacitor C_1.

In the second phase, C_2 is connected to the amplifier output, a charge transfer takes place between C_1 and C_2, and the output of the input amplifier is set to the voltage V_x, which is used to charge the capacitors C_4 and C_5 of the next stages.

During the third phase, the charge on C_2 is transferred onto C_3, while the charge on C_5 is maintained and the MSB is resolved by the comparator. Note that the charge transfer is not affected by C_1 because one of its input nodes is floating.

For the first phase of the determination of the next bit, the capacitor C_1 is charged to V_{REF} and the charge produced by V_x on C_5 is transferred onto C_2.

In the second phase, C_2 acts as a feedback capacitor and its initial charge is combined with the ones on C_1 and C_3 to update the value of the residue signal, thereby resulting in the summation of the two charge contributions associated with the input voltage and due to the reference voltage. The derived residue is used to charge the capacitors C_4 and C_5 connected to the output of the input amplifier.

During the third phase, the state of the output bit is determined by the comparator and the charge on C_2 is transferred onto C_3.

These last three phases are then repeated for each of remaining bits to be resolved. The ratio-independent algorithmic ADC requires three clock phases for the determination of each bit; therefore an N-bit conversion is performed in $3N$ clock phases.

TABLE 4.11

Switch Control Signals for a Bipolar Conversion

	MSB	Bit b_k $(k > 1)$	Bit b_k
S_1	$\phi_S \cdot \phi_1$		$\phi_S \cdot \phi_1$
S_2		$\overline{\phi}_S(\overline{b}_{k-1}\phi_1 + b_{k-1} \cdot \phi_2)$	$\overline{\phi}_S(\overline{b}_{k-1}\phi_1 + b_{k-1} \cdot \phi_2)$
S_3	$\phi_S \cdot \phi_2$	$\overline{\phi}_S(\overline{b}_{k-1}\phi_2 + b_{k-1} \cdot \phi_1)$	$\phi_S \cdot \phi_2 + \overline{\phi}_S(\overline{b}_{k-1}\phi_2 + b_{k-1} \cdot \phi_1)$
S_4	$\phi_S(\phi_1 + \phi_3)$	$\overline{\phi}_S \cdot \phi_3$	$\phi_S(\phi_1 + \phi_3) + \overline{\phi}_S \cdot \phi_3$
S_5		$\overline{\phi}_S \cdot \phi_1$	$\overline{\phi}_S \cdot \phi_1$

The clock signals for a bipolar conversion can be generated according to the Boolean logic equations of Table 4.11. In the case of a unipolar conversion with the input signal varying between 0 and V_{REF}, the converter switching scheme should be slightly modified. It is necessary to exchange the control signals for switches S_2 and S_3 during the determination of the MSB. As a result, the first residue will be of the form $V_{x(1)} = V_i - V_{REF}$.

The accuracy of the algorithmic ADC is affected by the nonlinearities due to charge injections of switches, and offset voltages of the amplifier and comparator. The amplifier can be designed to have enough gain and speed such that the deviations due to the finite gain and settling time are greatly reduced. But, the remaining uncompensated component imperfections limit the achievable resolution to about 10 bits.

4.1.9 Time-interleaved ADC

A solution for the design of data converters operating with a speed beyond the fundamental technological limit can consist of interleaving, in time, more than one ADC.

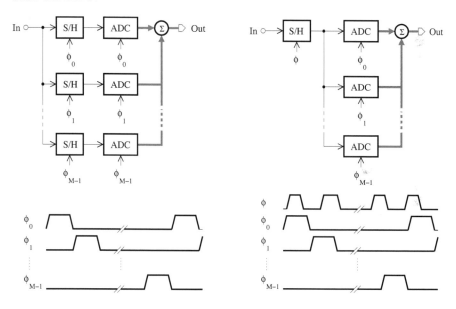

FIGURE 4.65
Time-interleaved ADC with identical parallel channels.

FIGURE 4.66
Time-interleaved ADC with a single-input S/H circuit.

One approach to control the sampling operation in time-interleaved structures is to use several clock signals, as shown in Figure 4.65, where M is the number of parallel channels. Ideally, the effective sampling rate of the resulting structure is increased by M times in comparison to the one of a single ADC. However, due to the difference in the delays between successive sampling instants, dynamic distortions can be observed in the spectrum of the output signal. This results in a degradation of the overall ADC performance. A way to reduce the timing mismatches is to use the structure of Figure 4.66, where the input signal is sampled at a high rate and then multiplexed over M parallel channels to be processed by ADCs. The drawback of this approach is the limited input bandwidth due to the increased capacitive loading of the input sampling stage.

Let $v_i(t_n)$ denote the sequences obtained by sampling the band-limited input signal at the instants t_n given by

$$t_n = (n + \alpha_m)T \qquad (4.98)$$

where $T = 1/f_s$ is the sampling period, α_m ($m = 0, 1, \cdots, M-1$) is the timing offset measured in percentage of the sampling period T, and $n = kM + m$.

By relying on the definition

$$\widehat{\mathbf{V}}_i(\omega) = \sum_{n=-\infty}^{+\infty} v_i(t_n)e^{-j\omega nT} \qquad (4.99)$$

to compute the signal spectrum as if the signal has been sampled uniformly, the digital output spectrum of the converter processing a sine wave will exhibit line spectra whose magnitudes are frequency dependent.

The spectrum of the nonuniformly sampled signal [38] can be computed as

$$\widehat{\mathbf{V}}_i(\omega) = \sum_{n=-\infty}^{+\infty} v_i(t_n)e^{-j\omega t_n} \qquad (4.100)$$

That is,

$$\widehat{\mathbf{V}}_i(\omega) = \frac{1}{T} \sum_{k=-\infty}^{+\infty} A(k)\mathbf{V}_i\left(\omega - k\frac{2\pi}{MT}\right) \qquad (4.101)$$

where

$$A(k) = \frac{1}{M} \sum_{m=0}^{M-1} e^{-jk\alpha_m(2\pi/M)}e^{-jkm(2\pi/M)} \qquad (4.102)$$

and $\mathbf{V}_i(\omega)$ is the Fourier transform of $v_i(t)$.

For $\alpha_m = 0$, the signal spectrum is reduced to

$$\widehat{\mathbf{V}}_i(\omega) = \frac{1}{T} \sum_{k=-\infty}^{+\infty} \mathbf{V}_i\left(\omega - k\frac{2\pi}{MT}\right) \qquad (4.103)$$

which corresponds to a uniformly sampled signal.

The signal spectrum, $\widehat{\mathbf{V}}_i(\omega)$, can be derived as follows:

$$\widehat{\mathbf{V}}_i(\omega) = \sum_{n=-\infty}^{+\infty} v_i(t_n)e^{-j\omega t_n} \tag{4.104}$$

$$= \sum_{k=-\infty}^{+\infty} \sum_{m=0}^{M-1} v_i((kM+m+\alpha_m)T)e^{-j\omega(kM+m+\alpha_m)T} \tag{4.105}$$

Using

$$v_i((kM+m+\alpha_m)T) = \frac{1}{2\pi}\int_{-\infty}^{+\infty} V_i(\Omega)e^{-j\Omega(kM+m+\alpha_m)T}d\Omega \tag{4.106}$$

and permuting the order of the first summation and integration, we obtain

$$\widehat{\mathbf{V}}_i(\omega) = \sum_{m=0}^{M-1}\frac{1}{2\pi}\int_{-\infty}^{+\infty} V_i(\Omega)\left[\sum_{k=-\infty}^{+\infty} e^{-j(\Omega-\omega)kMT}\right]e^{-j(\Omega-\omega)(m+\alpha_m)T}d\Omega \tag{4.107}$$

Because

$$\sum_{k=-\infty}^{+\infty} e^{-j(\Omega-\omega)kMT} = \sum_{k=-\infty}^{+\infty}\frac{2\pi}{MT}\delta\left(\Omega-\omega+k\frac{2\pi}{MT}\right) \tag{4.108}$$

it can be shown that

$$\widehat{\mathbf{V}}_i(\omega) = \sum_{k=-\infty}^{+\infty}\sum_{m=0}^{M-1}\frac{1}{MT}\int_{-\infty}^{+\infty} V_i(\Omega)\delta\left(\Omega-\omega+k\frac{2\pi}{MT}\right)e^{-j(\Omega-\omega)(m+\alpha_m)T}d\Omega \tag{4.109}$$

Hence,

$$\widehat{\mathbf{V}}_i(\omega) = \frac{1}{MT}\sum_{k=-\infty}^{+\infty}\sum_{m=0}^{M-1} V_i\left(\omega-k\frac{2\pi}{MT}\right)e^{-jk(2\pi/MT)(m+\alpha_m)T} \tag{4.110}$$

and finally,

$$\widehat{\mathbf{V}}_i(\omega) = \frac{1}{T}\sum_{k=-\infty}^{+\infty} A(k)V_i\left(\omega-k\frac{2\pi}{MT}\right) \tag{4.111}$$

where

$$A(k) = \frac{1}{M}\sum_{m=0}^{M-1} e^{-jk\alpha_m(2\pi/M)}e^{-jkm(2\pi/M)} \tag{4.112}$$

Classical time-interleaved ADC structures can be affected by the following sources of error.

- Timing skew errors

Let us consider a sinusoidal input signal of the following form:

$$v_i(t) = \sin(\omega_i t) \tag{4.113}$$

The Fourier transform of $v_i(t)$ is given by

$$\mathbf{V}_i(\omega) = j\pi[\delta(\omega + \omega_i) - \delta(\omega - \omega_i)] \tag{4.114}$$

With the assumption that $\omega_i = 2\pi f_i$, where f_i is the frequency, the substitution of Equation (4.114) into (4.101) leads to the next equation,

$$\widehat{\mathbf{V}}_i(\omega) = \frac{2\pi}{T} \sum_{k=-\infty}^{+\infty} \left[A(k)\delta\left(\omega + \omega_i - k\frac{2\pi}{MT}\right) + B(k)\delta\left(\omega - \omega_i - k\frac{2\pi}{MT}\right) \right] \tag{4.115}$$

where

$$A(k) = -\frac{1}{2jM} \sum_{m=0}^{M-1} e^{j\alpha_m(2\pi f_i/f_s)} e^{-jkm(2\pi/M)} \tag{4.116}$$

and

$$B(k) = \frac{1}{2jM} \sum_{m=0}^{M-1} e^{-j\alpha_m(2\pi f_i/f_s)} e^{-jkm(2\pi/M)} \tag{4.117}$$

Note that $A(k) = B^*(M - k)$, where $*$ denotes the notation for the complex conjugate. Due to timing errors between the ADC clock signals, pairs of line spectra centered at $\pm f_i + mf_s/M$ ($m = 1, 2, \cdots, M - 1$) appear in the output spectrum. The corresponding magnitudes are given by $|A(k)|$ and $|B(k)|$, respectively.

Practical time-interleaved ADCs exhibit a clock skew error of a few picoseconds. That is, the value of α_m computed from the discrete Fourier transform of the converter output signal is used to control programmable delays with picosecond resolution or clock signal generators. However, the major drawback of this approach is the high complexity of the digital hardware needed for the algorithm implementation [39]. In order to address the problem of timing skew mismatches, the generation of clock signals can be controlled by a delay-locked loop. Another alternative can consist of using the structure of Figure 4.66. By not resetting between samples, the full-speed single sample and hold (S/H) at the front-end provides subsequent circuit sections the whole clock period to operate on the held signal and eliminates in this way the timing skew errors.

• Gain and offset dispersions

The distortions due to the gain and offset dispersions can be modeled by assuming an input sinusoidal input signal of the form

$$v_i(t) = A_m \sin(\omega_i t) + V_m \tag{4.118}$$

where $m = 0, 1, \cdots, M - 1$. The corresponding Fourier transform is given by

$$\mathbf{V}_i(\omega) = j\pi A_m [\delta(\omega + \omega_i) - \delta(\omega - \omega_i)] + 2\pi V_m \delta(\omega) \tag{4.119}$$

Substituting Equation (4.119) into (4.101), the output spectrum can be written as

$$\widehat{\mathbf{V}}_i(\omega) = \frac{2\pi}{T} \sum_{k=-\infty}^{+\infty} A(k) \left[\delta\left(\omega + \omega_i - k\frac{2\pi}{MT}\right) + \delta\left(\omega - \omega_i - k\frac{2\pi}{MT}\right) \right]$$
$$+ \frac{2\pi}{T} \sum_{k=-\infty}^{+\infty} V(k)\delta\left(\omega - k\frac{2\pi}{MT}\right) \tag{4.120}$$

where

$$A(k) = -\frac{1}{2jM} \sum_{m=0}^{M-1} A_m e^{-jkm(2\pi/M)} \tag{4.121}$$

and

$$V(k) = \frac{1}{M} \sum_{m=0}^{M-1} V_m e^{-jkm(2\pi/M)} \tag{4.122}$$

The gain error results in sidebands centered at $\pm f_i + m f_s/M$. However, the components in each pair of line spectra have the same magnitude, $|A(k)|$, in contrast to distortions caused by clock skew mismatches.

The dispersion of the offset among the channels gives rise to distortions which can be observed in the frequency domain as tones at each path sampling frequency, f_s/M, and its integer multiples. The magnitude of these spectral lines is determined by $|V(k)|$.

Because the distortion power of gain and offset errors is not frequency dependent, it can then be compensated using appropriate circuit calibration techniques [40, 41].

In high-resolution ADCs, the thermal noise appears to be the most important nonideality. Specifically, the increase in the number of bits implies a reduction in the noise level, which can be achieved by augmenting the component sizes and equivalently the power consumption.

4.2 Summary

In practice, ADCs operating at the Nyquist rate are difficult to implement and may require a high power consumption, especially for high resolutions. Nyquist ADCs exhibit a quantization noise, which is uniformly spread from 0 to approximately half the sampling rate or clock signal frequency, and individually convert each input signal sample into a digital output code. They can be categorized into SAR, integrating, flash, sub-ranging, and pipelined architectures. In general, the achievable resolution and speed are limited by various noise contributions, and component and timing mismatches. For a given application, the ADC design is determined by the trade-off that can be achieved between the resolution, speed, chip area, and power consumption.

Flash ADCs can provide a resolution ranging from 5 to 9 bits at the sampling rate, while counting ADCs can offer a resolution of 10 to 20 bits at a speed 2^N times smaller than the sampling rate and with a latency equal to the product of 2^N and the clock period, where N is the number of bits. On the other hand, pipelined ADCs can achieve a resolution of 10 to 14 bits by operating at the sampling rate, while SAR ADCs can provide a resolution of 8 to 16 bits at a frequency equal to the sampling rate divided by the resulting number of bits. SAR and pipelined ADCs exhibit an identical latency equal to the product of the resolution in bits and the clock period, but they can be designed to meet the requirement of low power consumption.

4.3 Circuit design assessment

1. **Buffer amplifier for data converter interfacing**
    – Consider a buffer amplifier with the response to a step input given by
    $$v_0(t) = V_m[1 - \exp(-t/\tau)] \qquad (4.123)$$

    where V_m is the maximum amplitude of the output signal and τ is the time constant. In an application requiring a resolution of N bits, the amplifier should be designed such that the condition $[v_0(t) - V_m]/V_m \leq 1/2^{N+1}$ is satisfied for $t = t_s = 1/(2f_s)$, where f_s is the sampling frequency.

    Determine the time constant τ.

    – In the case where the input signal is a sinusoid of the form

    $$v_i(t) = V_{FS} \sin 2\pi f_B t \qquad (4.124)$$

    where V_{FS} and f_B represent the signal full-scale amplitude and bandwidth frequency, respectively, find the maximum aperture timing error $t_a = \Delta V/[(dv_i(t)/dt)|_{t=0}]$, assuming that the resulting amplitude error $|\Delta V|$ is to be less than a half of the least-significant bit (LSB) and $LSB = V_{FS}/2^N$.

2. **Quantizer model**

In analog-to-digital converters, each sample of the input signal is quantized to fit a finite resolution. This quantization process can be modeled using a characteristic and error function, as illustrated in Figures 4.67(a) and (b) in the case of midtread (a) and midrise (b) quantizers, respectively.

Assuming that the quantization error, $e_Q = \hat{x} - x$, is a stationary process and uncorrelated with the input signal, x, it can be seen that the probability density of the quantization error is uniformly distributed between $-\Delta/2$ and $\Delta/2$.

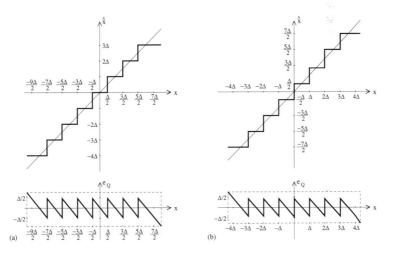

FIGURE 4.67

Characteristics and errors of (a) midtread and (b) midrise quantizers.

For both quantizers, show that

$$E(e_Q) = \int_{-\Delta/2}^{\Delta/2} e_Q p(e_Q) de_Q = 0 \qquad (4.125)$$

and

$$\sigma_Q^2 = E(e_Q^2) = \int_{-\Delta/2}^{\Delta/2} e_Q^2 p(e_Q) de_Q = \frac{\Delta^2}{12} \qquad (4.126)$$

where $p(e_Q)$ is the probability density of the quantization error and Δ is the quantizer step size.

3. **A model for the sampling jitter estimation**

The output of an S/H circuit, y_k, can be computed as

$$y_k = x(kT + \delta_k) + n_k \quad k \in \mathbb{Z} \tag{4.127}$$

where $x(t)$ is the input signal, T is the sampling period, n_k is the additive sampling noise, and δ_k denotes the error due to the deviation in the sampling instant.

Let

$$x(t) = A\cos(\omega_0 t + \phi) \tag{4.128}$$

be a sinusoid signal with the amplitude A, the initial phase ϕ, and the angular frequency ω_0.

Using the assumption $x(kT + \delta_k) \simeq x(kT) + \delta_k x'(kT)$, where x' represents the first derivative of x, verify that:

$$y_k \simeq A\cos(\omega_0 kT + \phi) + \epsilon_k \tag{4.129}$$

where $\epsilon_k = -A\omega_0 \delta_k \sin(\omega_0 kT + \phi) + n_k$.

Show that the variance of the error term ϵ_k can be written as

$$E[\epsilon_k^2] = E[A^2 \omega_0^2 \delta_k^2 \sin^2(\omega_0 kT + \phi) + n_k^2] \tag{4.130}$$

$$= \frac{A^2 \omega_0^2}{2} \sigma_\delta^2 - \frac{A^2 \omega_0^2}{2} \sigma_\delta^2 \cos(2\omega_0 kT + 2\phi) + \sigma_n^2 \tag{4.131}$$

where $\sigma_\delta^2 = E[\delta_k^2]$ and $\sigma_n^2 = E[n_k^2]$.

Propose a procedure based on the Fourier transform for the computation of the variances σ_δ^2 and σ_n^2.

4. **Switched-capacitor SAR ADC**

The circuit architecture shown in Fig 4.68 [42] achieves the conversion of bipolar signals into a digital code, the MSB of which indicates the polarity. The control signals S_j $(j = 1, 2, \cdots, 14)$ are defined in Table 4.12, $C_1 = C_2$, and $C_5 = C_6$. The clock signal ϕ_S determines the sampling and conversion phases. The sign of the input signal is detected during the on-state of ϕ_P. The bit b_0 will be set either to 1 if $V_i \geq 0$ or to 0 if the input signal is negative. Its value is maintained at the output of one latch during the next conversion period fixed by ϕ_H.

Let k denote the conversion cycle. The sampled analog input signal, $V_i(k)$, which is stored on C_1, is compared to the DAC output, $V(k)$, available on C_2. The sign of the threshold voltage generated by the DAC is similar to the one of the input signal. Show that the voltage $V_c(k)$ at the inverting input node of the comparator is given by

$$V_c(k) = \frac{-C_1 V_i(k) + C_2 V(k)}{C_1 + C_2 + C_p} + V_{off} \tag{4.132}$$

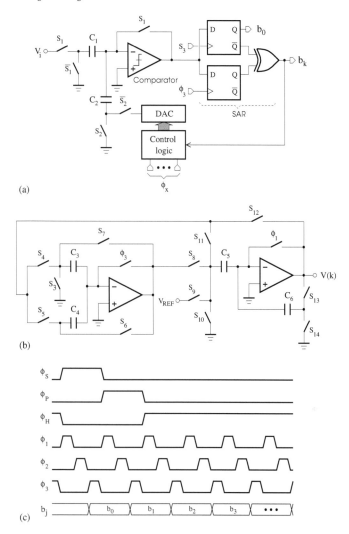

(a)

(b)

(c)

FIGURE 4.68
(a) Block diagram of a SAR ADC, $x = S, P, H, 1, 2, 3$; (b) SC DAC; (c) clock
and digital signal waveforms.

where C_p is the parasitic capacitance at the comparator input node
and V_{off} is the comparator *dc* offset voltage.

Propose a gate-level implementation of the control circuit.

The DAC is realized using an S/H and an amplifier circuit with the

TABLE 4.12
Digital Signals for the SAR ADC Switch Control

$S_1 : \phi_S\phi_1$	$S_2 : \phi_S + \phi_1\phi_P$	$S_3 : \phi_S\phi_3$
$S_4 : b_k\overline{\phi_S}\phi_3$	$S_5 : \phi_S\phi_3 + \overline{b_k}\,\overline{\phi_S}\phi_3$	$S_6 : b_k\phi_1\phi_H$
$S_7 : \overline{b_k}\phi_1\phi_H$	$S_8 : \phi_1\phi_H$	$S_9 : \phi_S\phi_2$
$S_{10} : \phi_S\phi_1 + \overline{b_k}\phi_1\phi_P$	$S_{11} : \overline{b_k}\overline{\phi_S}\phi_3$	$S_{12} : \phi_S\phi_3 + \overline{\phi_S}(\phi_2 + \phi3)$
$S_{13} : \phi_S\phi_2 + \overline{b_k}\phi_S\phi_3 + \overline{\phi_S}(\phi_2 + \phi3)$		$S_{14} : \phi_S\phi_1 + b_k\phi_1\phi_P$

gain of $1/2$. Its output signal can be written as

$$V(k) = -(-1)^{b_0} V_{REF} \left(2^{-k} + \sum_{j=0}^{k-1} b_j 2^{-j} \right) \qquad (4.133)$$

Depending on the value of b_0, the positive or negative charge due to V_{REF} and stored on C_5 is transferred onto C_4. Compare the results of the theoretical analysis of the DAC circuit to SPICE simulations.

Analyze the dependence of the converter resolution to the amplifier finite gain and mismatching between C_5 and C_6.

5. **SAR ADC**
   Consider the 4-bit SAR ADC shown in Figure 4.69. A dummy capacitor C is used in addition to the binary weighted capacitor array, so that the total capacitance of the SC DAC can be equal to 2^4C, allowing a binary division to be performed when the states of the SAR output bits are successively changed.

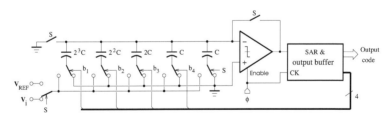

FIGURE 4.69
Circuit diagram of a 4-bit SAR ADC.

Use the equivalent model of the SC DAC shown in Figure 4.70 and verify the following statements:

(a) At the end of the sampling phase, the charge stored on the lower plates of the capacitors is $Q = -16CV_i$.

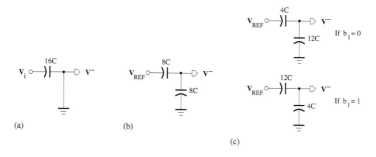

FIGURE 4.70

Equivalent model of the SC DAC: (a) sampling, (b) MSB or b_1, (c) b_2.

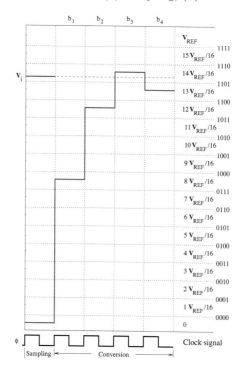

FIGURE 4.71

An example waveform of the 4-bit SAR ADC.

(b) During the determination of b_1 (MSB),

$$Q = -16CV_i = 8C(V^- - V_{REF}) + 8CV^-$$

and

$$V^- = -V_i + V_{REF}/2$$

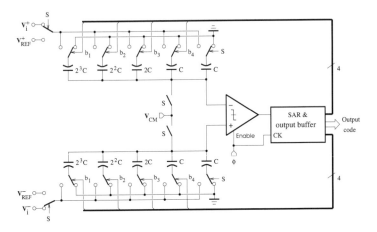

FIGURE 4.72
Circuit diagram of a 4-bit SAR ADC with differential SC DACs.

If $V_i > V_{REF}/2$, then $V^- < 0$ and the comparator output goes high, as a result b_1 is set to 1. On the other hand, if $V_i < V_{REF}/2$, then $V^- > 0$ and the comparator output goes low, b_1 is then set to 0.

(c) During the determination of b_2, the voltage at the comparator input is given by

$$V^- = \begin{cases} -V_i + V_{REF}/4 & \text{if } b_1 = 0 \\ -V_i + 3V_{REF}/4 & \text{if } b_1 = 1 \end{cases}$$

The conversion process then continues until all the remaining bits are generated.

Verify that the input voltage of the comparator can generally be written as

$$V^- = -V_i + V_{REF}\left(\frac{b_1}{2} + \frac{b_2}{4} + \frac{b_3}{8} + \frac{b_4}{16}\right) \tag{4.134}$$

Verify that the 4-bit SAR ADC operation, as described by the waveforms of Figure 4.71, corresponds to the output code 1101.

A 4-bit SAR ADC can also be implemented using differential SC DACs, as shown in Figure 4.72.

Analyze this circuit and show that the comparator input voltages are given by

$$V^+ = -V_i^- + V_{CM} + V_{REF}^- \left(\frac{b_1}{2} + \frac{b_2}{4} + \frac{b_3}{8} + \frac{b_4}{16} \right) \qquad (4.135)$$

$$V^- = -V_i^+ + V_{CM} + V_{REF}^+ \left(\frac{b_1}{2} + \frac{b_2}{4} + \frac{b_3}{8} + \frac{b_4}{16} \right) \qquad (4.136)$$

Deduce the expressions of the voltages V^+ and V^- in the specific case where $V_{REF}^+ = V_{CM} + V_{REF}/2$, $V_{REF}^- = V_{CM} - V_{REF}/2$, and $V_{CM} = V_{REF}/2$.

6. **Nyquist data converter analysis**
   • Consider the DAC depicted in Figure 4.73, which consists of a binary weighted capacitor array.

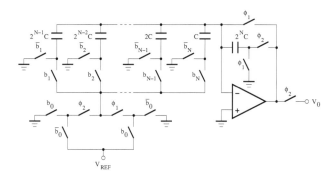

FIGURE 4.73
Block diagram of a charge redistribution DAC.

In the ideal case, show that the output voltage is of the form

$$V_0 = (-1)^{b_0} V_{REF} \sum_{k=1}^{N} 2^{-k} b_k \qquad (4.137)$$

where b_0 is the sign bit, b_k $(k = 1, 2, \cdots, N)$ represents the magnitude bit, and V_{REF} is the reference voltage.

Verify that the total capacitance required to achieve a resolution of N bits is $C_T = (2^{N+1} - 1)C$.

Assuming that the saturation level of the amplifier output voltage is αV_{DD}, where $\alpha = 0.8$ and $V_{DD} = 2.5$ V, solve the equation $|V_0| \leq \alpha V_{DD}$ with $C = 1$ pF to determine the maximum value of N.

• For the flash ADC of Figure 4.74, verify that the numbers of comparators and resistors required to achieve a resolution of N bits are $2^N + 1$ and 2^N, respectively.

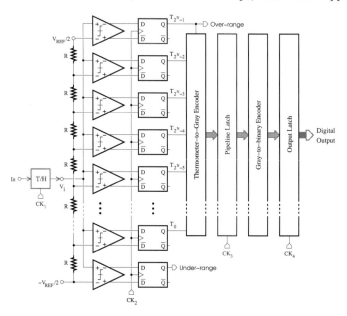

FIGURE 4.74
Block diagram of a flash ADC.

Let the differential nonlinearity (DNL) be defined as the difference between an actual step width and the ideal value of 1 least significant bit (LSB). For each code T_k $(k = 0, 1, \cdots, 2^N - 1)$, the DNL is given by

$$\text{DNL}_k = \frac{\triangle_k}{V_{LSB}} - 1 \qquad (4.138)$$

where $V_{LSB} = V_{REF}/2^N$, and $\triangle_k$ is the actual step size associated with the code T_k.

Assuming that the differential input voltage of each comparator is of the form $V^+ - V^- + V_{off}$, where V_{off} is the offset voltage, and V^+ and V^- are the voltage levels applied to the noninverting and inverting node, respectively, show that

$$\text{DNL}_k = \begin{cases} V_{off} & \text{if} \quad k = 2^N - 1 \\ 0 & \text{otherwise.} \end{cases} \qquad (4.139)$$

For $V_{off} = 10$ mV and $V_{REF} = 2.5$ V, determine the maximum achievable resolution, N_{max}, by solving the equation $\text{DNL}_k \leq V_{LSB}/2$, which guarantees an effective number of bits equal to N_{max}.

7. **Flash ADC**

A thermometer-to-binary encoder generates a binary output code representing the number of ones on the inputs. Its implementation can then be based on a ones-counter. This last approach has the advantage of providing a global bubble error suppression.

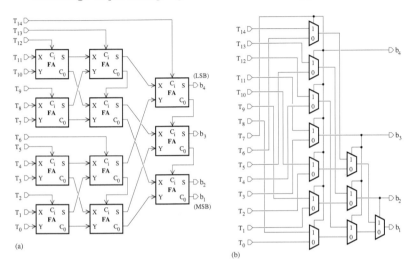

FIGURE 4.75

Tree encoder based on ones-counter for a 4-bit flash ADC: implementations using (a) full adders and (b) 2-to-1 multiplexers.

Analyze the circuits of Figure 4.75 to show that they realize a thermometer-to-binary encoding.

8. **Two-step ADC**

A two-step ADC can be implemented without an interstage gain stage, as shown in Figure 4.76. It is composed of an input stage that performs a coarse conversion, and an output stage that realizes a fine conversion.

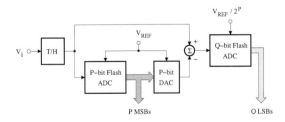

FIGURE 4.76

Two-step ADC without an interstage gain stage.

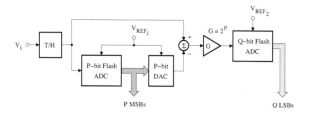

FIGURE 4.77
Two-step ADC with an interstage gain stage.

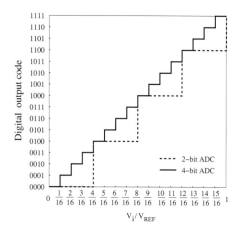

FIGURE 4.78
Transfer characteristic of the 4-bit two-step ADC.

Assuming that the 2-bit bipolar flash ADC is implemented using a ladder of 4 identical resistors R, 3 comparators, and a thermometer-to-binary decoder, complete the encoding table presented in Table 4.13.

TABLE 4.13
Truth Table of the 2-Bit Bipolar Flash ADC

	Comparator outputs			Binary code	
Input voltage range	T_2	T_1	T_0	b_1	b_2
$V_{REF}/2 < V_i \leq V_{REF}$	1	1	1	...	...
$0 < V_i \leq V_{REF}/2$	0	1	1	...	...
$-V_{REF}/2 < V_i \leq 0$	0	0	1	...	...
$-V_{REF} < V_i \leq -V_{REF}/2$	0	0	0	...	...

To relax comparator specifications, a two-step ADC can also be implemented with an interstage gain stage, as shown in Figure 4.77. It is considered to generate a residue of the form:

$$V_r = (G \pm \triangle G)\left(V_i - V_{REF} \sum_{k=0}^{P-1} b_{k+1} 2^{-k-1}\right) \tag{4.140}$$

where $G = 2^P$ and $\triangle G$ represents the gain error.

Assuming that $P = Q = 2$ bits, determine the residue V_r when the digital output code is 1000 and 0111.

Deduce that the difference between the residues when the digital output code is 1000 and 0111 can be written as

$$\triangle V_r = (G \pm \triangle G)\frac{V_{REF}}{4} \tag{4.141}$$

Determine $\triangle G$ so that $\triangle V_r$ remains less than or equal to one LSB, where

$$1\,\text{LSB} = \frac{V_{REF_2}}{4} = \frac{V_{REF_1}}{4 \times 4} = \frac{V_{REF}}{16} \tag{4.142}$$

and $V_{REF_1} = V_{REF}$.

Verify that the characteristic of the two-step ADC can be represented as shown in Figure 4.78.

9. **Pipeline ADC**

A pipeline ADC of Figure 4.79(a) consists of a track-and-hold (T/H) circuit, two 1.5-bit stages, a 2-bit stage, and a digital adder. For $k = 1, 2$, the residues (or outputs of the first and second stages) can be obtained as follows

$$V_k = \begin{cases} 2V_{k-1} + V_{REF} & \text{if} \quad V_{k-1} < -V_{REF}/4 \\ 2V_{k-1} & \text{if} \quad -V_{REF}/4 < V_{k-1} < V_{REF}/4 \\ 2V_{k-1} - V_{REF} & \text{if} \quad V_{k-1} > V_{REF}/4 \end{cases} \tag{4.143}$$

where $V_0 = V_i$ and $V_{REF} = 0.5$ V, and the corresponding digital output code is 00, 01, and 10, respectively.

Verify that the residues, V_1 and V_2, can be represented, as shown in Figure 4.79(b), where V_i/V_{REF} is the ratio of the input voltage to the reference voltage.

Verify that the diagram of Figure 4.80 effectively illustrates the operation of the pipeline ADC when the input voltage, V_i, is assumed to be equal to 0.1 V.

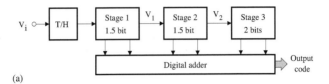

(a)

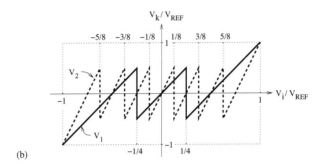

(b)

FIGURE 4.79
Pipeline ADC: (a) block diagram; (b) representation of residues V_1 and V_2.

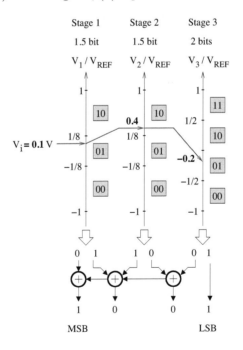

FIGURE 4.80
Illustration of the pipeline ADC operation.

10. SNR degradation due to clock skew errors

The analysis (see Subsection 4.1.9) of timing skew errors between the ADC clock signals of time-interleaved converters shows that pairs of line spectra centered around the frequencies $\pm f_i + m f_s/M$ $(m = 1, 2, \cdots, M-1)$ appear in the output spectrum. The corresponding magnitudes are given by $|A(k)|$ and $|B(k)|$, respectively, and

$$A(k) = -\frac{1}{2jM} \sum_{m=0}^{M-1} e^{j\alpha_m (2\pi f_i/f_s)} e^{-jkm(2\pi/M)} \qquad (4.144)$$

$$B(k) = \frac{1}{2jM} \sum_{m=0}^{M-1} e^{-j\alpha_m (2\pi f_i/f_s)} e^{-jkm(2\pi/M)} \qquad (4.145)$$

where f_i is the frequency of the input sine wave, f_s is the sampling frequency, and α_m is the relative error in the sampling instants with respect to the clock signal period.

Verify that the SNR due to clock skew errors is given by

$$SNR = 10\log_{10}\left(\frac{P_i}{P_\eta}\right) \qquad (4.146)$$

where the noise power is provided by the formula

$$P_\eta = P - P_i \qquad (4.147)$$

$$P_i = |A(0)|^2 + |B(0)|^2 \qquad (4.148)$$

is the power of the input signal, and by using Parseval's relation,[2] the output signal power is estimated as

$$P = P_A + P_B = \frac{1}{2} \qquad (4.149)$$

with

$$P_A = \sum_{k=0}^{M-1} |A(k)|^2 = \frac{1}{4} \qquad (4.150)$$

$$P_B = \sum_{k=0}^{M-1} |B(k)|^2 = \frac{1}{4} \qquad (4.151)$$

Note that $A(k)$ and $B(k)$ can be considered the discrete Fourier transform of the sequences $-(1/2jM)e^{j\alpha_m(2\pi f_i/f_s)}$ and $(1/2jM)e^{-j\alpha_m(2\pi f_i/f_s)}$, respectively.

---

[2] Let $x(n)$ be an N-point sequence, and $X(k)$ its discrete Fourier transform. The next equation,

$$\sum_{n=0}^{N-1} |x(n)|^2 = \frac{1}{N} \sum_{k=0}^{N-1} |X(k)|^2$$

is known as Parseval's relation.

Bibliography

[1] H. T. Russel, Jr., "An improved successive-approximation register design for use in A/D converters," *IEEE Trans. on Circuits and Systems*, vol. 25, pp. 550–554, July 1978.

[2] C. K. Yuen, "Another design of the successive approximation register for A/D converters," *Proc. of the IEEE*, vol. 67, pp. 873–874, May 1979.

[3] J. L. McCreary and P. R. Gray, "All-MOS charge redistribution analog-to-digital conversion techniques—Part I," *IEEE J. of Solid-State Circuits*, vol. 10, pp. 371–379, Dec. 1975.

[4] H.-S. Lee and D. A. Hodges, "Self-calibration technique for A/D converters," *IEEE Trans. on Circuits and Systems*, vol. 30, pp. 188–190, March 1983.

[5] Y. S. Yee, L. M. Terman, and L. G. Heller, "A two-stage weighted capacitor network for D/A-A/D conversion," *IEEE Journal of Solid-State Circuits*, vol. SC-14, pp. 778–781, Aug. 1979.

[6] L. L. Evans, "High speed integrating analog-to-digital converter," U.S. Patent 4,395,701, filed March 25, 1980; issued July 26, 1983.

[7] B. W. Phillips, "Bipolar dual-ramp analog-to-digital converter," U.S. Patent 3,906,486, filed July 2, 1973; issued September 16, 1975.

[8] N. H. Strong, "Instantaneous gain changing analog to digital converter," U.S. Patent 4,588,983, filed June 17, 1985; issued May 13, 1986.

[9] B. J. Rodgers and C. R. Thurber, "A monolithic $\pm 5\frac{1}{2}$-digit BiMOS A/D converter," *IEEE J. of Solid-State Circuits*, vol. 24, pp. 617–626, Jun. 1989.

[10] C. W. Mangelsdorf, "A 400-MHz input flash converter with error correction," *IEEE J. of Solid-State Circuits*, vol. 25, pp. 184–191, Feb. 1990.

[11] C. L. Portmann and T. H. Y. Meng, "Power-efficient metastability error reduction in CMOS flash A/D converters," *IEEE J. of Solid-State Circuits*, vol. 25, pp. 1132–1140, Aug. 1996.

[12] K. Kattmann and J. Barrow, "A technique for reducing differential non-linearity errors in flash A/D converters," *1991 IEEE ISSCC Digest of Technical Papers*, pp. 170–171, Feb. 1991.

[13] M. Choi and A. A. Abidi, "A 6-b 1.3-Gsample/s A/D converter in 0.35-μm CMOS," *IEEE J. of Solid-State Circuits*, vol. 36, pp. 1847–1858, Dec. 2001.

[14] P. C. S. Scholtens and M. Vertregt, "A 6-b 1.6-Gsample/s flash ADC in 0.18-μm CMOS using averaging termination," *IEEE J. of Solid-State Circuits*, vol. 27, pp. 1599–1609, Dec. 2002.

[15] X. Jiang and M.-C. F. Chang, "A 1-GHz signal bandwidth 6-bit CMOS ADC with power-efficient averaging," *IEEE J. of Solid-State Circuits*, vol. 40, pp. 532–535, Feb. 2005.

[16] R. E. J. Van de Grift, I. W. J. M. Rutten, and M. Van der Veen, "An 8-bit video ADC incorporating folding and interpolation techniques," *IEEE J. of Solid-State Circuits*, vol. 22, pp. 944–953, Dec. 1987.

[17] R. J. van de Plassche and P. Baltus, "An 8-bit 100-MHz full-Nyquist analog-to-digital converter," *IEEE J. of Solid-State Circuits*, vol. 23, pp. 1334–1344, Dec. 1988.

[18] M. P. Flynn and D. J. Allstot, "CMOS folding A/D converter with current-mode interpolation," *IEEE J. of Solid-State Circuits*, vol. 31, pp. 1248–1257, Sept. 1996.

[19] A. G. W. Venes and R. J. van de Plassche, "An 80-MHz, 8-b CMOS folding A/D converter with distributed track-and-hold preprocessing," *IEEE J. of Solid-State Circuits*, vol. 31, pp. 1846–1853, Dec. 1996.

[20] S. Limotyrakis, K. Nam, and B. A. Wooley, "Analysis and simulation of distortion in folding and interpolating A/D converters," *IEEE Trans. on Circuits and Systems–II*, vol. 49, pp. 161–169, March 2002.

[21] Y. Li and E. Sánchez-Sinencio, "A wide input bandwidth 7-bit 300-Msamples/s folding and current-mode interpolating ADC," *IEEE J. of Solid-State Circuits*, vol. 38, pp. 1405–1410, Aug. 2003.

[22] T. C. Verster, "A method to increase the accuracy of fast serial-parallel analog-to-digital converters," *IEEE Trans. on Electronic Computers*, EC-13, pp. 471–473, 1964.

[23] S. H. Lewis, H. S. Fetterman, G. F. Gross, Jr., R. Ramachandran, and T. R. Viswanathan, "A 10-b 20-Msample/s analog-to-digital converter," *IEEE J. of Solid-State Circuits*, vol. 27, pp. 351–358, March 1992.

[24] T. Cho and P. R. Gray, "A 10-b, 20-Msample/s, 35-mW pipeline A/D converter," *IEEE J. of Solid-State Circuits*, vol. 30, pp. 166–172, March 1995.

[25] A. M. Abo and P. R. Gray, "A 1.5-V, 10-bit, 14.3-MS/s CMOS pipeline analog-to-digital converter," *IEEE J. of Solid-State Circuits*, vol. 34, pp. 599–606, May 1999.

[26] D. W. Cline and P. R. Gray, "A power optimized 13-b 5-Msamples/s pipelined analog-to-digital converter in 1.2-μm CMOS," *IEEE J. of Solid-State Circuits*, vol. 31, pp. 294–303, March 1996.

[27] B.-S. Song, M. F. Tompsett, and K. R. Lakshmikumar, "A 12-bit 1-Msample/s capacitor error-averaging pipeline A/D converter," *IEEE J. of Solid-State Circuits*, vol. 23, pp. 1324–1333, Dec. 1988.

[28] Y. Chiu, P. R. Gray, and B. Nikolić, "A 14-b 12-MS/s CMOS pipeline ADC with over 100-dB SFDR," *IEEE J. of Solid-State Circuits*, vol. 39, pp. 2139–2151, Dec. 2004.

[29] E. Siragusa and I. Galton, "A digitally enhanced 1.8-V 15-bit 40-MSample/s CMOS pipelined ADC," *IEEE J. of Solid-State Circuits*, vol. 39, pp. 2126–2138, Dec. 2004.

[30] M. Daito, H. Matsui, M. Ueda, and K. Iizuka, "A 14-bit 20-MS/s pipelined ADC with digital distortion calibration," *IEEE J. of Solid-State Circuits*, vol. 41, pp. 2417–2423, Nov. 2006.

[31] W. (W.) Yang, D. Kelly, I. Mehr, M. T. Sayuk, and L. Singer, "A 3-V 340-mW 14-b 75-Msample/s CMOS ADC with 85-dB SFDR at Nyquist input," *IEEE J. of Solid-State Circuits*, vol. 36, pp. 1931–1936, Dec. 2001.

[32] R. H. McCharles, V. A. Saletore, W. C. Black. Jr., and D. A. Hodges, "An algorithmic analog-to-digital converter," *1977 IEEE ISSCC Digest of Technical Papers*, section IX, pp. 96–97, Feb. 1977.

[33] C.-C. Lee, "Switched-capacitor circuit analog-to-digital converter," U.S. Patent 4,529,965, filed May 3, 1983; issued July 16, 1985.

[34] P. W. Li, M. Chin, P. R. Gray, and R. Castello, "A ratio-independent algorithmic analog-to-digital conversion technique," *IEEE J. of Solid-State Circuits*, vol. 19, pp. 828–836, Dec. 1984.

[35] K. Nagaraj, "Efficient circuit configurations for algorithmic analog to digital converters," *IEEE Trans. on Circuits and Systems–II*, vol. 40, pp. 777–785, Dec. 1993.

[36] H. Onodera, T. Tateishi, and K. Tamaru, "A cyclic A/D converter that does not require ratio-matched components," *IEEE J. of Solid-State Circuits*, vol. 23, pp. 152–158, Feb. 1988.

[37] H. Matsumoto and K. Watanabe, "Improved switched-capacitor algorithmic analogue-to-digital convertor," *Electronic Letters*, vol. 21, pp. 430–431, March 1985.

[38] Y.-C. Jenq, "Perfect reconstruction of digital spectrum from nonuniformly sampled signals," *IEEE Trans. on Instrum. Meas.*, vol. 46, pp. 649–652, June 1997.

[39] Y.-C. Jenq, "Digital spectra of nonuniformly sampled signals: Fundamentals and high-speed waveform digitizers," *IEEE Trans. on Instrum. Meas.*, vol. 37, pp. 245–251, June 1988.

[40] T. Ndjountche and R. Unbehauen, "Adaptive calibration techniques for time-interleaved ADCs," *Electronics Letters*, vol. 37, pp. 412–414, March 2001.

[41] T. Ndjountche, F.-L. Luo, and R. Unbehauen, "A high-frequency double-sampling second-order delta-sigma modulator," *IEEE Trans. on Circuits and Systems–II*, vol. 52, pp. 841–845, Dec. 2005.

[42] S. Ogawa and K. Watanabe, "A switched-capacitor successive-approximation A/D converter," *IEEE Trans. on Instrum. and Meas.*, vol. 42, pp. 847–853, Aug. 1993.

5

Delta-Sigma Data Converters

CONTENTS

In comparison with other analog-to-digital converters (ADCs) and digital-to-analog converters (DACs), delta-sigma ($\Delta\Sigma$) data converters (or generally oversampling data converters) exhibit a reduced sensitivity to analog component matching. $\Delta\Sigma$ converters are usually the best choice for applications requiring a resolution greater than 20 bits. In these converters, the input signal can be sampled at a rate much greater than the Nyquist frequency (i.e., twice the bandwidth or highest frequency of the signal being sampled), and the quantization noise is shaped by the modulator to be low in the signal band and high in the out-band spectrum. The specifications of either the analog anti-aliasing filter for the analog-to-digital conversion or the smoothing filter for the digital-to-analog conversion are then relaxed due to the signal oversampling, and the remaining out-band noise is attenuated by a filter. The actual reduction of in-band noise power level depends on the modulator structure and the oversampling ratio (OSR) and a high resolution is achieved with a penalty in speed, as the modulator hardware has to operate at the oversampling rate, and an increased complexity of the filter hardware. Oversampling data converters then present a trade-off between speed and resolution.

$\Delta\Sigma$ modulators can be exploited in the implementation of calibration stages required to improve the linearity of Nyquist converters. Built-in self-test is another application for analog signal synthesis where $\Delta\Sigma$ modulators are ideally suited.

5.1 Delta-sigma analog-to-digital converter

$\Delta\Sigma$ ADCs can be implemented using either continuous-time (CT) or discrete-time (DT) filters, as shown in Figures 5.1 and 5.2. The signal sampling is performed at the input node in the DT case by an S/H circuit, while it is implemented after the filtering in the CT structure. In both types, the system consists of a $\Delta\Sigma$ modulator followed by a digital decimator. A filter, a quantizer, and a DAC are used as building blocks of the modulator, which is based on an output-feedback structure. Note that even modulators based on a CT filter possess an equivalent in the DT domain, where the converter design is generally achieved.

During the data conversion, the modulator feedback forces the average value of the quantized signal to follow one of the input signals and the quantization noise in the signal band is attenuated. A decimation filter is required to eliminate the out-of-band noise and to reduce the sampling rate of the modulator output signal.

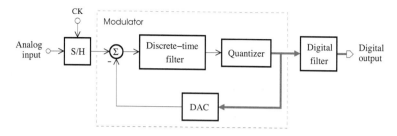

FIGURE 5.1
Block diagram of a DT $\Delta\Sigma$ ADC.

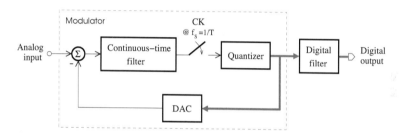

FIGURE 5.2
Block diagram of a CT $\Delta\Sigma$ ADC.

5.1.1 Time domain behavior

At each processing step, a $\Delta\Sigma$ modulator generates a digital estimation of the signal, which is subtracted from the actual sample of the input signal and the digital conversion of the resulting sequence is achieved such that the output and input signals tend to be equal on average.

In the general case, a single-stage modulator can be described in the time domain using the following equations

$$x(n) = \sum_{j=0}^{J_S} h_S(j)s(n-j) + \sum_{j=1}^{J_Q} h_Q(j)e_Q(n-j) \tag{5.1}$$

$$e_Q(n) = y(n) - x(n) \tag{5.2}$$

$$y(n) = Q[x(n)] \tag{5.3}$$

where h_S and h_Q are the impulse responses of the signal transfer function (STF) and quantization noise transfer function (QNTF), respectively; x denotes the output state of the modulator filter, J_S and J_Q represent the length of the STF and QNTF, respectively; and Q is the equivalent piecewise-constant function of the quantizer. The modulator will be considered *stable* for a given input and initial conditions if the signal samples x are bounded and the quantizer is not overloaded.

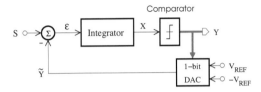

FIGURE 5.3
Block diagram of a first-order $\Delta\Sigma$ modulator.

In the special case of the first-order $\Delta\Sigma$ modulator shown in Figure 5.3, the sum ϵ of the input signal s and the output $\tilde{y}$ of a 1-bit feedback DAC is applied to an integrator, whose output x is connected to the input of a comparator delivering the digital sequence y available at the modulator output, which is then used to drive the 1-bit feedback DAC. Assuming that the supply voltages of the comparator are V_{DD} and $-V_{SS}$, and the DAC reference voltages are $\pm V_{REF}$, the modulator can be described using the next time-domain equations,

$$x(n) = x(n-1) + \epsilon(n-1) \tag{5.4}$$
$$y(n) = Q[x(n)] \tag{5.5}$$
$$\epsilon(n) = s(n) - \tilde{y}(n) \tag{5.6}$$

where

$$Q[x(n)] = \begin{cases} H & \text{if} \quad x(n) \geq 0 \\ L & \text{if} \quad x(n) < 0 \end{cases} \tag{5.7}$$

and

$$\tilde{y}(n) = \begin{cases} V_{REF} & \text{if} \quad y(n) = 1 \\ -V_{REF} & \text{if} \quad y(n) = 0 \end{cases} \tag{5.8}$$

At the modulator output, a logic high state, H, corresponds to a voltage level of about V_{DD} and a logic low state, L, is represented by approximately $-V_{SS}$.

Let a dc input signal of 0.25 V be applied to the modulator and the reference voltages of the DAC be ± 1 V. The state and output sequences of the modulator are given in Table 5.1, where the initial conditions are specified in the row associated with $n = 0$. It can be observed that the state and output sequences for $n \in [2, 9]$ are periodically repeated starting from $n = 10$. For the first-order modulator, the quantization noise is correlated with the input signal and appears not to be entirely random. The allowed input range is from V_{REF} to $-V_{REF}$, resulting in a

TABLE 5.1

State and Output Sequences of the First-Order Modulator

n	$s(n)$	$x(n)$	$y(n)$	$\tilde{y}(n)$	$\epsilon(n)$
0	—	0.10	H	1	−1.00
1	0.25	−0.90	L	−1	1.25
2	**0.25**	**0.35**	**H**	**1**	**−0.75**
3	0.25	−0.40	L	−1	1.25
4	0.25	0.85	H	1	−0.75
5	0.25	0.10	H	1	−0.75
6	0.25	−0.65	L	−1	1.25
7	0.25	0.60	H	1	−0.75
8	0.25	−0.15	L	−1	1.25
9	0.25	1.10	H	1	−0.75
10	**0.25**	**0.35**	**H**	**1**	**−0.75**
11	0.25	−0.40	L	−1	1.25
12	0.25	0.85	H	1	−0.75
13	0.25	0.10	H	1	−0.75
14	0.25	−0.65	L	−1	1.25
15	0.25	0.60	H	1	−0.75

converter full-scale range (FSR) of $2V_{REF}$, or say 2 V. A 0.25-V input signal is 1.25 V above the lower −1-V limit of the FSR, that is, the input represents $(1.25/2) \times 100 = 62.5\%$ of the FSR. By averaging the first 8 samples of the output sequence, the number of bits at the high state is 5, leading to the H-state density given by $(5/8) \times 100 = 62.5\%$.

In practice, the modulator output is decoded using a digital low-pass filter that averages every given number of samples, which can be increased to improve the overall resolution. For input signals around the mid-scale, the H-state density is about 50% in the modulator output sequence. An increase in the input signal toward the higher limit of the FSR results in an augmentation of the H-state density, while a decrease in the input signal toward the lower limit of the FSR induces a reduction in the H-state density.

Waveforms of a first-order modulator with a sine wave input signal are illustrated in Figure 5.4. The oversampling ratio is assumed to be 64. In general, it can be increased to improve the conversion resolution.

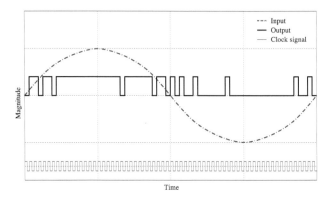

FIGURE 5.4
Waveforms of a first-order modulator with a sine wave input.

Due to oversampling, a $\Delta\Sigma$ modulator makes use of the available speed to exchange the resolution in time for that in amplitude.

5.1.2   Linear model of a discrete-time modulator

A linear model of a discrete-time modulator can be obtained based on the assumptions that different feed-ins to the filter are used by the input and feedback signals, and the quantization is done with an additive error. With reference to Figure 5.5, the output signal of the modulator can be computed as

$$Y(z) = H_S(z)S(z) + H_Q(z)E_Q(z) \qquad (5.9)$$

where

$$H_S(z) = \frac{Y(z)}{S(z)} = \frac{qH(z)}{1 + qH(z)} \qquad (5.10)$$

$$H_Q(z) = \frac{Y(z)}{E_Q(z)} = \frac{1}{1 + qH(z)} \qquad (5.11)$$

and H represents the z-domain transfer function of the loop filter, H_S denotes the signal transfer function (STF), H_Q is the quantization noise transfer function (QNTF), E_Q represents the quantization error, and q $(q > 0)$ is the quantizer gain. To simplify the analysis, the modulator is generally modeled by replacing the quantizer with a unity-gain element followed by an additive noise source. Hence,

$$H_S(z) = 1 - H_Q(z) \qquad (5.12)$$

and the QNTF determines the modulator performance and stability. It should be emphasized that a modulator realized with real components is a nonlinear system and the coupling between the signal and quantization noise is neglected in the above description.

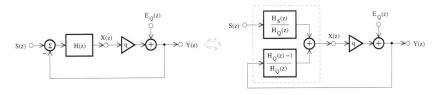

FIGURE 5.5
Block diagram of the DT $\Delta\Sigma$ modulator linear model.

$\Delta\Sigma$ modulators are generally described by the order of the loop filter, the characteristic of which determines the shape of the noise spectrum. Lowpass filters are generally used to meet the desired resolution in audio applications, while modulators based on a bandpass filter are preferred for the digitization of high-frequency band-limited signals in telecommunication systems. In the z-domain, an arbitrary L-th-order lowpass modulator can be transformed into a bandpass modulator of order $2L$ using the transformation

$$z^{-1} \rightarrow -z^{-2} \tag{5.13}$$

In this way, the zeros of the QNTF are shifted from dc to $f_s/4$, where f_s is the sampling frequency.

The resolution achievable with a single-bit $\Delta\Sigma$ ADC is limited and can generally be improved using converters based on high-order filters [4] or multibit quantizers [12].

5.1.3 Modulator dynamic range

Consider a $\Delta\Sigma$ modulator with order L operating with the oversampling ratio $OSR = f_s/f_N = f_s/(2f_{max})$, where f_s is the sampling frequency, f_N is the Nyquist frequency, and f_{max} is the highest spectral component present in the input signal. The quantization noise is shaped by a transfer function of the form

$$H_Q(z) = (1 - z^{-1})^L \tag{5.14}$$

In the frequency domain, we have

$$H_Q(jf) = \left[1 - \exp\left(-j2\pi\frac{f}{f_s}\right)\right]^L = \left[2j\sin\left(\pi\frac{f}{f_s}\right)\exp\left(-j\pi\frac{f}{f_s}\right)\right]^L \tag{5.15}$$

and

$$|H_Q(jf)| = \left[2\sin\left(\pi\frac{f}{f_s}\right)\right]^L \tag{5.16}$$

Figure 5.6 shows the plot of H_Q magnitudes for $L = 1, 2, 3, 4, 5$. The quantization noise suppression over the low-frequency signal band is improved as the value of L, or equivalently the modulator order, is increased.

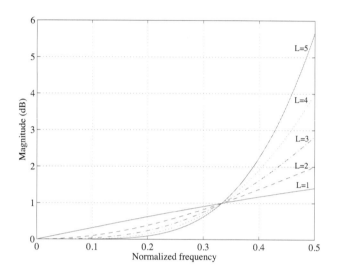

FIGURE 5.6
Plot of H_Q magnitudes for $L = 1, 2, 3, 4, 5$.

Let $\triangle$ be the quantizer step size. Assuming that the quantization error, e_Q, is evenly distributed between $-\triangle/2$ and $\triangle/2$, the mean value of e_Q is zero and the probability density of e_Q can be expressed in the form

$$
p(e_Q) = \begin{cases} \dfrac{1}{\triangle} & \text{if } e_Q \in [-\triangle/2, \triangle/2] \\ 0 & \text{otherwise.} \end{cases} \tag{5.17}
$$

The variance of the quantization noise, σ_Q^2, is then given by

$$
\sigma_Q^2 = \int_{-\infty}^{\infty} e_Q^2 p(e_Q) \mathrm{d}e_Q = \frac{1}{\triangle} \int_{-\triangle/2}^{\triangle/2} e_Q^2 \mathrm{d}e_Q = \frac{\triangle^2}{12} \tag{5.18}
$$

Generally, the input signal is sampled at the frequency f_s, and the spectral density of the quantization noise to be filtered is supposed to remain constant between 0 and $f_s/2$. That is,

$$
\sigma_Q^2 = \int_0^{f_s/2} p_i \mathrm{d}f = p_i \int_0^{f_s/2} \mathrm{d}f = p_i(f_s/2) \tag{5.19}
$$

or, equivalently,

$$
p_i = \sigma_Q^2 \frac{1}{f_s/2} = \frac{\triangle^2}{12} \frac{2}{f_s} \tag{5.20}
$$

The spectral density of the quantization noise at the modulator output can be written as

$$p_0 = |H_Q(jf)|^2 p_i = \frac{2^{2L} \triangle^2}{6 f_s} \sin^{2L}\left(\pi \frac{f}{f_s}\right) \qquad (5.21)$$

The power of the quantization noise in the Nyquist frequency range is given by

$$P_Q = \int_0^{f_N/2} p_0(f) \mathrm{d}f \qquad (5.22)$$

Because the value of the oversampling ratio $OSR = f_s/f_N$ is generally high, we have $f_N \ll f_s$. As a consequence, we have $0 \le f \le f_N/2 \ll f_s$ and $\sin(\pi f/f_s) \simeq \pi f/f_s$. Hence,

$$P_Q \simeq \frac{2^{2L} \triangle^2}{6 f_s} \int_0^{f_N/2} \left(\pi \frac{f}{f_s}\right)^{2L} \mathrm{d}f \qquad (5.23)$$

Finally, we obtain

$$P_Q \simeq \frac{\triangle^2}{12} \frac{\pi^{2L}}{2L+1} \left(\frac{1}{OSR}\right)^{2L+1} \qquad (5.24)$$

In the case of a rounding quantizer, each input sample is assigned to the nearest quantization level. The quantization error is then limited to the range of $-\triangle/2$ to $\triangle/2$, and

$$\triangle = \frac{\text{FSR}}{2^B - 1} \qquad (5.25)$$

where B is the number of bits of the quantizer. In the case of a sinusoidal signal with a peak-to-peak amplitude equal to the quantizer full-scale range, FSR, the average power is

$$P_S = \sigma_S^2 = \mathrm{E}[s^2(n)] = \frac{\text{FSR}^2}{8} \qquad (5.26)$$

The dynamic range (DR) of the modulator can then be expressed as

$$DR^2 = \frac{P_S}{P_Q}$$

$$= \frac{\left(\dfrac{\text{FSR}}{2\sqrt{2}}\right)^2}{\dfrac{\text{FSR}^2}{12(2^B-1)^2} \dfrac{\pi^{2L}}{2L+1} \left(\dfrac{1}{OSR}\right)^{2L+1}} = \frac{3}{2} \frac{2L+1}{\pi^{2L}} (2^B - 1)^2 \, OSR^{2L+1}$$

$$\qquad (5.27)$$

or in decibels,

$$DR(\text{in dB}) = 10 \log_{10}\left(\frac{3}{2} \frac{2L+1}{\pi^{2L}} (2^B - 1)^2\right) + 10(2L+1) \log_{10} OSR \quad (5.28)$$

Thus, the dynamic range, DR, is a function of the filter order and the over-sampling ratio, OSR. The multiplication of the OSR by a factor of 2 results in an increase in the DR on the order of $3(2L+1)$ dB, or equivalently, $(L+1/2)$ bits of resolution. Note that the dynamic range given by Equation (5.28) may be considered an upper bound because it is based on a linear model of the modulator. Furthermore, the stability requirements of modulators with an order equal to or greater than 2 are only met by using design techniques or structures that can constrain the modulator dynamic range well below this upper bound.

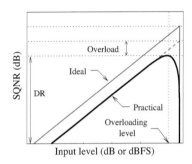

FIGURE 5.7
Curve of the signal-to-quantization noise ratio (SQNR) versus the input level.

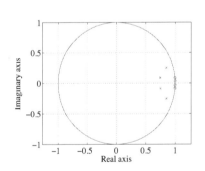

FIGURE 5.8
Pole-zero plot of the NTF.

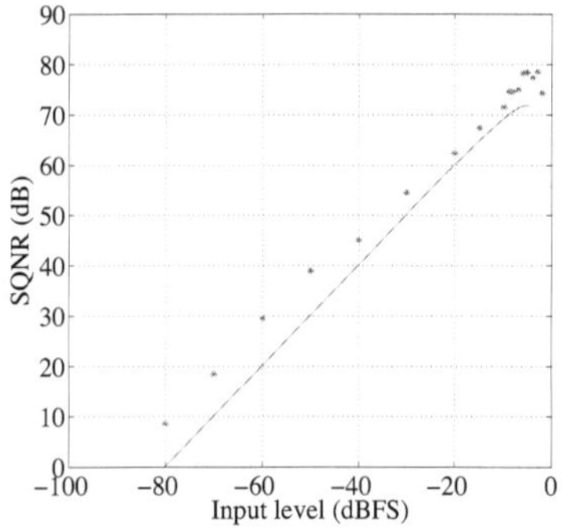

FIGURE 5.9
SQNR curve of the $\Delta\Sigma$ modulator.

The curve of the signal-to-quantization noise ratio (SQNR) versus the input level is depicted in Figure 5.7. A linear scaling effect is observed between the modulator SQNRs obtained by relying on an ideal model or a practical chip. Furthermore, a premature clipping occurs in the practical SQNR at high input levels because the slew rate of active components appears to be limited. Note that the input level can be evaluated in dB, or in dBFS (decibels relative to full scale), provided the output spectrum is normalized so that a full-scale sine wave can appear at 0 dB.

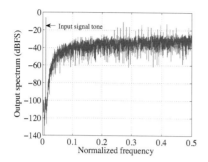

FIGURE 5.10

Output power spectrum of a fourth-order lowpass $\Delta\Sigma$ modulator.

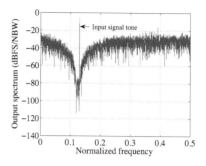

FIGURE 5.11

Output power spectrum of a fourth-order bandpass $\Delta\Sigma$ modulator.

A typical way to synthesize a $\Delta\Sigma$ modulator is to select the integrator type and signal paths such that all zeros of the noise transfer function are located on the unit circle, or equivalently at $z = 1$. This approach has the advantage of reducing the complexity of the modulator architecture, but the achievable SNR and DR may be limited.

Given a modulator order, the above performance characteristics can be improved by selecting a noise transfer function with zeros on the unit circle and poles inside the unit circle [5,6], as shown in Figure 5.8. The poles can be chosen to be identical to the ones of a filter approximation function, such as Butterworth or Chebyshev polynomial, while the zeros are optimally placed on the unit circle. In this case, the noise transfer function is of the form $H_Q(z) = N(z)/D(z)$, where N and D are two polynomials, and it is assumed that the realizability constraint, $\lim_{z\to\infty} H_Q(z) = 1$, and stability requirement are met. The resulting SQNR curve is depicted in Figure 5.9. The fourth-order lowpass modulator considered here exhibits a maximum signal-to-noise ratio of about 80 dB at an oversampling ratio of 32.

The DR formula is still valid in the case of $2L$-th bandpass modulators obtained by applying the dc-to-$f_s/4$ transformation to L-th lowpass modulator prototypes. For lowpass modulators, the STF is designed as a lowpass filtering function and the QNTF is chosen as a highpass function, while, for bandpass modulators, the STF is a bandpass function and the QNTF is a band-reject or notch function. The output spectra of fourth-order lowpass and bandpass modulators are depicted in Figures 5.10 and 5.11, respectively. The output spectrum of the bandpass modulator is determined relative to the noise power bandwidth (NBW).

5.1.4   Continuous-time modulator

Continuous-time (CT) modulators are more suitable for high-frequency applications. However, they are more sensitive to clock jitter than   their SC counterparts.

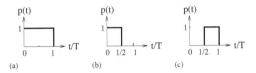

FIGURE 5.12
DAC waveforms: (a) NRZ, (b) RZ, (c) HRZ.

In CT modulators, the sampling of the signal takes place in the loop and the stability must be analyzed in the DT domain. The design of the modulator filter then results in the equivalent transfer function $H(z)$. Note that the unstable behavior will be observed when the poles are located far away from the zeros even for input signals with a low amplitude, while the performance is compromised by placing the poles very close to the zeros. Let us assume that the operation of a 1-bit quantizer can be described by the following expression,

$$p(t) = \begin{cases} 1 & \text{if } p_1 T \le t \le p_2 T \\ 0 & \text{otherwise,} \end{cases} \qquad (5.29)$$

where $p_1, p_2 \in [0, 1]$ and T is the sampling period. In the s-domain, this corresponds to the zero-order hold pulse transfer function given by

$$P(s) = (e^{-s p_1 T} - e^{-s p_2 T})/s \qquad (5.30)$$

The type of the DAC pulse, which is determined by the values of p_1 and p_2, is nonreturn-to-zero (NRZ) for $p_1 = 0$ and $p_2 = 1$, return-to-zero (RZ) for $p_1 = 0$ and $p_2 = 1/2$, and half-clock-period delayed return-to-zero (HRZ) for $p_1 = 1/2$ and $p_2 = 1$ (see Figure 5.12).

An NRZ DAC is characterized by an output signal that remains constant throughout the whole period of the clock signal. It can easily be implemented. However, the mismatch between the rise and fall times of the output waveform causes inter-symbol interference that limits the linearity of the NRZ DAC. When this data-dependent transient nonlinearity is translated into even-order harmonic distortions, it can be canceled by fully differential circuits.

A RZ (or HRZ) DAC exhibits an output signal that is active only during a half of the clock signal period. To deliver the same amount of charge per clock signal period as an NRZ DAC, its output signal magnitude should be two times greater than that of the NRZ DAC. It can still operate linearly even in the presence of a mismatch between the rise and fall times of the output waveform. However, the RZ (or HRZ) DAC is more sensitive to clock jitter than the NRZ DAC because it requires twice as many output signal transitions.

CT modulators are very sensitive to clock jitter because the amount of feedback charge generated by the DAC is a function of the clock signal pulse width. The use of various other (exponentially decaying, sine-shaped) DAC pulses with a duty cycle shorter than the clock period is one of the approaches to reduce the sensitivity of CT modulators to clock jitter. The clock jitter effect can also be attenuated by increasing the DAC resolution or by lowering the oversampling ratio. But, due to the inherent mismatch between the DAC elements, multi-bit DACs generally require calibration.

The CT transfer function, $H(s)$, can be obtained using the impulse invariant transformation as follows,

$$\mathcal{Z}^{-1}\{H(z)\} = \mathcal{L}^{-1}\{P(s)\hat{H}(s)\}|_{t=nT} \tag{5.31}$$

where $\mathcal{Z}^{-1}$ and $\mathcal{L}^{-1}$ denote the inverses of the z-transform and Laplace transform, respectively; $H(z)$ is the transfer function of the DT filter; and $\hat{H}(s)$ represents the transfer function of the CT filter. This last equation can equivalently be written in the time domain as

$$h(nT) = [p(t) * \hat{h}(t)]\Big|_{t=nT} = \left(\int_{-\infty}^{\infty} p(\tau) * \hat{h}(t-\tau)d\tau\right)\Big|_{t=nT} \tag{5.32}$$

where $*$ represents the time convolution, $h(nT)$ is the impulse response of the DT filter, $p(t)$ is the impulse response of the DAC, and $\hat{h}(t)$ denotes the impulse response of the CT filter.

The classical approach used for the design of a CT modulator consists of first choosing the appropriate z-domain QNTF, $H_Q(z)$, that meets the required specifications and converting it to the DT loop transfer function $H(z) = (H_Q(z) - 1)/H_Q(z)$. The CT filter transfer function, $\hat{H}(s)$, is then obtained by solving the equation of the impulse invariant transformation with a symbolic math program or numerical methods. Whenever possible, the DT transfer function of the modulator can be decomposed into partial fractions of the form $H_i(z)$, $i = 1, 2$, and the equivalent CT function is derived using the results of Table 5.2, where f_s is the clock signal frequency [7]. Note that l'Hopital's rule was exploited to obtain the s-domain equivalent functions in the specific case where $z_k = 1$ (i.e., the poles are located at dc). The impulse invariant method has the advantage of resulting in circuits with a low complexity in comparison with other approaches.

In practice, the settling behavior of the DAC and quantizer are affected by the excess loop delay due to the nonzero switching time of transistors in the quantizer latch and DAC, and timing jitter.

It can be observed that in the CT structure, the clock jitter, which causes a variation in the width of DAC pulses, disturbs the sum of the input signal and quantization noise, because the sampling occurs at the quantizer rather than the input. In the DT case, only the input signal is affected. As a result,

TABLE 5.2
Impulse-Invariant Transformation of Functions with Single, Double, and Triple Poles

z-domain	s-domain
$H_1(z) = \dfrac{z^{-1}}{1 - z_k z^{-1}}$	$\hat{H}_1(s) = \begin{cases} \dfrac{s_k}{q_1(s - s_k)}, & \text{if } z_k \neq 1 \\[2ex] \dfrac{f_s}{(p_2 - p_1)s}, & \text{if } z_k = 1 \end{cases}$
$H_2(z) = \dfrac{z^{-2}}{(1 - z_k z^{-1})^2}$	$\hat{H}_2(s) = \begin{cases} \dfrac{(q_2 s_k + q_1 f_s)s - q_2 s_k^2}{z_k q_1^2 (s - s_k)^2}, \\[2ex] \qquad\qquad\qquad \text{if } z_k \neq 1 \\[2ex] \dfrac{-f_s\left(1 - \dfrac{p_1 + p_2}{2}\right)s + f_s^2}{(p_2 - p_1)s^2}, \\[2ex] \qquad\qquad\qquad \text{if } z_k = 1 \end{cases}$
$H_3(z) = \dfrac{z^{-3}}{(1 - z_k z^{-1})^3}$	$\hat{H}_3(s) = \begin{cases} \dfrac{r_2 f_s s^2 + r_1 f_s^2 s + r_0 f_s^3}{z_k^2 q_1^3 (s - s_k)^3}, \\[2ex] \qquad\qquad\qquad \text{if } z_k \neq 1 \\[2ex] \dfrac{r f_s s^2 - f_s^2\left(\dfrac{3}{2} - \dfrac{p_1 + p_2}{2}\right)s + f_s^3}{(p_2 - p_1)s^3}, \\[2ex] \qquad\qquad\qquad \text{if } z_k = 1 \end{cases}$

where
$f_s = 1/T$
$s_k = \ln(z_k)/T$
$q_1 = z_k^{1-p_1} - z_k^{1-p_2}$
$q_2 = (1 - p_2)z_k^{1-p_2} - (1 - p_1)z_k^{1-p_1}$
$r_0 = (q_4/2)s_k^3$
$r_1 = -q_4 s_k^2 + q_3 s_k + q_1^2$
$r_2 = (q_4/2)s_k - q_3$
$r = 1 + [p_1(p_1 - 9) + p_2(p_2 - 9) + 4p_1 p_2]/12$
$q_3 = (3/2 - p_1)(z_k^{1-p_1})^2 + (3/2 - p_2)(z_k^{1-p_2})^2 + (p_1 + p_2 - 3)z_k^{1-p_1}z_k^{1-p_2}$
and
$q_4 = (1 - p_1)(2 - p_1)(z_k^{1-p_1})^2 + (1 - p_2)(2 - p_2)(z_k^{1-p_2})^2$
$\qquad + [p_1(p_1 + 3) + p_2(p_2 + 3) - 4(1 + p_1 p_2)]z_k^{1-p_1}z_k^{1-p_2}$

the signal-to-noise ratio (SNR) of CT modulators is more severely affected by the timing jitter in the quantizer clock than the SNR of the equivalent DT versions. It should be noted that modulators with an NRZ DAC are less sensitive to the clock jitter than the one with RZ or HRZ DACs.

Furthermore, the performance of CT $\Delta\Sigma$ modulators can be affected by the so-called excess delay, which is required by the quantizer to update its output. As a result, the DAC pulse can extend beyond T (or the clock period end) and the order of the equivalent DT loop filter is now one unit higher than the CT filter order. In general, solutions at the circuit level (appropriate selection of the DAC pulse, feedback coefficient tuning, use of extra feedback paths) can be used for the compensation.

5.1.5 Lowpass delta-sigma modulator

Lowpass $\Delta\Sigma$ modulators are based on discrete-time integrators with a delay, and whose transfer function is

$$I(z) = \frac{z^{-1}}{1 - z^{-1}} \tag{5.33}$$

and a comparator. Note that the term $1/(1 - z^{-1})$ is generally realized by a switched-capacitor integrator, but the delay z^{-1} introduced in the transfer function numerator can be implemented using appropriate clock signals at the integrator input and output, or at the integrator input and to drive the quantizer.

5.1.5.1 Single-stage modulator with a 1-bit quantizer

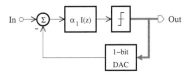

FIGURE 5.13
Block diagram of a first-order modulator.

The block diagram of a first-order modulator is shown in Figure 5.13. It consists of an integrator, a 1-bit quantizer or comparator, and a 1-bit DAC. With the assumption that $\alpha_1 = 1$, the STF and QNTF are given by

$$H_S(z) = z^{-1} \quad \text{and} \quad H_Q(z) = 1 - z^{-1} \tag{5.34}$$

This structure has a large dynamic range, and is simple and less sensitive to the component nonidealities. However, the quantization noise can be signal dependent and not statistically uncorrelated with the input signal as it is usually assumed. As a consequence, single-frequency tones appear in the modulator output spectrum for slowly varying input signals. This effect can be prevented by whitening the quantization noise through *dithering*. It consists of adding a pseudo-random sequence, which is independent and uncorrelated with the input signal, at the quantizer input. The transfer function from the dither input to the modulator output must be proportional to the one of the quantization noise. Furthermore, the magnitude of the dither signal should be chosen so that the quantizer cannot overload. The increase of the number of state variables (integrator input and output) can also help to reduce tones by preventing the formation of a repeating bit pattern at the modulator output.

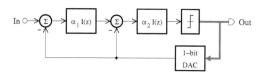

FIGURE 5.14
Block diagram of a second-order modulator.

The modulator shown in Figure 5.14 achieves a second-order shaping of the quantization noise. It uses two integrators in the filter loop. In comparison to the first-order structure, the number of internal states is increased and the occurrence likelihood of spectral tones is reduced. Here for $\alpha_2/2 = 2\alpha_1 = 1$, the STF and QNTF can be written in the form

$$H_S(z) = z^{-2} \qquad \text{and} \qquad H_Q(z) = (1 - z^{-1})^2 \qquad (5.35)$$

Here, the QNTF exhibits two zeros at dc and can provide an improved attenuation of the noise in the baseband compared to the one of the first-order modulator.

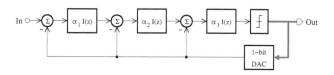

FIGURE 5.15
Block diagram of a third-order modulator.

The third-order modulator of Figure 5.15 is derived by adding an integrator stage and a feedback path to the second-order structure. The STF and QNTF can be, respectively, expressed as

$$H_S(z) = \frac{\prod_{i=1}^{3} \alpha_i [I(z)]^3}{1 + \sum_{i=1}^{3} \prod_{j=i}^{3} \alpha_j [I(z)]^{3-i+1}} \tag{5.36}$$

$$= \frac{\alpha_1 \alpha_2 \alpha_3 z^{-3}}{D(z)} \tag{5.37}$$

and

$$H_Q(z) = \frac{1}{1 + \sum_{i=1}^{3} \prod_{j=i}^{3} \alpha_j [I(z)]^{3-i+1}} \tag{5.38}$$

$$= \frac{(1 - z^{-1})^3}{D(z)} \tag{5.39}$$

where

$$D(z) = 1 + (\alpha_3 - 3)z^{-1} + [\alpha_3(\alpha_2 - 2) + 3]z^{-2} + [\alpha_3(\alpha_2 \alpha_1 - \alpha_2 + 1) - 1]z^{-3} \tag{5.40}$$

With the following set of coefficients, $\alpha_1 = 0.2$ and $\alpha_2 = \alpha_3 = 0.5$, the expected SNR, DR, premature overloading are 80 dB, 86 dB, and 0.55 (with reference to 1, ideally), respectively, at an OSR of 64.

Due to the nonlinear nature of $\Delta\Sigma$ modulators, the stability can depend on characteristics such as the input signal level or initial conditions. Simulations show that the use of a multi-bit quantizer, which can better accommodate large signals than a single-bit quantizer, is necessary to stabilize a single-loop modulator with the QNTF of the form $H_Q(z) = (1 - z^{-1})^L$, where $L > 2$. An L-th-order $\Delta\Sigma$ modulator will then be stable if the quantizer possesses $B \geq L + 1$ bits of resolution. However, higher-order single-loop modulators with a single-bit quantizer can be made stable by matching the QNTF to a more general highpass or band-reject transfer function. This is generally achieved either using the Butterworth or inverse-Chebyshev filter approximations to find the QNTF and then suitable zeros are added to the numerator in order to improve the attenuation of the baseband quantization noise or by numerically finding the poles and zeros of the QNTF, which can provide a more effective shaping of the baseband quantization noise out of the band of interest, with the help of computer-aided design tools. It may then be necessary to increase the complexity of the loop filter structure, as shown in Figures 5.16 and 5.17. These third-order modulator structures are based on single-stage topologies with feedforward and feedback paths and can allow the QNTF zeros to be spread over the signal bandwidth instead of being all placed at dc.

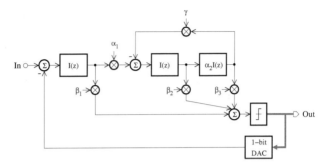

FIGURE 5.16
Block diagram of the third-order modulator with a feedforward summation and local resonator feedback.

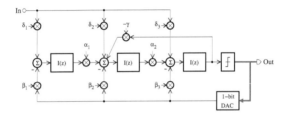

FIGURE 5.17
Block diagram of the third-order modulator with feedforward input paths, distributed feedbacks, and a local resonator feedback.

Single-bit modulators with an order equal to or greater than three possess a dynamic range lower than the one predicted by Equation (5.28) due to the attenuation required in the signal path in order to meet the loop stability condition. The design objective is to find the coefficient combination that provides the maximal dynamic range, while maintaining the modulator stability.

Most of the methods for generating the loop coefficients from modulator specifications focus on synthesizing a QNTF based on a filtering function. Because a delay-free loop around a quantizer is not implementable, the associated QNTF should have the property that $\lim_{z \to \infty} H_Q(z) = 1$. Let the noise power gain (NPG) be defined as

$$NPG = \frac{1}{\pi} \int_0^\pi (|H_Q(e^{j\omega})|)^2 d\omega \qquad (5.41)$$

The modulator must be designed such that the NPG limitation is satisfied. That is,

$$\text{NPG}_{min} \leq \text{NPG} \leq \text{NPG}_{max} \qquad (5.42)$$

where NPG_{min} is determined by the acceptable level of in-band tones, and

NPG_{max} depends on the stability requirement that is affected by the modulator order and the maximum power of the input dc signal. To take into account the effect of coefficient variations due to component imperfections, the upper bound of the NPG must be selected with a safety margin from the instability border. The resulting NPG is generally a function of the in-band noise suppression, the OSR, and the modulator order.

As a design example with a maximum stable input signal of -6 dB, a bandwidth of 20 kHz, an OSR of 64, a SNR of 89 dB, and DR of 92 dB, the coefficients of the third-order modulator shown in Figure 5.16 can be obtained as: $\alpha_1 = 0.598$, $\alpha_2 = 0.709$, $\alpha_3 = 0.196$, $\beta_1 = 1.543$, $\beta_2 = 0.895$, $\beta_3 = 0.782$, and $\gamma = 0.013$.

In the case of the third-order modulator shown in Figure 5.17, an improved DR of 95 dB is achieved with the following set of coefficients: $\alpha_1 = 0.2273$, $\alpha_2 = 0.2972$, $\alpha_3 = 0.7060$, $\beta_1 = 1$, $\beta_2 = 1.2915$, $\beta_3 = 0.9332$, $\delta_1 = 1$, $\delta_2 = \delta_3 = 0$, and $\gamma = 0.0086$.

Note that γ is the most sensitive coefficient for SNR performance because it determines the locations of the NTF zeros. The input frequency used for SNR and DR calculations is $f_s/1024$ and f_s is the sampling frequency.

5.1.5.2 Dithering

In general, the time-domain output waveform of $\Delta\Sigma$ modulators can be affected by idle tones or pattern noise, which appears as periodic impulses whose peak levels are much greater than their rms values. This behavior, which is due to the correlation between the quantizer error and the dc level of the input signal, is undesirable, especially in audio applications. A solution consists of using dithering, which can be realized by adding a pseudo-random sequence to the quantizer input. As a result, the quantization noise is made almost independent of the modulator input signal and asymptotically white in some cases.

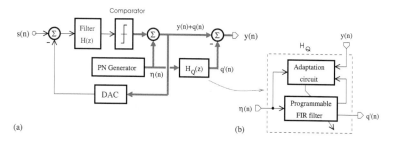

FIGURE 5.18
(a) Block diagram of a modulator with dithering; (b) adaptive filter-based realization of $H_Q(z)$.

The block diagram of a modulator with dithering is shown in Figure 5.18(a). Here, the dithering signal, which is a digital sequence provided by a pseudo-random number generator, can be added to the comparator output [8], instead of the input. Because it is shaped by the modulator in the same way as the quantization noise, its cancelation at the modulator output is realized using a filter section with the transfer function, H_Q, and a subtractor. The noise transfer function, H_Q, is determined by the specifications of the loop filter, and is implemented using either a conventional digital filter or an adaptive filter [9], as illustrated in Figure 5.18(b). This latter approach can provide a better tracking of the modulator response.

5.1.5.3 Design examples

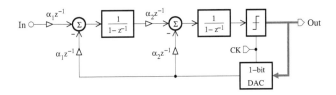

FIGURE 5.19
Block diagram of a second-order lowpass DT modulator.

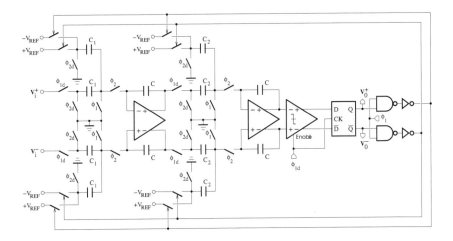

FIGURE 5.20
Circuit diagram of a second-order lowpass DT modulator.

The block diagram of a second-order lowpass modulator is shown in Figure 5.19, where $\alpha_1 = 1/2$, $\alpha_2 = 2$. This modulator is implemented as a fully differential circuit depicted in Figure 5.20, where the comparator is assumed to have an enable (or comparison) phase and a reset phase. It is based on a switched-capacitor (SC) integrator and operates with nonoverlapping two-phase clock signals. Phase 1 includes ϕ_1 and ϕ_{1d}, while ϕ_2 and ϕ_{2d} constitute phase 2. The comparator can provide erroneous decisions due to the fact that its decision time generally increases for input signals with a low magnitude. It is then followed by a flip-flop, which can reduce the bit error due to the metastability. The switching of the reference voltages is controlled by a digital circuit, which consists of NAND gates and inverters with a buffer function. Simulation results show that the dynamic range of the modulator increases by 15 dB for every doubling of the OSR.

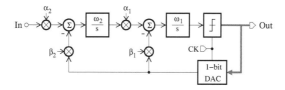

FIGURE 5.21
Block diagram of a second-order lowpass CT modulator.

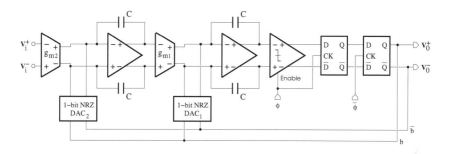

FIGURE 5.22
Circuit diagram of a second-order lowpass CT modulator.

The block diagram of a second-order lowpass modulator, which uses a continuous-time filter, is shown in Figure 5.21, where $\alpha_1 = \alpha_2 = 1$, $\beta_1 = 1.5$, $\beta_2 = 1$, and $\omega_1 = \omega_2 = 1$. The implementation of this modulator shown in Figure 5.22 is based on g_m-C operational amplifier integrators. The output signal of the filter is quantized by the latched comparator and then processed by the D flip-flops, which drive the two 1-bit switched-current DACs used in the feedback path.

Figure 5.23 shows the principle, waveforms, and circuit diagram of a differential 1-bit NRZ DAC. The DAC output current remains constant over a full period of the clock signal.

Because $\hat{H}(s)$ cannot uniquely be determined by the impulse invariant transformation, a suitable CT filter prototype is generally used to solve Equation (5.31). For the filter used in the

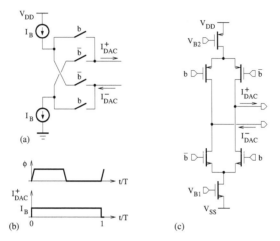

FIGURE 5.23
Differential 1-bit NRZ DAC: (a) principle, (b) waveforms, (c) circuit diagram.

modulator of Figure 5.22, it can be assumed that the DAC pulse
is of the NRZ type and the filter has the following transfer function:

$$\hat{H}(s) = \frac{\beta_1}{s} + \frac{\beta_2}{s^2} \tag{5.43}$$

Here, the equivalent DT transfer function can be obtained as

$$H(z) = \mathcal{Z}\left\{\mathcal{L}^{-1}\left[P(s)\hat{H}(s)\right]_{t=nT}\right\} \tag{5.44}$$

Because the s-transform of the DAC pulse is given by

$$P(s) = \frac{1 - e^{-sT}}{s} \tag{5.45}$$

we have

$$H(z) = (1 - z^{-1})\mathcal{Z}\left\{\mathcal{L}^{-1}\left[\frac{\hat{H}(s)}{s}\right]_{t=nT}\right\} \tag{5.46}$$

Noting that

$$\mathcal{Z}\left\{\mathcal{L}^{-1}\left[\frac{1}{s^2}\right]_{t=nT}\right\} = \frac{Tz^{-1}}{(1 - z^{-1})^2} \tag{5.47}$$

$$\mathcal{Z}\left\{\mathcal{L}^{-1}\left[\frac{1}{s^3}\right]_{t=nT}\right\} = \frac{T^2z^{-1}(1 + z^{-1})}{(1 - z^{-1})^3} \tag{5.48}$$

and $T = 1$, the equivalent z-domain transfer function can be expressed as

$$H(z) = \frac{(2\beta_1 + \beta_2)z^{-1} + (-2\beta_1 + \beta_2)z^{-2}}{2(1 - z^{-1})^2} \quad (5.49)$$

In the case where $\beta_1 = 1.5$ and $\beta_2 = 1$, we can obtain

$$H(z) = \frac{2z^{-1} - z^{-2}}{(1 - z^{-1})^2} \quad (5.50)$$

The CT modulator of Figure 5.21 is then equivalent to a second-order DT modulator with the QNTF given by

$$H_Q(z) = \frac{1}{1 + H(z)} = (1 - z^{-1})^2 \quad (5.51)$$

Note that the design of a CT modulator can be based on other types of DAC pulses. In order to obtain the output at any time between two consecutive sampling instants, the expression of the equivalent DT transfer function should be rewritten as

$$H(z) = \mathcal{Z}_m \left\{ \mathcal{L}^{-1} \left[P(s)\hat{H}(s) \right]_{t=nT} \right\} \quad (5.52)$$

where $\mathcal{Z}_m$ denotes the modified z-transform.

In the analysis of the excess loop delay, the transfer function provided by the modified z-transform, which is based on the assumption that the delay occurs at the output of the CT filter, is similar but not identical to the one obtained by solving the impulse invariant transformation equation in the time domain, where the fact that the delay occurs prior to the DAC pulse, as it is the case in practical circuit, can be taken into account.

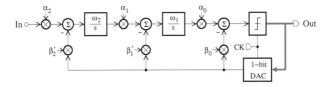

FIGURE 5.24
Block diagram of a second-order lowpass CT modulator with an extra feedback path.

$$\alpha_0 = \alpha_1 = \alpha_2 = 1, \, w_1 = w_2 = 1.$$

One approach to compensate for the excess loop delay consists of adding extra feedback paths to the modulator. In the modulator block diagram shown in Figure 5.24, the additional path connects the DAC output to the summer inserted between the last integrator and the quantizer. The transfer function of the CT filter now becomes

$$\hat{H}(s) = \beta_0 + \frac{\beta_1'}{s} + \frac{\beta_2'}{s^2} \tag{5.53}$$

Considering an excess loop delay equal to τ, the z-domain equivalent transfer function of the filter can be obtained as

$$H_\tau(z) = \mathcal{Z}\left\{\mathcal{L}^{-1}\left[P_\tau(s)\hat{H}(s)\right]_{t=nT}\right\} \tag{5.54}$$

where T is the clock signal period, and the Laplace transform of the DAC pulse, $P_\tau(s)$, is given by

$$P_\tau(s) = \int_\tau^{T+\tau} e^{-st}dt \tag{5.55}$$

$$= \frac{e^{-s\tau} - e^{-s(T+\tau)}}{s} \tag{5.56}$$

$$= \frac{e^{-s\tau} - e^{-sT}}{s} + e^{-sT}\frac{1 - e^{-s\tau}}{s} \tag{5.57}$$

Using the following impulse invariant transformations

$$\mathcal{Z}\left\{\mathcal{L}^{-1}\left[\frac{P_\tau(s)}{s}\right]_{t=nT}\right\} = \frac{(1-\tau)z^{-1}}{1-z^{-1}} + z^{-1}\frac{\tau z^{-1}}{1-z^{-1}} \tag{5.58}$$

and

$$\mathcal{Z}\left\{\mathcal{L}^{-1}\left[\frac{P_\tau(s)}{s^2}\right]_{t=nT}\right\} = \frac{(1-\tau)^2 z^{-1} + (1-\tau^2)z^{-2}}{2(1-z^{-1})^2}$$

$$+ z^{-1}\frac{\tau(2-\tau)z^{-1} + \tau^2 z^{-2}}{2(1-z^{-1})^2} \tag{5.59}$$

it can be shown that

$$H_\tau(z) = \frac{az^{-1} + bz^{-2} + cz^{-3}}{2(1-z^{-1})^2} \tag{5.60}$$

where

$$a = 2\beta_0 + 2\beta_1'(1-\tau) + \beta_2'(1-\tau)^2 \tag{5.61}$$

$$b = -4\beta_0 - 2\beta_1'(1-2\tau) + \beta_2'(1+2\tau-2\tau^2) \tag{5.62}$$

and

$$c = 2\beta_0 - 2\beta_1'\tau + \beta_2'\tau^2 \tag{5.63}$$

The compensation for the excess loop delay is achieved provided the transfer functions $H(z)$ and $H_\tau(z)$ are matched. Setting Equation (5.49) equal to (5.60) yields

$$\beta_0 = \beta_1 \tau + \beta_2 \tau^2 / 2 \tag{5.64}$$

$$\beta_1' = \beta_1 + \beta_2 \tau \tag{5.65}$$

and

$$\beta_2' = \beta_2 \tag{5.66}$$

The aforementioned method for the derivation of the feedback coefficients may become cumbersome and impractical due the nonideal characteristics (parasitic poles and zeros) of amplifiers used in the integrator design. As a consequence, numerical techniques and behavioral simulations are often used in practice.

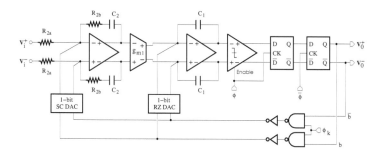

FIGURE 5.25
Circuit diagram of a second-order lowpass CT modulator.

A CT modulator can also be implemented using RZ DACs. Let us consider the following second-order DT signal transfer function,

$$H(z) = \frac{z^{-1}(2 - z^{-1})}{(1 - z^{-1})^2} \tag{5.67}$$

$$= H_2(z) + H_1(z) \tag{5.68}$$

where $H_1(z) = 2/(z - 1)$ and $H_2(z) = 1/(z - 1)^2$. Using a switched-capacitor with a series resistor (SCR) DAC [10, 11] and a switched-current RZ DAC, the equivalent CT transfer function can be obtained by making the quantizer inputs of DT and CT modulator prototypes equal at the sampling instants, nT. In the z-domain, for $j = 1, 2$, this translates into the equation to be solved for the CT transfer function,

$$\mathcal{Z}^{-1}\{H_j(z)\} = \mathcal{L}^{-1}\{P_{H_j}(s)\hat{H}_j(s)\}|_{t=nT} \tag{5.69}$$

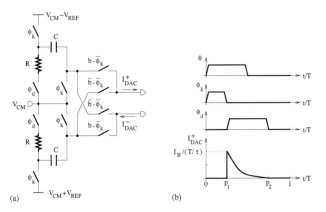

FIGURE 5.26
Circuit diagram and waveforms of a 1-bit SCR DAC.

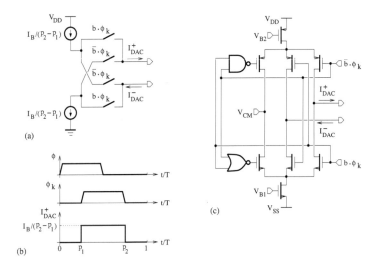

FIGURE 5.27
Principle, waveforms, and circuit diagram of a 1-bit RZ DAC.

where

$$P_{H_2}(s) = \frac{e^{-sp_1 T}\left(1 - e^{-(s+1/\tau_d)(p_2 - p_1)T}\right)}{s + 1/\tau_d} \tag{5.70}$$

$$P_{H_1}(s) = \frac{e^{-sp_1 T} - e^{-sp_2 T}}{s} \tag{5.71}$$

and τ_d is the discharging time constant of the DAC. In the time domain, by making the impulse responses identical at the sampling instants, the equivalent equation is derived as,

$$h(nT) = p_{h_j}(t) * h_j(t)|_{t=nT} = \int_{-\infty}^{+\infty} p_{h_j}(\tau) h_j(t - \tau) d\tau \bigg|_{t=nT} \qquad (5.72)$$

where

$$p_{h_2}(t) = \begin{cases} e^{-(t-p_1 T)/\tau_d} & p_1 T \le t < p_2 T \\ 0 & \text{otherwise} \end{cases} \qquad (5.73)$$

$$p_{h_1}(t) = \begin{cases} 1 & p_1 T \le t < p_2 T \\ 0 & \text{otherwise.} \end{cases} \qquad (5.74)$$

The CT transfer function of the loop filter can be obtained as

$$\hat{H}(s) = \hat{H}_2(s) + \hat{H}_1(s) \qquad (5.75)$$

$$= \frac{r_{21}s + r_{20}}{s^2} + \frac{2r_{10}}{s} \qquad (5.76)$$

where

$$r_{10} = \frac{1}{(p_2 - p_1)T} \qquad (5.77)$$

$$r_{20} = \frac{1}{\tau_d(1 - e^{-(p_2 - p_1)T/\tau_d})T} \qquad (5.78)$$

$$r_{21} = \frac{p_1 + \tau_d/T - 1 - (p_2 + \tau_d/T - 1)e^{-(p_2 - p_1)T/\tau_d}}{\tau_d(1 - e^{-(p_2 - p_1)T/\tau_d})^2} \qquad (5.79)$$

The resulting second-order CT modulator can be implemented as shown in Figure 5.25. The loop filter consists of a feedback RC integrator followed by a $g_m C$-OA integrator.

The circuit diagram and waveforms of a 1-bit SCR DAC are represented in Figure 5.26, where ϕ is the clock signal. During the phase ϕ_k, that lasts for a quarter of the clock signal period, the capacitors C are charged to a potential difference of $2V_{REF}$. Afterward, depending on the input data bit b (comparator output), the capacitors C are discharged through the resistors R that are switched on during the phase ϕ_d, whose duration is one-half of the clock signal period.

Figure 5.27 shows the principle, waveforms, and circuit diagram of a differential 1-bit RZ DAC. Here, ϕ_k lasts for a quarter of the clock signal period. Depending on the input data bit, the DAC can deliver a current of the form, $I_{DAC}^+ = bI_B/(p_2 - p_1)$, where $b = \pm 1$. During the return-to-zero phase, either the transistor switch driven by the NAND gate or the one driven by the NOR gate is closed so that the DAC current can be directed to the path connected to the common-mode voltage, V_{CM}. In this way, the current source transistors are prevented from going into the non-linear region, hence relaxing the settling requirements.

Note that by choosing the peak current equal to $2I_B$ for the RZ DAC and I_BT/τ for the SCR DAC, each DAC can deliver the same total charge per clock signal period if $\tau \ll T/2$.

In general, by operating with symmetric signals, differential architectures have the advantage of reducing the inter-symbol interference effects caused by unequal rise and fall times of the DAC pulses.

5.1.5.4   Modulator architectures with a multi-bit quantizer

A multi-bit $\Delta\Sigma$ modulator is realized by using a multi-bit quantizer (or say, a B-bit ADC) and a DAC. The main advantage of multi-bit modulators is the increase in the dynamic range by about $20 \log_{10}(2^B - 1)$ dB compared to that of modulators with a single-bit quantizer. This is due to the fact that the power of the quantization noise is proportional to the square of the quantizer step size, which is reduced by increasing the number of quantization levels in the converter range. The oversampling ratio required to achieve a given conversion resolution can then be reduced. In the design of high-order ($L > 2$), single-loop modulators, the use of a multi-bit quantizer helps prevent the instability due to quantizer overload and observed in most single-bit structures. In the case of low-order ($L \leq 2$) modulators, an improved attenuation of tones can be expected because the quantization noise is more randomly distributed as the number of bits of the quantizer is increased.

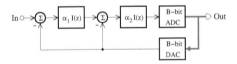

FIGURE 5.28
Block diagram of a multi-bit second-order modulator.

The block diagram of a multi-bit second-order modulator is depicted in Figure 5.28, where $\alpha_1 = 1/2$, $\alpha_2 = 2$, and $I(z) = z^{-1}/(1 - z^{-1})$. The multi-bit DAC used in the feedback loop can be implemented using either current-steering circuits, wherein transistor matching is essential to obtain a high linearity, or charge-redistribution circuits, in which capacitor matching is required. By modeling the quantizer as an additive source with the quantization noise E_Q and the multi-bit DAC as an additive source with the nonlinearity error E_D, the modulator output, Y, is given by

$$Y(z) = z^{-2}S(z) + (1 - z^{-1})^2 E_Q(z) - E_D(z) \qquad (5.80)$$

where S denotes the input signal. Hence, the performance of the multi-bit modulator is limited by the nonlinearity of the B-bit internal DAC, which results in distortions directly added to the input signal. Modulator designs were reported using digital correction techniques [12] or dynamic element matching

(DEM) methods such as data-weighted averaging (DWA) to reduce the effect of the DAC mismatch errors that can be introduced in the signal baseband.

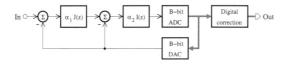

FIGURE 5.29
Block diagram of a multi-bit second-order modulator with digital calibration.

The general architecture used for the digital calibration is illustrated by the block diagram shown in Figure 5.29. During the calibration, the system is configured to allow the estimation of DAC errors that are stored in memory and subsequently used for correction.

The internal ADC is generally of the flash type and its hardware complexity increases exponentially with the bit resolution. That is, a suitable choice for the number of bits, B, is on the order of 4. The block diagram of the digital calibration required to meet the accuracy of 16 bits using a second-order modulator with an OSR of 128 is shown in Figure 5.30 [12]. It is based on the length truncation of data stored in a random access memory (RAM). In the worst case, the 4-bit internal DAC is assumed to exhibit a linearity of 9 bits. The modulator output is transferred to the address lines for the selection of the corresponding RAM word with the length of 10 bits (1 sign bit + 9 bits). This latter is reduced to 3 bits by a first-order digital $\Delta\Sigma$ modulator. The compressed RAM word, $\mathbf{e}$, is then added to $\mathbf{x}$, which is a 10-bit delayed version of the modulator output. The modulator output sequences form the four MSBs, the fifth MSB is set to 1 to assign the positive sign of $\mathbf{x}$ to the addition result for any value of $\mathbf{e}$, and the remaining bits are zero. The 10-to-4 bit truncation of $\mathbf{s}$ is then achieved using the structure shown in Figure 5.31. Note that the characteristic of the truncator is determined by the signal resolution predicted by simulations to be at least 18 bits. The scheme depicted in Figure 5.32 is used to store the conversion errors of the 4-bit DAC in the RAM. The analog equivalent of the digital input code generated by a 4-bit counter is applied to the initial multi-bit $\Delta\Sigma$ modulator operating as a single-bit converter. The decimation stage is based on counters, which can provide an 18-bit word for each digital input code, which is held for 2^{18} clock periods. The error data to be stored in the RAM is computed as the difference between the converted and the original input code. The overall calibration requires $2^4 \times 2^{18}$ clock periods and is achieved off-line. However, with a DAC structure that can operate with multiple inputs and outputs, such as the resistor-string converter, the modulator can be duplicated to allow a background calibration.

An alternative technique used to mitigate the effect of component mismatch in multi-bit DACs is dynamic element matching (DEM), which consists of using an algorithm to assign randomly the DAC unit elements to the

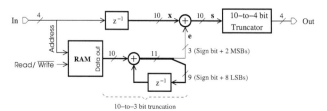

FIGURE 5.30
Block diagram of the digital calibration. (From [12], ©1993 IEEE.)

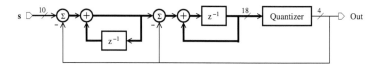

FIGURE 5.31
Block diagram of the modulator used for the 10-to-4 bit truncation.

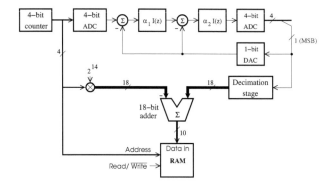

FIGURE 5.32
Scheme for the storage of the DAC error data in the RAM. (From [12], ©1993 IEEE.)

code being converted. In this way, the linearity error, which is generated in the case where some mismatched elements are more frequently selected than others over a given time period, is modulated at frequencies outside the signal band. The DEM technique can be implemented using a shifter controlled by a suitable selection logic. However, the clock frequency of the resulting modulator can be limited by the time delay introduced in the feedback path by this additional logic.

In comparison with single-bit topologies, multi-bit modulators offer a better performance (an increase of the dynamic range by 6 dB per additional bit as a result of the reduced quantization noise, and an improved attenuation of tones because the randomness assumption of the quantization noise is better satisfied as the number of quantization levels is increased) without increasing the OSR, but they can be limited by the stringent linearity requirement placed on the feedback DAC.

5.1.5.5 Cascaded modulator

The use of a high-order filter structure or a multi-bit internal quantizer can be adopted to improve the dynamic range of a modulator without increasing the oversampling ratio. However, each of these design solutions is known to be limited by potential shortcomings. High-order single-loop modulators may become prone to instability due to quantizer overload caused by large signals or the integrator initial conditions. The performance of modulators with a multi-bit quantizer is affected by the nonlinearity of the internal DAC and the increasing loading of amplifiers. A suitable design alternative can consist of performing high-order filtering through a cascade of low-order structures to ensure modulator stability and using a multi-bit quantizer only in the final stage, whose noise cancelation logic also attenuates the nonlinearity of the multi-bit DAC in the signal baseband [13]. The resulting implementation is known as a multistage or cascaded modulator. Here, cascaded modulators are realized using only first-order and second-order structures, which feature relaxed stability criteria. In this way, the dynamic range of the cascaded modulator can be larger than the one of single-loop structures, provided an adequate matching is achieved between the loop coefficients.

The input signal of the first stage is S and the subsequent stages are fed by a signal, which is either the inverted version of the quantization noise generated by the previous stage or the output of the last integrator in the previous stage, and which can be computed from the following equations:

$$E_{Qi}(z) = Y_i(z) - \frac{X_i(z)}{\alpha_i} \quad \text{or} \quad X_i(z) = \alpha_i(Y_i(z) - E_{Qi}(z)) \qquad i = 1, 2, 3$$
$$(5.81)$$

for the first-order modulator, and

$$E_{Q1}(z) = Y_1(z) - \frac{X_2(z)}{\alpha_1 \alpha_2} \quad \text{or} \quad X_2(z) = \alpha_1 \alpha_2(Y_1(z) - E_{Q1}(z)) \quad (5.82)$$

in the case of the second-order modulator. Here, E_{Q1}, E_{Q2}, and E_{Q3} are the quantization noises; Y_1, Y_2, and Y_3 denote the modulator outputs; X_1, X_2, and X_3 represent the integrator outputs; and α_1, α_2, and α_3 are scaling coefficients. The purpose of the scaling process is to maximize the overload level by using all the available signal swing at the output of each integrator without clipping. To keep the modulator output independent of the integrator coefficients, the output of the last integrator in a given stage is multiplied by a factor proportional to the inverse of the product of all the integrator coefficients of that stage before being summed at the input of the next stage.

Second-order modulator

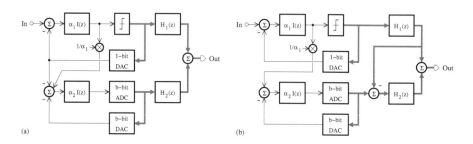

FIGURE 5.33
Block diagrams of 1-1 cascaded lowpass modulators.

The second-order architectures depicted in Figure 5.33 require two first-order modulator stages. It can be assumed that $\alpha_1 = \alpha_2 = 1$.

• Second-order, 1-1 cascaded lowpass modulator (a)
In the modulator structure of Figure 5.33(a), the quantization noise of the first stage is estimated, inverted, and applied to the next stage. The linear analysis in the z-domain of the modulator yields the following equations,

$$Y_1(z) = z^{-1}S(z) + (1 - z^{-1})E_{Q1}(z) \tag{5.83}$$

$$Y_2(z) = -z^{-1}E_{Q1}(z) + (1 - z^{-1})E_{Q2}(z) \tag{5.84}$$

and

$$Y(z) = H_1(z)Y_1(z) + H_2(z)Y_2(z) \tag{5.85}$$

where Y_1 and Y_2 denote the outputs of the first and second stages, respectively, and E_{Q1} and E_{Q2} represent the quantization noises of the first and second stages, respectively.

• Second-order, 1-1 cascaded lowpass modulator (b)
For the implementation illustrated by the block diagram of Figure 5.33(b), the output of the first integrator is connected to the input of the second modulator stage. The following expressions can be derived:

$$Y_1(z) = z^{-1}S(z) + (1 - z^{-1})E_{Q1}(z) \tag{5.86}$$

$$Y_2(z) = z^{-1}X_1(z) + (1 - z^{-1})E_{Q2}(z) \tag{5.87}$$

and

$$Y(z) = H_1(z)Y_1(z) + H_2(z)[Y_2(z) - H_1(z)Y_1(z)] \tag{5.88}$$

where Y_1 and Y_2 denote the outputs of the first and second stages, respectively, and E_{Q1} and E_{Q2} represent the quantization noises of the first and second stages, respectively. The output, X_1, of the integrator in the first stage can be obtained as

$$X_1(z) = z^{-1}(S(z) - E_{Q1}(z)) \tag{5.89}$$

and it can then be found that

$$Y_2(z) = z^{-2}(S(z) - E_{Q1}(z)) + (1 - z^{-1})E_{Q2}(z) \tag{5.90}$$

By choosing the transfer functions, H_1 and H_2, of the digital cancelation logics for the modulators of Figure 5.33 as

$$H_1(z) = z^{-1} \quad \text{and} \quad H_2(z) = 1 - z^{-1} \tag{5.91}$$

the overall output should ideally exhibit only the quantization noise of the last modulator stage. Hence,

$$Y(z) = H_S(z)S(z) + H_Q(z)E_{Q2}(z) \tag{5.92}$$

where the STF and QNTF are, respectively, of the form

$$H_S(z) = z^{-2} \quad \text{and} \quad H_Q(z) = (1 - z^{-1})^2 \tag{5.93}$$

Third-order modulator

A third-order lowpass modulator can be implemented by the 1-1-1 cascaded or 2-1 cascaded structures as shown in Figure 5.34, where $\alpha_1 = \alpha_2 = \alpha_3 = 1$, or Figure 5.35, where $\alpha_1 = 1/2$, $\alpha_2 = 2$, $\alpha_3 = 1$.

- Third-order, 1-1-1 cascaded lowpass modulator (a)

With reference to the modulator of Figure 5.34(a), we have

$$Y_1(z) = z^{-1}S(z) + (1 - z^{-1})E_{Q1}(z) \tag{5.94}$$
$$Y_2(z) = -z^{-1}E_{Q1}(z) + (1 - z^{-1})E_{Q2}(z) \tag{5.95}$$
$$Y_3(z) = -z^{-1}E_{Q2}(z) + (1 - z^{-1})E_{Q3}(z) \tag{5.96}$$

and

$$Y(z) = H_1(z)Y_1(z) + H_2(z)Y_2(z) + H_3(z)Y_3(z) \tag{5.97}$$

where Y_1, Y_2, and Y_3 are the outputs of the first, second, and third stages, respectively; and E_{Q1}, E_{Q2}, and E_{Q3} represent the quantization noises of the first, second, and third stages, respectively. To remove the quantization noises

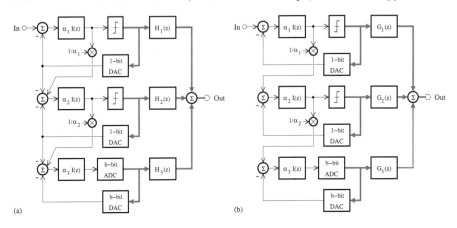

(a)   (b)

FIGURE 5.34
Block diagrams of 1-1-1 cascaded lowpass modulators.

of the two first stages from the modulator output, it is necessary to use digital circuit sections with the transfer functions

$$H_1(z) = z^{-2} \qquad H_2(z) = z^{-1}(1 - z^{-1}) \qquad \text{and} \qquad H_3(z) = (1 - z^{-1})^2$$
$$(5.98)$$

• Third-order, 1-1-1 cascaded lowpass modulator (b)
In the case of the structure depicted in Figure 5.34(b), the following expressions can be derived:

$$Y_1(z) = z^{-1}S(z) + (1 - z^{-1})E_{Q1}(z) \tag{5.99}$$

$$Y_2(z) = z^{-1}X_1(z) + (1 - z^{-1})E_{Q2}(z) \tag{5.100}$$

$$Y_3(z) = z^{-1}X_2(z) + (1 - z^{-1})E_{Q3}(z) \tag{5.101}$$

and

$$Y(z) = G_1(z)Y_1(z) + G_2(z)Y_2(z) + G_3(z)Y_3(z) \tag{5.102}$$

where Y_1, Y_2, and Y_3 are the outputs of the first, second, and third stages, respectively; and E_{Q1}, E_{Q2}, and E_{Q3} represent the quantization noises of the first, second, and third stages, respectively. The output of the first integrator can be computed as

$$X_1(z) = z^{-1}(S(z) - E_{Q1}(z)) \tag{5.103}$$

and Y_2 becomes

$$Y_2(z) = z^{-2}(S(z) - E_{Q1}(z)) + (1 - z^{-1})E_{Q2}(z) \tag{5.104}$$

while the output of the second integrator is obtained as

$$X_2(z) = z^{-2}(S(z) - E_{Q1}(z)) - z^{-1}E_{Q2} \tag{5.105}$$

and Y_3 can take the form

$$Y_3(z) = z^{-3}(S(z) - E_{Q1}(z)) - z^{-2}E_{Q2}(z) + (1 - z^{-1})E_{Q3}(z) \qquad (5.106)$$

Here, the cancelation of the quantization noise of the first previous stages is achieved with the following transfer functions:

$$G_1(z) = z^{-3} \qquad G_2(z) = z^{-2}(1 - z^{-1}) \qquad \text{and} \qquad G_3(z) = (1 - z^{-1})^2 \qquad (5.107)$$

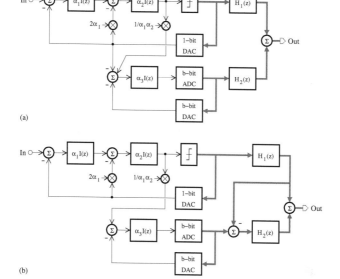

FIGURE 5.35
Block diagrams of 2-1 cascaded lowpass modulators.

- Third-order, 2-1 cascaded lowpass modulator (a)
The modulator structure shown in Figure 5.35(a) can be described by

$$Y_1(z) = z^{-2}S(z) + (1 - z^{-1})^2 E_{Q1}(z) \qquad (5.108)$$
$$Y_2(z) = -z^{-1}E_{Q1}(z) + (1 - z^{-1})E_{Q2}(z) \qquad (5.109)$$

and

$$Y(z) = H_1(z)Y_1(z) + H_2(z)Y_2(z) \qquad (5.110)$$

where Y_1 and Y_2 are the outputs of the first and second stages, respectively; and E_{Q1} and E_{Q2} represent the quantization noises of the first and second stages, respectively.

• Third-order, 2-1 cascaded lowpass modulator (b)

In the case of the modulator structure of Figure 5.35(b), we have

$$Y_1(z) = z^{-2}S(z) + (1 - z^{-1})^2 E_{Q1}(z) \tag{5.111}$$

$$Y_2(z) = z^{-1}X_2(z) + (1 - z^{-1})E_{Q2}(z) \tag{5.112}$$

and

$$Y(z) = H_1(z)Y_1(z) + H_2(z)[Y_2(z) - H_1(z)Y_1(z)] \tag{5.113}$$

where Y_1 and Y_2 are the outputs of the first and second stages, respectively; and E_{Q1} and E_{Q2} represent the quantization noises of the first and second stages, respectively. The output of the second integrator in the first stage, X_2, is given by

$$X_2(z) = z^{-2}S(z) + z^{-1}(z^{-1} - 2)E_{Q1}(z) \tag{5.114}$$

and Y_2 takes the form

$$Y_2(z) = z^{-3}S(z) + z^{-2}(z^{-1} - 2)E_{Q1}(z) + (1 - z^{-1})E_{Q2}(z) \tag{5.115}$$

In the cases of the 2-1 cascaded structures depicted in Figure 5.35, the transfer functions of the cancelation logic are given by

$$H_1(z) = z^{-1} \quad \text{and} \quad H_2(z) = (1 - z^{-1})^2 \tag{5.116}$$

and the overall output of the modulator can be written as

$$Y(z) = H_S(z)S(z) + H_Q(z)E_{Q3}(z) \tag{5.117}$$

where the STF and QNTF are given by

$$H_S(z) = z^{-3} \quad \text{and} \quad H_Q(z) = (1 - z^{-1})^3 \tag{5.118}$$

Ideally, the 2-1 and 1-1-1 cascaded modulators can be designed to realize the same STF and QNTF. However, because the cancelation of the quantization noise generated by the first stage is achieved after the second-order shaping in the 2-1 cascaded modulator, the required component matching is more relaxed than in the 1-1-1 cascaded structure, which is based on first-order stages.

Fourth-order modulator

The block diagrams of fourth-order lowpass modulators are depicted in Figure 5.36, where $\alpha_1 = 1/2$, $\alpha_2 = 2$, and $\alpha_3 = \alpha_4 = 1$, and Figure 5.37, where $\alpha_1 = \alpha_3 = 1/2$ and $\alpha_2 = \alpha_4 = 2$.

• Fourth-order, 2-1-1 cascaded lowpass modulator (a)

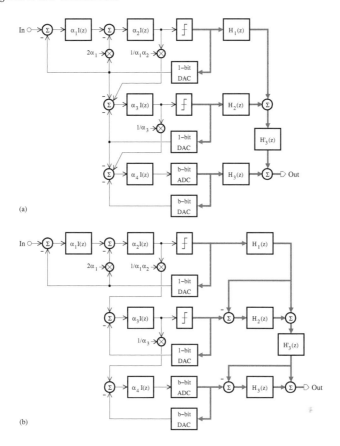

FIGURE 5.36
Block diagrams of 2-1-1 cascaded lowpass modulators.

With reference to the 2-1-1 cascaded modulator structure shown in Figure 5.36(a), we can obtain

$$Y_1(z) = z^{-2}S(z) + (1 - z^{-1})^2 E_{Q1}(z) \tag{5.119}$$

$$Y_2(z) = -z^{-1}E_{Q1}(z) + (1 - z^{-1})E_{Q2}(z) \tag{5.120}$$

$$Y_3(z) = -z^{-1}E_{Q2}(z) + (1 - z^{-1})E_{Q3}(z) \tag{5.121}$$

and

$$Y(z) = H_3'(z)[H_1(z)Y_1(z) + H_2(z)Y_2(z)] + H_3(z)Y_3(z) \tag{5.122}$$

where Y_1, Y_2, and Y_3 are the outputs of the first, second, and third stages, respectively; and E_{Q1}, E_{Q2}, and E_{Q3} represent the quantization noises of the first, second, and third stages, respectively.

• Fourth-order, 2-1-1 cascaded lowpass modulator (b)

For the 2-1-1 cascaded modulator structure depicted in Figure 5.36(b), it can be shown that

$$Y_1(z) = z^{-2}S(z) + (1 - z^{-1})^2 E_{Q1}(z) \tag{5.123}$$

$$Y_2(z) = z^{-1}X_2(z) + (1 - z^{-1})E_{Q2}(z) \tag{5.124}$$

$$Y_3(z) = z^{-1}X_3(z) + (1 - z^{-1})E_{Q3}(z) \tag{5.125}$$

and

$$Y(z) = H_3(z)[Y_3(z) - Y'(z)] + Y'(z) \tag{5.126}$$

where

$$Y'(z) = H_3'(z)\{H_1(z)Y_1(z) + H_2(z)[Y_2(z) - H_1(z)Y_1(z)]\} \tag{5.127}$$

Here, Y_1, Y_2, and Y_3 are the outputs of the first, second, and third stages, respectively; and E_1, E_2, and E_3 represent the quantization noises of the first, second, and third stages, respectively. The output, X_2, of the second integrator in the first stage is

$$X_2(z) = z^{-2}S(z) + z^{-1}(z^{-1} - 2)E_{Q1}(z) \tag{5.128}$$

and Y_2 can be written as

$$Y_2(z) = z^{-3}S(z) + z^{-2}(z^{-1} - 2)E_{Q1}(z) + (1 - z^{-1})E_{Q2}(z) \tag{5.129}$$

For the output, X_3, of the integrator in the second stage, the following expression is obtained

$$X_3(z) = z^{-3}S(z) + z^{-2}(z^{-1} - 2)E_{Q1}(z) - z^{-1}E_{Q2}(z) \tag{5.130}$$

and

$$Y_3(z) = z^{-4}S(z) + z^{-3}(z^{-1} - 2)E_{Q1}(z) - z^{-2}E_{Q2}(z) + (1 - z^{-1})E_{Q3}(z) \tag{5.131}$$

The transfer functions of the cancelation logic for the 2-1-1 cascaded structures shown in Figure 5.36 are derived such that the quantization noises of the first two stages should be removed from the overall output of the modulator. That is,

$$H_1(z) = H_3'(z) = z^{-1}, \quad H_2(z) = (1 - z^{-1})^2 \quad \text{and} \quad H_3(z) = (1 - z^{-1})^3 \tag{5.132}$$

• Fourth-order, 2-2 cascaded lowpass modulator (a)

The analysis of the 2-2 cascaded modulator structure shown in Figure 5.37(a) yields

$$Y_1(z) = z^{-2}S(z) + (1 - z^{-1})^2 E_{Q1}(z) \tag{5.133}$$

$$Y_2(z) = z^{-2}E_{Q1}(z) + (1 - z^{-1})^2 E_{Q2}(z) \tag{5.134}$$

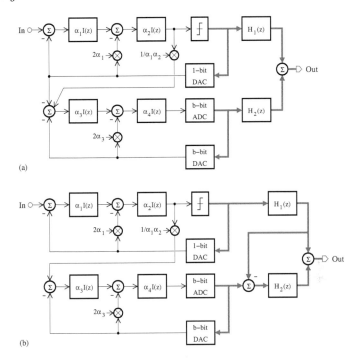

FIGURE 5.37
Block diagrams of 2-2 cascaded lowpass modulators.

and

$$Y(z) = H_1(z)Y_1(z) + H_2(z)Y_2(z) \tag{5.135}$$

where Y_1 and Y_2 are the outputs of the first and second stages, respectively; and E_{Q1} and E_{Q2} represent the quantization noises of the first and second stages, respectively.

• Fourth-order, 2-2 cascaded lowpass modulator (b)
The 2-2 cascaded modulator structure of Figure 5.37(b) can be characterized by equations of the form

$$Y_1(z) = z^{-2}S(z) + (1 - z^{-1})^2 E_{Q1}(z) \tag{5.136}$$
$$Y_2(z) = z^{-2}X_2(z) + (1 - z^{-1})^2 E_{Q2}(z) \tag{5.137}$$

and

$$Y(z) = H_2(z)[Y_2(z) - H_1(z)Y_1(z)] + H_1(z)Y_1(z) \tag{5.138}$$

where Y_1 and Y_2 are the outputs of the first and second stages, respectively;

and E_{Q1} and E_{Q2} represent the quantization noises of the first and second stages, respectively. By expressing the output, X_2, of the second integrator in the first stage as

$$X_2(z) = z^{-2}S(z) + z^{-1}(z^{-1} - 2)E_{Q1}(z) \tag{5.139}$$

we obtain

$$Y_2(z) = z^{-4}S(z) + z^{-3}(z^{-1} - 2)E_{Q1}(z) + (1 - z^{-1})^2 E_{Q2}(z) \tag{5.140}$$

For the 2-2 cascaded modulators of Figure 5.37, the digital cancelation logic suppresses the quantization noise of the first stage, provided that

$$H_1(z) = z^{-2} \quad \text{and} \quad H_2(z) = (1 - z^{-1})^2 \tag{5.141}$$

The output of the above fourth-order modulator is of the form

$$Y(z) = H_S(z)S(z) + H_Q(z)E_{Q3}(z) \tag{5.142}$$

where the STF and QNTF are given by

$$H_S(z) = z^{-4} \quad \text{and} \quad H_Q(z) = (1 - z^{-1})^4 \tag{5.143}$$

In practice, the cancelation of the quantization noise due to the first modulator stages is limited by the matching level achievable between the loop gains.

5.1.5.6 Effect of the multi-bit DAC nonlinearity

Given an L-th-order cascaded modulator with k stages, the nonlinearity of the multi-bit DAC used in the last stage can be modeled as an additive noise characterized by the z-domain function, E_D. The linear analysis yields a more general equation of the modulator output, Y, as follows,

$$Y(z) = H_S(z)S(z) + H_Q(z)E_{Qk}(z) + H_D(z)E_D(z) \tag{5.144}$$

where S denotes the input signal, E_{Qk} is the quantization noise of the last stage, and STF and QNTF are, respectively, given by

$$H_S(z) = z^{-L} \quad \text{and} \quad H_Q(z) = (1 - z^{-1})^L \tag{5.145}$$

Here, the DAC nonlinearity is attenuated in the signal baseband by the transfer function

$$H_D(z) = H_k(z) = (1 - z^{-1})^\eta \tag{5.146}$$

where η represents the number of integrators used in the stages 1 to $k - 1$. This can be understood by observing that the output of the multi-bit DAC is actually fed back to the input of the last stage, but it can be considered a signal shaped by an η-th-order transfer function from the perspective of the overall modulator. Note that the 2-1-1 and 2-2 cascaded modulators realize the same STF and QNTF, but the errors due to the feedback multi-bit DAC are shaped by third- and second-order transfer functions, respectively. As a result, the 2-1-1 cascaded modulator features better attenuation of the DAC nonlinearities in the baseband than does the 2-2 cascaded structure.

5.1.5.7 Quantization noise-shaping and inter-stage coefficient scaling

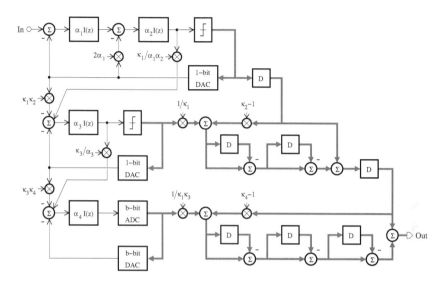

FIGURE 5.38
Block diagram of a 2-1-1 cascaded lowpass modulator with scaling coefficients.

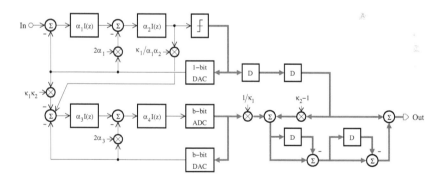

FIGURE 5.39
Block diagram of a 2-2 cascaded lowpass modulator with scaling coefficients.

To avoid clipping at high levels of the quantization noise, it may be necessary to use additional inter-stage scaling coefficients in cascaded modulators as shown in the block diagram of Figures 5.38 and 5.39. The 2-1-1 cascaded modulator uses two inter-stage gains, κ_1 and κ_3, and the DAC feedback signal is scaled by κ_2 and κ_4; while for the 2-2 cascaded structure, κ_1 denotes the inter-stage gain and κ_2 is the scaling coefficient of the DAC feedback signal.

The QNTF can be obtained as

$$H_Q(z) = \frac{(1 - z^{-1})^4}{\kappa} \tag{5.147}$$

where κ is, respectively, equal to $\kappa_1 \kappa_3$ and κ_1 for the 2-1-1 and 2-2 cascaded modulators. The design objective is to find the modulator coefficients that yield the maximum dynamic range. The level of the signal transferred from one stage to the next is set to avoid a premature overload by appropriately selecting the values of the coefficients κ_1, κ_2, κ_3, and κ_4. The multiplication of the last stage output with a factor $1/\kappa$ that is greater than 1 leads to a decrease in the modulator resolution by $\log_2(1/\kappa)$ bits. Thus, a trade-off must be made between the minimization of the quantization noise in the baseband and the achievable improvement of the overload condition due to the $1/\kappa$ scaling effect. To simplify the implementation of the digital circuit for the noise cancelation, the factor $1/\kappa$ is generally chosen as a power of 2. Note that a feedback path is added to the cancelation logics for each of the coefficients κ_2 and κ_4, that is different from unity.

5.1.6   Bandpass delta-sigma modulator

For a given technology, bandpass $\Delta\Sigma$ modulators are generally dedicated for the analog-to-digital conversion of signals with a higher frequency than the one supported by lowpass modulators. This is due to the fact that the bandwidth frequency of a bandpass modulator is limited to $f_s/(2\,\text{OSR})$, where OSR is the oversampling ratio and f_s is the sampling frequency, instead of the signal frequency as is the case for a lowpass modulator, thus making possible the conversion of signals with frequencies up to $f_s/2$. Bandpass modulators can then find applications in the digitalization of intermediate frequency signals in wireless receivers.

The key parameters in the design of $\Delta\Sigma$ modulators for a specified signal-to-noise ratio (SNR) and dynamic range (DR) are the OSR, the order or structure of the loop filter, and the quantizer resolution. A bandpass modulator can be designed to have the passband center frequency located anywhere between 0 and $f_s/2$. However, the class of bandpass modulators with the passband centered around $f_s/4$ seems to exhibit some advantages. It can easily be derived from lowpass prototypes using the z^{-1} to $-z^{-2}$ transformation. The resulting bandpass modulator can be implemented using building blocks with a reduced complexity because it is based on second-order resonators with the transfer function given by

$$R(z) = I(\hat{z})|_{\hat{z}^{-1}=-z^{-2}} = \left.\frac{\hat{z}^{-1}}{1 - \hat{z}^{-1}}\right|_{\hat{z}^{-1}=-z^{-2}} = -\frac{z^{-2}}{1 + z^{-2}} \tag{5.148}$$

The function R is characterized by a resonance occurring at the frequency $f_s/4$ due to the pair of complex poles located at $z = \pm j$.

In the general case, the transformation of the lowpass prototype into a bandpass modulator can be achieved using the following z-variable substitution,

$$z^{-1} \rightarrow -z^{-1} \frac{z^{-1} - \alpha}{1 - \alpha z^{-1}} \qquad (5.149)$$

where $\alpha = \cos 2\pi(f_0/f_s)$ and f_0 represents the desired center frequency. Assuming that $f_0 = f_s/4$, the above expression is reduced to the transformation, $z^{-1} \rightarrow -z^{-2}$. However, this general approach has the inconvenience of not preserving the dynamic properties of the lowpass prototype and may result in a circuit implementation with increased complexity.

5.1.6.1 Single-loop bandpass delta-sigma modulator

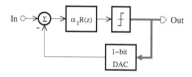

FIGURE 5.40
Block diagram of a second-order bandpass modulator.

The second-order bandpass modulator shown in Figure 5.40 is derived by performing the dc-to-$f_s/4$ transformation on a first-order lowpass prototype. Assuming that $\alpha_1 = 1$, the STF and QNTF of the bandpass modulator are, respectively, given by

$$H_S(z) = -z^{-2} \qquad \text{and} \qquad H_Q(z) = 1 + z^{-2} \qquad (5.150)$$

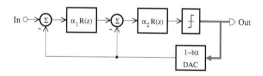

FIGURE 5.41
Block diagram of a fourth-order bandpass modulator.

The fourth-order bandpass modulator depicted in Figure 5.41 is obtained from the single-loop second-order lowpass modulator. In the case where we have $\alpha_1 = 1/2$ and $\alpha_2 = 2$, the STF and QNTF can be expressed as

$$H_S(z) = z^{-4} \qquad \text{and} \qquad H_Q(z) = (1 + z^{-2})^2 \qquad (5.151)$$

5.1.6.2　Cascaded bandpass delta-sigma modulator

The SNR can be increased using a higher OSR or higher-order filter. However, the power consumption due to the settling requirements of amplifiers becomes a constraint in modulators with a high value of OSR. Because higher-order modulators based on a single loop may be prone to instability, an alternative is to use cascaded structures, provided that the mismatch between the analog and digital transfer functions is maintained at an acceptable level.

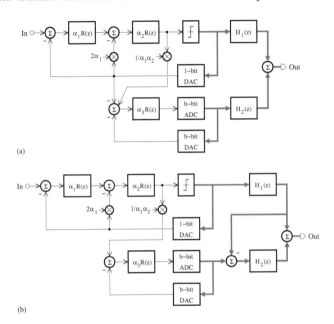

(a)

(b)

FIGURE 5.42
Block diagrams of 4-2 cascaded bandpass modulators.

The block diagrams of 4-2 cascaded bandpass modulators, as shown in Figure 5.42, are derived from the 2-1 cascaded lowpass modulators. To proceed further, it can be assumed that $\alpha_1 = 1/2$, $\alpha_2 = 2$, and $\alpha_3 = 1$. With the transfer functions of the cancelation logic being given by

$$H_1(z) = z^{-2} \quad \text{and} \quad H_3(z) = (1 + z^{-2})^2 \tag{5.152}$$

the overall output of the bandpass modulators can be expressed as

$$Y(z) = H_S(z)S(z) + H_Q(z)E_2(z) \tag{5.153}$$

where S is the input signal, E_2 is the quantization noise of the second stage, and the STF and QNTF are given by

$$H_S(z) = -z^{-6} \quad \text{and} \quad H_Q(z) = (1 + z^{-2})^3 \tag{5.154}$$

For the design of a bandpass modulator based on the z^{-1} to $-z^{-2}$ transformation, the zeros ($z = \pm j$) of the QNTF are located at $f_s/4$ instead of dc (or say, $z = 1$), as is the case for lowpass modulators. In addition, this transformation has the advantage of preserving the stability property of the lowpass prototype.

5.1.6.3   Design examples

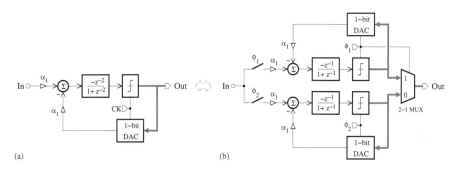

FIGURE 5.43
Block diagram of a second-order bandpass DT modulator.

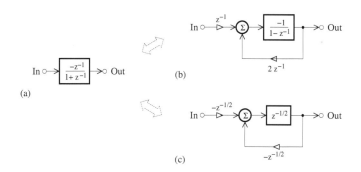

FIGURE 5.44
Block diagrams of alternative realizations of the transfer function, $H(z) = -z^{-1}/(1 + z^{-1})$.

The block diagram of a DT, second-order bandpass modulator is illustrated in Figure 5.43(a). It is derived from a first-order lowpass filter prototype using the z^{-1} to $-z^{-2}$ transformation. As a result, the integrator is replaced by a resonator with the transfer function $-z^{-2}/(1 + z^{-2})$, which can be implemented using various SC topologies. The resonator implementation using a loop of two undamped SC integrators is generally plagued by transfer function errors due to the finite dc gain and bandwidth of the amplifier. The bandpass modulator is then preferably realized using a two-path structure, as depicted in

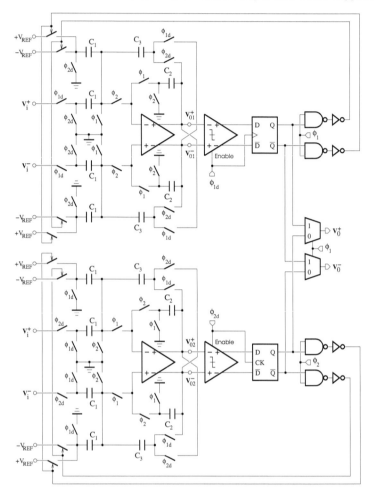

FIGURE 5.45
Circuit diagram of the second-order bandpass DT modulator based on a first-order highpass filter.

Figure 5.43(b), where ϕ_1 and ϕ_2 represent both nonoverlapping clock phases, and each path is clocked in a time-interleaved fashion. The z-domain transfer function of the filter is obtained as

$$H(z) = R(z^2) = -\frac{z^{-1}}{1 + z^{-1}} \qquad (5.155)$$

It should be noted that the effective sampling frequency of the overall modulator is two times the one of a single path, as implied by the variable z^2. However, without a careful circuit and layout technique, the modulator dy-

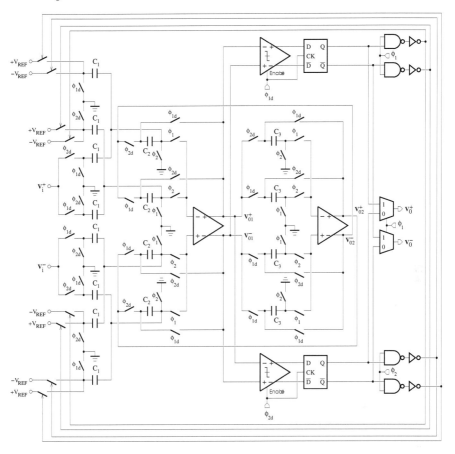

FIGURE 5.46
Circuit diagram of the second-order bandpass DT modulator based on a half-delay cell.

namic range may be limited by path mismatches, which appear as spurious frequency components in the output spectrum.

Depending on the filter topologies, the accuracy of the transfer function pole location can be limited by capacitor mismatches. Figure 5.44 shows the block diagram of various structures that can be used to precisely realize the highpass transfer function, $H(z)$.

Based on the filter block diagram of Figure 5.44(b), the circuit diagram of the modulator was realized as shown in Figure 5.45. During the sampling phase, the capacitors C_1 are connected to the input and reference voltages, respectively; the charge stored on capacitors C_2 and C_3 are proportional to the output voltage and the inverse of the output voltage, respectively; and the capacitors C_2 are included in the feedback path around the amplifier. During the integrating phase, the capacitors C_3 are switched to the amplifier feedback path and a charge transfer takes place between the capacitors C_1 and C_3; the capacitors C_2 become a load connected to the output voltage. That is,

$$C_3 V_{0j}(n) = C_1[V_i(n-1) + V_{REF}(n-1)] - C_3 V_{0j}(n-1) \quad j = 1, 2 \quad (5.156)$$

and

$$V_{0j}(z) = H(z)[V_i(z) + V_{REF}(z)] \tag{5.157}$$

where

$$H(z) = \pm \frac{C_1}{C_3} \frac{z^{-1}}{1 + z^{-1}} \tag{5.158}$$

The transfer function $H(z)$ was derived with the assumption that the clock phasing at the filter input and at the level of the comparator latch is such that a delay of one clock period exists from the input to output (leading to the term z^{-1} in the numerator). Note that the sign of the transfer function can be modified by simply reversing the positive and negative input nodes of the differential implementation. Furthermore, the signal is sampled during the first clock signal and the charge transfer occurs during the second clock phase for one path; whereas for the other path, we have the signal sampling on the second phase and the charge transfer on the first phase. Because the capacitors C_3 are used in the amplifier feedback path and to invert the polarity of the output voltage, the numerator of the filter transfer function is not affected by capacitor mismatches and the pole location can remain on the unit circle. The value of C_2 is not critical for the z-domain design, but it must be chosen so that a low level of the thermal noise and an adequate settling of the amplifier during the sampling phase can be achieved.

The discrete-time filter in the double-sampling bandpass modulator can also be implemented using two half-delay cells, as depicted in Figure 5.44(c). This approach is selected for the SC implementation of the modulator shown in Figure 5.46. The sampling of the signal during one clock phase and the charge transfer during the next clock phase required for the realization of the half delay are performed by the same capacitor, or say C_2 and C_3 for the first and second stages, respectively. As a result, the pole of the filter transfer function remains insensitive to capacitor mismatches. Then, we can write

$$C_1[V_i(n-1/2) + V_{REF}(n-1/2)] = C_2[V_{01}(n) + V_{02}(n-1/2)] \tag{5.159}$$
$$C_3[V_{02}(n-1/2) - V_{01}(n-1)] = 0 \tag{5.160}$$

and

$$V_{02}(z) = H(z)[V_i(z) + V_{REF}(z)] \tag{5.161}$$

where

$$H(z) = \pm \frac{C_1}{C_2} \frac{z^{-1}}{1 + z^{-1}} \tag{5.162}$$

The term z^{-1} included in the numerator of the $H(z)$ transfer function is due to the fact that there is a delay of one clock period between the sampling instants at the filter input and at the level of the comparator latch. The size of capacitors C_1 is determined by the dc gain of the filter, while C_2 and C_3 can be unit size capacitors.

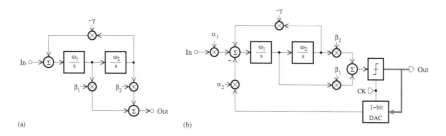

FIGURE 5.47
Block diagrams of (a) a CT resonator and (b) a second-order bandpass CT modulator.

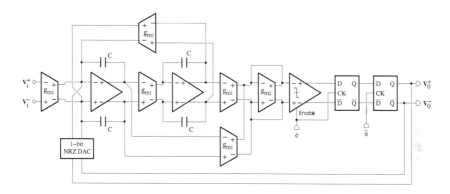

FIGURE 5.48
Circuit diagram of a second-order bandpass CT modulator.

The CT bandpass modulator to be designed is based on a second-order resonator. Let X be the input signal of the CT resonator shown in Figure 5.47(a), and X_1 and X_2 the output variables of the first and second integrators, respectively. With the resonator output signal given by

$$Y(s) = \beta_1 X_1(s) + \beta_2 X_2(s) \tag{5.163}$$

where

$$X_1(s) = \frac{s\omega_1}{s^2 + \gamma\omega_1\omega_2} X(s) \tag{5.164}$$

and

$$X_2(s) = \frac{\omega_1\omega_2}{s^2 + \gamma\omega_1\omega_2}X(s) \tag{5.165}$$

the transfer function is obtained as

$$\hat{H}(s) = \frac{Y(s)}{X(s)} = \frac{\beta_1\omega_1 s + \beta_2\omega_1\omega_2}{s^2 + \gamma\omega_1\omega_2} \tag{5.166}$$

where β_1, β_2, ω_1, ω_2, and γ are real coefficients.

For a second-order CT modulator designed with the loop coefficients $\beta_1 = -1/2$, $\beta_2 = -1/2$, $\gamma = 1$, and $\omega_1 = \omega_2 = \omega_0$, the transfer function of the resonator is given by

$$\hat{H}(s) = -\frac{(\omega_0/2)s + \omega_0^2/2}{s^2 + \omega_0^2} \tag{5.167}$$

In the case of an NRZ DAC pulse, the equivalent z-domain transfer function can be derived as

$$H(z) = (1 - z^{-1})\mathcal{Z}\left\{\mathcal{L}^{-1}\left[\frac{\hat{H}(s)}{s}\right]_{t=nT}\right\} \tag{5.168}$$

The partial-fraction expansion of the function $\hat{H}(s)/s$ takes the form

$$\frac{\hat{H}(s)}{s} = -\frac{1/2}{s} - \frac{-(1/2)s + \omega_0/2}{s^2 + \omega_0^2} \tag{5.169}$$

Because

$$\mathcal{Z}\left\{\mathcal{L}^{-1}\left[\frac{1}{s}\right]_{t=nT}\right\} = \frac{1}{1 - z^{-1}} \tag{5.170}$$

$$\mathcal{Z}\left\{\mathcal{L}^{-1}\left[\frac{s}{s^2 + \omega_0^2}\right]_{t=nT}\right\} = \frac{1 - z^{-1}\cos\omega_0 T}{1 - 2z^{-1}\cos\omega_0 T + z^{-2}} \tag{5.171}$$

$$\mathcal{Z}\left\{\mathcal{L}^{-1}\left[\frac{\omega_0}{s^2 + \omega_0^2}\right]_{t=nT}\right\} = \frac{z^{-1}\sin\omega_0 T}{1 - 2z^{-1}\cos\omega_0 T + z^{-2}} \tag{5.172}$$

and the center frequency of the modulator is located at $f_s/4$, that is,

$$\omega_0 T = \frac{\pi}{2} \tag{5.173}$$

where T is the period of the clock signal, the transfer function $H(z)$ is reduced to

$$H(z) = -\frac{1-z^{-1}}{2}\left(\frac{1}{1-z^{-1}} - \frac{1}{1+z^{-2}} + \frac{z^{-1}}{1+z^{-2}}\right) \qquad (5.174)$$

Finally, the DT version of the filter transfer function is obtained as

$$H(z) = -\frac{z^{-1}}{1+z^{-2}} \qquad (5.175)$$

Note that different CT filters can be derived, depending on the numerator implementation of the DT transfer function due to the occurrence of the signal sampling inside the modulator loop. For the function $H(z)$ given by Equation (5.175), it is assumed that the digital section of the modulator exhibits a delay of one clock period, or equivalently z^{-1}. However, it is also possible to consider that the digital stage between the comparator and the DAC has no delay, and the DT transfer function of the filter is

$$H(z) = -\frac{z^{-2}}{1+z^{-2}} \qquad (5.176)$$

Especially, we need to have the set of coefficients, $\beta_1 = 1/2$, $\beta_2 = -1/2$, $\gamma = 1$, and $\omega_1 = \omega_2 = \omega_0$, yielding the CT resonator transfer function

$$\hat{H}(s) = -\frac{-(\omega_0/2)s + \omega_0^2/2}{s^2 + \omega_0^2} \qquad (5.177)$$

The resonator can be implemented with g_m-C operational amplifier integrators, a local feedback, and two feedforward gain stages. In this way, the coefficients of the transfer function can be related to the values of capacitances and transconductances.

The modulator design can start by choosing a second-order z-domain QNTF given by

$$H_Q(z) = 1 + z^{-2} \qquad (5.178)$$

The loop transfer function is derived as

$$H(z) = \frac{1 - H_Q(z)}{H_Q(z)} = -\frac{z^{-2}}{1+z^{-2}} \qquad (5.179)$$

To remove the delay introduced by the digital section, here z^{-1}, the loop transfer function is multiplied by $1/z^{-1}$. This results in the transfer function of the DT filter, which is then converted to the equivalent CT filter of the form

$$\hat{H}(s) = -\frac{(\omega_0/2)s + \omega_0^2/2}{s^2 + \omega_0^2} \tag{5.180}$$

where $\omega_0 = \pi/(2T)$ and T is the period of the clock signal. This conversion is achieved using the impulse invariant transform and assuming a DAC pulse of the NRZ type. Figure 5.47(b) shows the block diagram of the bandpass modulator. The same value is assigned to the coefficients α_1, α_2, and γ, which can be used to scale the signal level at the input of the first integrator, while $\beta_1 = -1/2$, $\beta_2 = -1/2$, $\gamma = 1$, and $\omega_1 = \omega_2 = \omega_0$. The modulator implementation with g_m-C operational amplifier circuits is illustrated by the circuit diagram shown in Figure 5.48, where $g_{m2} = 2g_{m1}$ and $\omega_0 = g_{m1}/C$.

Unlike DT modulators using an SC filter, whose coefficients are related to capacitor ratios, CT modulators based on a g_m-C operational amplifier filter have coefficients that are determined by the absolute values of transconductances and capacitors. A matching on the order of 1% is achievable between g_m/C values. However, the absolute g_m/C value can be subject to fluctuations of about 20% due to CMOS IC process variations. A solution may consist of using an on-chip tuning circuit or a calibration stage based on a programmable transconductor or capacitor array.

5.1.7 DT modulator synthesis

Given the specifications (OSR, signal-to-noise ratio, SNR, dynamic range, DR) of a DT (lowpass) modulator, the (high-pass) QNTF, $H_Q(z)$, can be derived by relying on traditional filter approximation functions, such as Butterworth and Chebyshev frequency responses, or using numerical methods. Then, the loop filter transfer function is computed using $H(z) = -1 + 1/H_Q(z)$, and simulations are used to estimate the corresponding peak SNR and DR. If the specifications are not met, the synthesis procedure should be repeated with a higher order or cutoff frequency for the QNTF. For the transfer function realizability, it is necessary that $H_Q(\infty)$ must be equal to one.

In general, conventional synthesis approaches result in the optimal design only for certain performance criteria.

Numerical methods are used to first determine the optimized QNTF zeroes as solutions of the equations obtained by setting the first derivative of the in-band power spectral density to zero, and then to iteratively find the best placement, but not necessarily optimal, for the QNTF poles. QNTFs of the Butterworth type have a very high dc loop gain or noise suppression at lower frequencies, but less noise suppression at higher frequencies. QNTFs based on the Chebyshev filter response feature a steeper roll-off. But they can also exhibit ripples either in the stopband or in the passband.

For the modulator stability, the closed-loop poles should be located inside the unit circle. But the modulator can also become unstable if the level of the quantizer input signal exceeds the quantizer range. This is due to the fact that the poles can move out of the unit circle as a result of a quantizer gain reduction caused by the increase of the input signal level.

Note that when the out-of-band gain of a QNTF is too high, the quantizer can saturate even for small input levels. For instance, a QNTF of the form, $(1 - z^{-1})^N$, exhibits a high out-of-band gain,[1] $||H_Q||_\infty$, that is given by 2^N at $z = -1$. In this case, the stability condition can be satisfied by increasing the quantizer resolution or by adding poles to the QNTF. As a rule-of-thumb, the modulator stability requires that $||H_Q||_\infty < 1.6$ (or 4.08 dB).

5.1.8 CT modulator synthesis

To reduce the number and complexity of feedback DACs, reconfigurable CT $\Delta\Sigma$ modulators, that find applications in multi-mode receivers, are generally designed using architectures with multiple feedforward paths and only one feedback path.

However, such architectures generally provide less attenuation of aliased signal components than architectures with multiple feedback paths. They also exhibit an undesirable out-of-band peaking in the STF. As a result, the dynamic range can be reduced due to the presence of strong out-of-band blockers, especially in wireless receivers. On the other hand, the use of multiple feedback paths can have the inconvenience of leading to an increase of integrator output swings. That is, low-gain integrators with large integration capacitors must be used to avoid signal clipping.

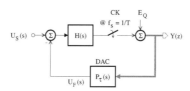

FIGURE 5.49
Equivalent linear model of a CT modulator.

Methods that are generally adopted to design CT modulators based on a loop filter with a single input path rely on the linear model shown in Figure 5.49.

Let $H_Q(z)$ be the QNTF obtained from the modulator specification. The transfer function of the corresponding DT loop filter can be obtained as:

$$H(z) = \frac{1 - H_Q(z)}{H_Q(z)} \qquad (5.181)$$

---

[1]For a system characterized by the z-domain transfer function $H_Q(z)$, we have: $||H_Q||_\infty = \max_{\theta \in [0,\pi]} |H_Q(e^{j\theta})|$. The $||H_Q||_\infty$ norm is infinite if the system has poles that are located on the unit circle.

Given the transfer function, $\hat{H}(s)$, of the CT loop filter derived from the selected modulator architecture, the impulse-invariant transformation is then used to map $\hat{H}(s)$ to $H_\tau(z)$, as follows:

$$H_\tau(z) = \mathcal{Z}\{\mathcal{L}^{-1}[\hat{H}(s)P_\tau(s)]_{t=nT}\} \tag{5.182}$$

where T denotes the period of the clock signal, $\mathcal{Z}$ is the z-transform, $\mathcal{L}^{-1}$ is the inverse Laplace transform, τ is the excess-loop delay, and $P_\tau(s)$ is the Laplace transform of the DAC pulse. In practice, this can be achieved using pre-computed mapping tables for NRZ, RZ, and HRZ DAC pulses or numerical methods.

Finally, the modulator coefficients are determined by matching $H_\tau(z)$ and $H(z)$.

In the aforementioned design method, all modulator coefficients are uniquely determined to achieve a target QNTF, and the STF is not directly taken into account.

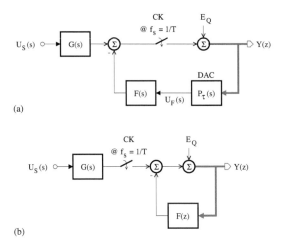

(a)

(b)

FIGURE 5.50
Equivalent linear models of a CT modulator.

The design of CT modulators with a peaking-free STF is achieved using alternative methods [14, 15] that can allow the simultaneous synthesis of the QNTF and the transfer function of the feedforward signal path. To this end, the linear models of the CT modulator can be derived as illustrated in Figure 5.50, where $G(s)$ denotes the feedforward transfer function, $F(s)$ denotes the feedback transfer function, $P_\tau(s)$ is the transfer function of the DAC pulse, and τ represents the excess-loop delay. In the z-domain, the QNTF can be obtained as,

$$H_Q(z) = \frac{Y(z)}{E_Q(z)} = \frac{1}{1 - F(z)} \tag{5.183}$$

Or equivalently, the DT feedback transfer function is given by

$$F(z) = \frac{H_Q(z) - 1}{H_Q(z)} \tag{5.184}$$

Considering the CT modulator as a linear system with two input variables and one output variable, the following state-space representation can be obtained:

$$s\mathbf{X}(s) = \mathbf{A}\mathbf{X}(s) + \mathbf{B}\mathbf{U}(s) \tag{5.185}$$
$$Y(s) = \mathbf{C}\mathbf{X}(s) + \mathbf{D}\mathbf{U}(s) \tag{5.186}$$

where $\mathbf{A}$ is the state matrix, $\mathbf{B}$ is the input matrix, $\mathbf{C}$ is the output matrix, $\mathbf{D}$ is the feedforward matrix, and $\mathbf{U}(s) = [U_F(s)\ U_S(s)]^T$, with $U_S(s)$ being the modulator input and $U_F(s)$ the DAC output. The transfer functions can be derived as:

$$\mathbf{T}(s) = \mathbf{C}(s\mathbf{I} - \mathbf{A})^{-1}\mathbf{B} + \mathbf{D} \tag{5.187}$$

where

$$\mathbf{T}(s) = \left[\frac{Y(s)}{U_F(s)}\ \frac{Y(s)}{U_S(s)} \right]^T \tag{5.188}$$

with $F(s) = Y(s)/U_F(s)$ and $G(s) = Y(s)/U_S(s)$.

The transfer functions, $F(s)$ and $G(s)$, have an identical denominator because the eigenvalues of the modulator state matrix are unique. Furthermore, the synthesis of a peaking-free STF with a monotonic roll-off frequency response requires that $G(s)$ be an all-pole transfer function.

Hence,

$$F(s) = \frac{Y(s)}{U_F(s)} = -\frac{\displaystyle\sum_{i=0}^{N-1} a_i s^i}{D(s)} \tag{5.189}$$

$$G(s) = \frac{Y(s)}{U_S(s)} = \frac{d_0}{D(s)} \tag{5.190}$$

where

$$D(s) = \sum_{i=0}^{N} b_i s^i \tag{5.191}$$

with a_i, b_i, and d_0 being constants that can be related to the modulator coefficients.

At dc, the signal gain should be equal to 1 (or 0 dB). Due to the fact that $F(z)|_{z=1} = F(s)|_{s=0}$, the analysis of the linear model leads to

$$\frac{Y(z)|_{z=1}}{U_S(s)|_{s=0}} = \frac{G(s)|_{s=0}}{1 - F(s)|_{s=0}} = \frac{d_0/b_0}{1 + a_0/b_0} = 1 \tag{5.192}$$

and
$$d_0 = a_0 + b_0 \tag{5.193}$$

Using numerical methods, the DT QNTF can be obtained from the modulator specifications. Based on (5.184), it is used to compute the DT feedback transfer function, $F(z)$, that can then be transformed into the CT transfer function $\hat{F}(s)$. Finally, the modulator coefficients are determined by matching $F(s)$ and $\hat{F}(s)$ and using (5.193).

5.1.9 Decimation filter

Once the modulator has transformed the analog samples into a low-resolution code with the frequency OSR $\cdot f_s$ much greater than two times the highest spectral component of the input signal, a digital filter is used to attenuate the out-of-band quantization noise, and high-frequency interferences that can be aliased into the passband. The decimation of the modulator output bit stream at the Nyquist rate is achieved by a lowpass filter followed by a down-sampler with the factor D equal to the OSR. The aliasing in the decimation process is avoided by pre-filtering the signal samples to eliminate the components above the frequency $f_s/(2D)$.

The SNR is simply increased when the resolution changes from the low resolution to B bits as a result of the down-sampling and filtering. However, it should be noted that there is no one-to-one correspondence between the input and output samples as is the case for some ADCs and each input sample value contributes to the whole train of output samples.

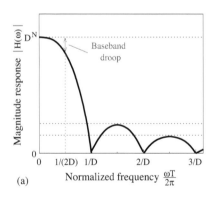
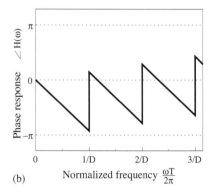

FIGURE 5.51
(a) Magnitude and (b) phase responses of a decimation filter with N sections.

The filter architecture should be chosen to have a linear-phase frequency response[2] and minimize the hardware complexity. These requirements can be

---

[2]A filter is assumed to have a linear phase if its transfer function can be written as

met by a moving-average filter [16], the transfer function of which is given by

$$G(z) = \sum_{i=0}^{D-1} z^{-i} = \begin{cases} D & \text{if } z = 1 \\ \dfrac{1 - z^{-D}}{1 - z^{-1}} & \text{otherwise,} \end{cases} \tag{5.194}$$

where $D = f_s/f_D$ is the decimation ratio and f_D is the decimation frequency. In the frequency domain, that is, for $z = e^{j\omega T}$, we have

$$G(\omega) = \left(\frac{e^{j\omega DT/2} - e^{-j\omega DT/2}}{e^{j\omega T/2} - e^{-j\omega T/2}} \right) \frac{e^{-j\omega DT/2}}{e^{-j\omega T/2}} \tag{5.195}$$

and

$$G(\omega) = \frac{\sin(\omega DT/2)}{\sin(\omega T/2)} e^{-j\omega(D-1)T/2} \tag{5.196}$$

where $T = 1/f_s$ and f_s is the sampling frequency. It was assumed that $\sin(x)\hat{=}(e^{jx} - e^{-jx})/2j$. The filter zeroes are located uniformly at multiples of the decimation frequency, f_s/D. The magnitude response is obtained as

$$|G(\omega)| = \begin{cases} D & \text{if } \omega = 0 \\ \left| \dfrac{\sin(\omega DT/2)}{\sin(\omega T/2)} \right| & \text{otherwise;} \end{cases} \tag{5.197}$$

and the phase response is given by

$$\angle G(\omega) = \begin{cases} -(D-1)\dfrac{\omega T}{2} & \text{if } G(\omega) \geq 0 \\ -(D-1)\dfrac{\omega T}{2} \pm \pi & \text{if } G(\omega) < 0 \end{cases} \tag{5.198}$$

Note that the phase response changes linearly with the frequency. The stop-band attenuation of a single filter section with the transfer function $G(z)$ is limited. Given a modulator with the order L, the required decimation filter should have a transfer function of the form [17]

$$H(\omega) = [G(\omega)]^N \tag{5.199}$$

where $N = L + 1$, in order to reduce the effect of an aliasing of the out-of band noise on the baseband signal. Figure 5.51 shows the frequency responses of a decimation filter with N sections. Generally, they will have $D - 1$ spectral zeros, $\lfloor D/2 \rfloor$ of which are located between 0 and $f_s/2$, where $\lfloor x \rfloor$ denotes the largest integer not greater than x. By sampling a signal at a rate of f_s/D, the baseband of interest should be restricted to the frequency range from 0

$G(\omega) = \alpha e^{-j\tau\omega} G_R(\omega)$, where α and τ are complex and real constants, respectively, and $G_R(\omega)$ is a real-valued function of ω.

to $f_b = f_s/(2D)$. The worst-case distortion occurs at the edge frequency of the baseband, which is characterized by $\omega_b = \pi(f_s/D)$ and where the transfer function magnitude of the decimation filter is

$$|H(\omega_b)| = [\sin(\pi/D)]^{-N} \tag{5.200}$$

Within the signal bandwidth, a decimation filter with N sections then exhibits a maximum attenuation of

$$\text{Droop} = -20N \log_{10}[\sin(\pi/D)] \quad \text{dB} \tag{5.201}$$

Generally, the signal components in the frequency range from $k(f_s/D) - f_b$ to $k(f_s/D) + f_b$, $k = 1, 2, \ldots, \lfloor D/2 \rfloor$ can alias back into the signal baseband. In practical applications, the useful signal bandwidth is then selected such that $f_b \ll f_s/(2D)$ to improve the filter attenuation in aliasing bands centered around multiples of the decimation frequency.

(a)

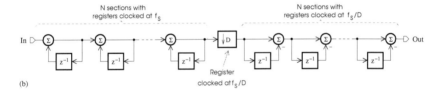

(b)

FIGURE 5.52
(a) Single-stage decimation filter and (b) its CIC filter-based implementation.

By moving the numerator of the filter transfer function after the rate change stage where the sampling rate can be reduced to f_s/D (see Figure 5.52(a)), the power consumption of the resulting structure is minimized. Figure 5.52(b) shows an implementation of the decimation filter, which consists of N integrators using shift registers clocked at the sampling rate f_s, a rate change stage that can be realized using a shift register operating at the decimation rate, f_s/D, and N differentiators based on shift registers with a clock signal of f_s/D. Such an implementation, known as a cascaded integrator-comb (CIC) filter,[3] has the advantage of requiring only registers and adders.

A simplistic analysis of a first-order modulator shows that the average value of the input signal is contained in the serial output bitstream. The decimation filter following the modulator reduces the sampling rate of the signal sequence and increases the signal resolution by averaging the input bitstream. In the case of a decimation by a factor D, this is achieved by transferring one out of every D signal samples to the output.

Let us assume that the modulator delivers a 1-bit output bitstream of the form

$$HLHLHHLH$$

where H and L denote the high and low logic states, respectively, to be downsampled by a factor 8. The output of the decimation filter is related to the H-state density, given by $5/8 = 0.625$, that is, the signal sequence with an overall of 8 states has 5 states at the logic high level. Because $0.625 = 1 \times 2^{-1} + 0 \times 2^{-2} + 1 \times 2^{-3}$, the input signal can be interpreted as a 3-bit number with the binary representation, 101. Here, the decimation process results in a reduction of the sampling rate by a factor of 8 and the conversion of the 1-bit serial input into a 3-bit number.

Data loss due to arithmetic overflow is avoided in the decimation filter by using a two's complement representation and registers with a sufficiently large length. If the input data of the decimation filter are encoded using a two's complement binary format with B_{in} bits, the magnitude of the output samples will be bounded by H_{max} satisfying the following equation,

$$H_{max} = 2^{B_{in}-1} \sum_{n=0}^{D-1} |h(n)| \tag{5.202}$$

Next, the transfer function of the decimation filter can be expressed as

$$H(z) = \sum_{n=0}^{D-1} h(n)z^{-n} = \left[\sum_{n=0}^{D-1} z^{-n} \right]^N \tag{5.203}$$

The decimation stage can be considered a finite-impulse response (FIR) filter producing an output word, which represents a weighted average of its most recent input samples.

Because the product of positive coefficient polynomials is a polynomial

---

[3]The decimation CIC filter is also called a *sinc* filter. This is due to the fact that the frequency magnitude response can be expressed in terms of *sinc* functions, where $sinc(x)$ is defined as $\sin(x)/x$.

with positive coefficients, all the coefficients $h(n)$ are positive, and

$$\sum_{n=0}^{D-1} |h(n)| = \sum_{n=0}^{D-1} h(n) = H(1) = D^N \qquad (5.204)$$

Assuming that B_{out} is the number of bits of the output data, we have

$$H_{max} \leq 2^{B_{out}} \qquad (5.205)$$

and

$$B_{out} = \lceil N \log_2(D) + B_{in} \rceil \qquad (5.206)$$

where $\lceil x \rceil$ is the smallest integer not less than x. By considering the least significant bit (LSB) to be the bit number zero, the most significant bit (MSB) number B_{max} in the output is given by

$$B_{max} = \lceil N \log_2(D) + B_{in} - 1 \rceil \qquad (5.207)$$

Although the overall dc gain of a CIC decimation filter with N sections is finite, individual integrators, whose gain is infinite at dc, can be subjected to numerical overflows. If two's complement arithmetic is used, and if the sum of more than two numbers is guaranteed not to overflow, then overflows in partial sums will be interpreted as either the most positive or most negative representable number, depending on its sign. The difference between any two successive samples computed by the following differentiator cancels out the overflow provided the data in all filter sections are represented with the same MSB position as the one of the output samples, as set by the word length, B_{max}.

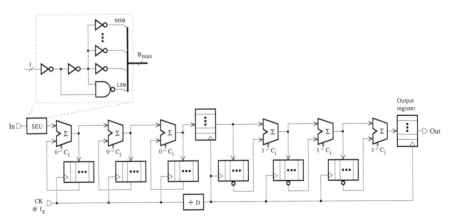

FIGURE 5.53
Circuit diagram of a CIC decimation filter with three sections.

The circuit diagram of a CIC decimation filter with three sections is illustrated in Figure 5.53. This structure can be used as a decimation stage for a single-bit second-order modulator. The sign extension unit (SEU) increases the resolution and converts the binary input data stream into the two's complement representation with B_{max} bits. This is achieved by converting a sample with a high logic level to 1 in two's complement and a sample with a low logic level to -1 in two's complement. The subtraction in the differentiator is realized using the adder carry-in node to add one to the sum of one input sequence and the complement obtained by inverting the bits of the other input sequence. A hard-wired logic realization of the CIC decimation filter only requires registers, adders, and a clock divider. However, other implementations may use a digital signal processor, which relies on multiply and accumulate, and address increment and decrement (or data shifting) operations.

The word length of registers and adders are related to the overall dc gain of the decimation filter and the input data word length. In applications with a decimation factor greater than 64, the number of bits required for the data representation can become excessively high as the number of filter sections increases, leading to a structure that is prohibitively difficult to implement. Because registers and adders are sized to have the same MSB, the word length can be scaled down by discarding the least significant bits (LSBs) to the bit position in any section that cannot grow beyond the LSB of the output word. The number of LSBs to be truncated or rounded from one filter section to another can be determined by assuming that the truncation error at the filter's output uniformly bounds the error incurred at the intermediate sections.

The use of truncation to reduce the word length in the N sections of a CIC decimation filter can result in a total of $2N + 1$ error sources. The errors associated with the truncation at integrator and differentiator inputs are labeled with j running from 1 to N and from $N+1$ to $2N$, respectively. The truncation of the data available at the output of the decimation filter produces an error source specified by $2N + 1$. If B_j represents the number of bits truncated at the j-th section, the truncation or rounding error has a uniform distribution with a width of

$$E_j = \begin{cases} 0 & \text{when using the full precision} \\ 2^{B_j} & \text{otherwise.} \end{cases} \tag{5.208}$$

The corresponding mean is $E_j/2$ in the case of truncation; otherwise it is zero, and the variance is, respectively, given by

$$\sigma_j^2 = \frac{E_j^2}{12} \tag{5.209}$$

The noise error introduced at the j-th section propagates through the filter. It can be verified that the overall mean at the filter output is a function of only the contributions of the first and last error sources. The statistical dispersion of all errors is better tracked by the variance, which is then used for the derivation of the design criteria. Assuming that the noise sources are mutually uncorrelated, the variance at the filter output due to the truncation at the j-th stage is

$$\sigma_{T_j}^2 = \sigma_j^2 F_j^2 \tag{5.210}$$

where

$$F_j^2 = \begin{cases} \sum_{n=0}^{(D-1)N+j-1} h_j^2(n) & j = 1, 2, \ldots, N \\ \sum_{n=0}^{2N+1-j} h_j^2(n) & j = N+1, N+2, \ldots, 2N \\ 1, & j = 2N+1 \end{cases} \tag{5.211}$$

and $h_j(n)$ is the impulse response from the j-th error source to the output. The overall variance at the filter output can be written as

$$\sigma_T^2 = \sum_{j=1}^{2N+1} \sigma_{T_j}^2 \tag{5.212}$$

In practice, it can be assumed that the variance from the first $2N$ error sources is at least as small as the variance of the last error source and the overall error is evenly spread out between these $2N$ sources. Hence,

$$\sigma_{T_j}^2 \leq \frac{1}{2N}\sigma_{T_{2N+1}}^2, \quad j = 1, 2, 3, \ldots, 2N, \tag{5.213}$$

where

$$\sigma_{T_j}^2 = \frac{1}{12}2^{2B_j}F_j^2 \tag{5.214}$$

For each filter section, the number of LSBs that can be thrown aside is then

$$B_j = \left\lfloor \left| \frac{1}{2}\log_2 \frac{6}{N} + \log_2 \sigma_{T_{2N+1}} - \log_2 F_j \right| \right\rfloor \tag{5.215}$$

for $j = 1, 2, 3, \ldots, 2N$. With the length of the output data being B_{out}, we can get the number of LSBs discarded at the output as follows,

$$B_{2N+1} = B_{max} - B_{out} + 1 \tag{5.216}$$

Note that the truncation of B_{2N+1} bits at the filter output (or rounding the filter output) will produce an output noise with

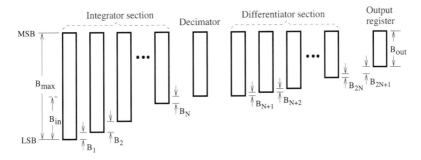

FIGURE 5.54
Register word lengths in a CIC decimation filter.

variance $2^{2B_{2N+1}}/12$. Figure 5.54 shows the distribution of the register word length in a CIC decimation filter.

Note that alternate decompositions of the transfer function, $H(z)$, can be exploited to arrive at other implementation structures of the CIC decimation filter.

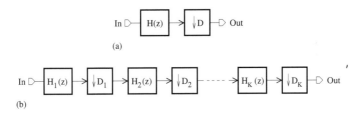

FIGURE 5.55
(a) Single-stage and (b) multistage decimation filters.

The decimation filter can be realized using a single-stage architecture as shown in Figure 5.55(a). With the FIR filter commonly used for the decimation, the required order (or length) is generally proportional to the sampling frequency and inversely proportional to the width of the transition band. Because the decimation filter should provide a narrow transition band, a hardware-efficient design then relies on a multistage architecture [18] as shown in Figure 5.55(b), where the earlier sections have a lower order than the latter ones. In this way, the input signal with a high sampling rate is processed by a filter that exhibits a large transition width, and therefore possesses a low order. For the following sections, both sampling frequency and transition width are reduced and an acceptable filter length can still be achieved.

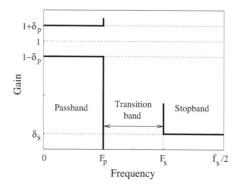

FIGURE 5.56
Lowpass filter specifications.

Let the lowpass filter required by a single-stage decimation be characterized by the specifications shown in Figure 5.56, where δ_p and δ_s are the passband and stopband ripples, respectively, and F_p and F_s denote the edge frequencies of the passband and stopband, respectively. Conventional design methods such as the Parks-McClellan algorithm and linear programming can be used to find the FIR filter length and coefficients such that the magnitude frequency response, $|H(\omega)|$, meets the following requirements:

$$||H(\omega)| - 1| \leq \delta_p \qquad \text{for} \quad 0 \leq \omega \leq \Omega_p \tag{5.217}$$

and

$$|H(\omega)| \leq \delta_s \qquad \text{for} \quad \Omega_s \leq \omega \leq \omega_s/2 \tag{5.218}$$

where $\Omega_p = 2\pi F_p$ and $\Omega_s = 2\pi F_s$.

With a multistage decimation filter [20], the overall decimation ratio D is factored into the product,

$$D = \prod_{k=1}^{K} D_k \tag{5.219}$$

where K is the number of stages, and each independent section within the structure decimates the signal by D_k, which is a positive integer. The k-th stage must have a ripple less than δ_p/K in the passband defined by

$$0 \leq f \leq F_p \tag{5.220}$$

where F_p is supposed to be the highest frequency in the original signal, and a ripple less than δ_s in the stopband specified by

$$F_k - \frac{F_K}{2} \leq f \leq \frac{F_{k-1}}{2}, \qquad k = 1, 2, \ldots, K \tag{5.221}$$

where F_k is the sampling frequency of the k-th stage. Hence,

$$F_k = \frac{F_{k-1}}{D_k} \qquad (5.222)$$

and $F_K = F_0/D$, where F_0 represents the input sampling frequency, Df_s, of the first stage. The passband ripple of the individual filters is selected with the aim of maintaining the overall passband ripple within the bounds set by δ_p. For each stage, the stopband ripple, and the edge frequencies of the passband and stopband are determined such that the effects of aliasing on the baseband spanning from 0 to $F_0/2D$ can be eliminated. Note that the transition band of the last stage is the same as the one of the single-stage architecture, but the sampling frequency is somewhat reduced.

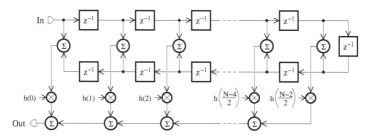

FIGURE 5.57
Direct form structure of a linear-phase FIR filter with N even.

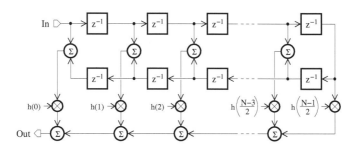

FIGURE 5.58
Direct form structure of a linear-phase FIR filter with N odd.

The decimation filter should be designed to provide sufficient attenuation of unwanted high-frequency signals that can be aliased into the baseband, and to feature a hardware-efficient structure. By increasing the stopband attenuation using more CIC filter sections, the passband droop is also increased. To compensate for the limitations of the CIC structure, a typical multistage decimation filter will then consist of a CIC filter, followed by half-band FIR filters and a FIR compensation filter.

An FIR filter can be described by the transfer function

$$H(z) = \sum_{n=0}^{N-1} h(n)z^{-n} \tag{5.223}$$

where N is the length of the filter and $h(n)$ denotes the filter coefficients. In the case of a lowpass FIR filter constrained to be a linear phase system, the coefficients must satisfy the condition

$$h(n) = h(N-1-n) \tag{5.224}$$

For N even, the filter transfer function can be decomposed as

$$H(z) = \sum_{n=0}^{\frac{N}{2}-1} h(n)z^{-n} + \sum_{n=N/2}^{N-1} h(n)z^{-n} \tag{5.225}$$

$$= \sum_{n=0}^{\frac{N-2}{2}} h(n)z^{-n} + \sum_{n=0}^{\frac{N-2}{2}} h(N-1-n)z^{-(N-1-n)} \tag{5.226}$$

$$= \sum_{n=0}^{\frac{N-2}{2}} h(n)\left[z^{-n} + z^{-(N-1-n)}\right] \tag{5.227}$$

and for N odd, we have

$$H(z) = \sum_{n=0}^{\frac{N-1}{2}-1} h(n)z^{-n} + h\left(\frac{N-1}{2}\right)z^{-(N-1)/2} + \sum_{n=\frac{N-1}{2}+1}^{N-1} h(n)z^{-n} \tag{5.228}$$

$$= \sum_{n=0}^{\frac{N-3}{2}} h(n)z^{-n} + h\left(\frac{N-1}{2}\right)z^{-(N-1)/2} + \sum_{n=0}^{\frac{N-3}{2}} h(N-1-n)z^{-(N-1-n)} \tag{5.229}$$

$$= \sum_{n=0}^{\frac{N-3}{2}} h(n)\left[z^{-n} + z^{-(N-1-n)}\right] + h\left(\frac{N-1}{2}\right)z^{-(N-1)/2} \tag{5.230}$$

As a result, the filter exhibits a symmetric $h(n)$ and the number of coefficients is reduced to either $N/2$ for N even or $(N+1)/2$ for N odd. Figures 5.57 and 5.58 show the direct form structures of a linear-phase FIR filter, where N is even and odd, respectively. By reversing the signal flow-graph of the direct form FIR filters while maintaining the same transfer function, the transpose form structures of Figures 5.59 and 5.60 can be derived in the cases, where N is even and odd, respectively. The main advantage of transpose architectures is that adders are naturally pipelined without introducing additional latency.

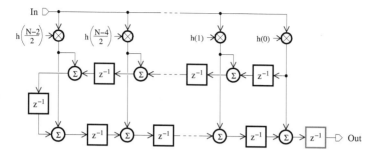

FIGURE 5.59
Transpose form structure of a linear-phase FIR filter with N even.

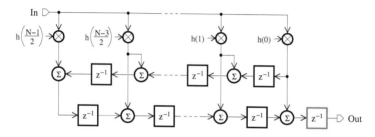

FIGURE 5.60
Transpose form structure of a linear-phase FIR filter with N odd.

Note that pipelining consists of adding latches or flip-flops between logic sections to reduce the critical path and increase the throughput of a system.

The block diagram of a two-fold decimation filter based on the direct form structure is depicted in Figure 5.61(a). This implementation features the computational complexity of the FIR filter and is not hardware efficient. An improvement can be obtained using a polyphase structure and then swapping the position of the filter and down-sampler, as shown in Figure 5.61(b). In general, for an FIR filter of length N, the next polyphase decomposition can be obtained [21]:

$$
H(z) = \begin{cases} \displaystyle\sum_{n=0}^{\frac{N-2}{2}} h(2n)z^{-2n} + z^{-1} \sum_{n=0}^{\frac{N-2}{2}} h(2n+1)z^{-2n} & \text{if } N \text{ even} \\ \displaystyle\sum_{n=0}^{\frac{N-1}{2}} h(2n)z^{-2n} + z^{-1} \sum_{n=0}^{\frac{N-3}{2}} h(2n+1)z^{-2n} & \text{if } N \text{ odd.} \end{cases}
\tag{5.231}
$$

Let

$$
h_i(n) = h(2n+i) \qquad i = 0,1
\tag{5.232}
$$

denote the n-th filter coefficient of the i-th polyphase component. The transfer

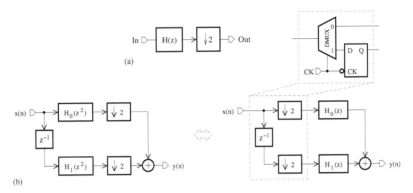

FIGURE 5.61
Block diagrams of two-fold decimation filters based on (a) direct form and (b)
polyphase structures.

function of the FIR filter can be written as

$$H(z) = H_0(z^2) + z^{-1}H_1(z^2) \qquad (5.233)$$

where

$$H_0(z) = \begin{cases} \sum_{n=0}^{\frac{N-2}{2}} h_0(n)z^{-n} & \text{if } N \text{ even} \\ \sum_{n=0}^{\frac{N-1}{2}} h_0(n)z^{-n} & \text{if } N \text{ odd} \end{cases} \qquad (5.234)$$

and

$$H_1(z) = \begin{cases} \sum_{n=0}^{\frac{N-2}{2}} h_1(n)z^{-n} & \text{if } N \text{ even} \\ \sum_{n=0}^{\frac{N-3}{2}} h_1(n)z^{-n} & \text{if } N \text{ odd.} \end{cases} \qquad (5.235)$$

Basically, the filter coefficients $h(n)$ were grouped into even- and odd-
numbered samples.

For N even, the number of multipliers and delay units is reduced by making
use of the fact that the filter coefficients of polyphase components exist in
mirror image pairs. This leads to the block diagrams of a linear-phase FIR
filter with a decimation factor of 2, as illustrated in Figures 5.62 and 5.63,
where $\lceil x \rceil$ denotes the smallest integer not less than x.

For N odd, the implementation of the decimation filter is made hardware
efficient by exploiting the coefficient symmetry. With $\lfloor x \rfloor$ being the largest
integer less than or equal to x, Figures 5.64 and 5.65 show the resulting block
diagrams of a linear-phase FIR filter with a decimation factor of 2.

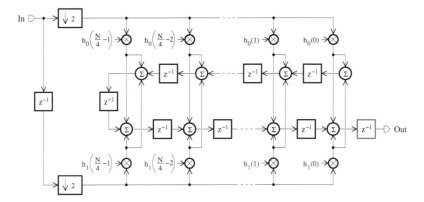

FIGURE 5.62
Block diagram of a linear-phase FIR filter with a decimation factor of 2 (N even and $N/2$ even).

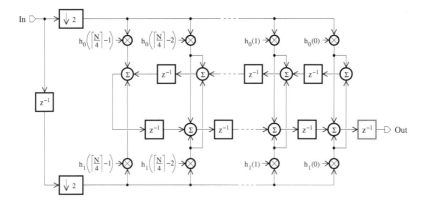

FIGURE 5.63
Block diagram of a linear-phase FIR filter with a decimation factor of 2 (N even and $N/2$ odd).

The impulse response of a half-band FIR filter [22, 23] is characterized by

$$h(n) = \begin{cases} \alpha & \text{if } n = (N-1)/2 \\ 0 & \text{if } n = 2k-1, \quad k \neq (N+1)/4 \quad k = 1, 2, \ldots, (N-1)/2 \\ h(2k) & \text{if } n = 2k, \quad k = 0, 1, 2, \ldots, (N-1)/2 \end{cases}$$

$$(5.236)$$

where α is usually $1/2$ and the filter length is of the form, $N = 4L - 1$, with L being an integer. All coefficients with n odd, except for $n = (N-1)/2$, are

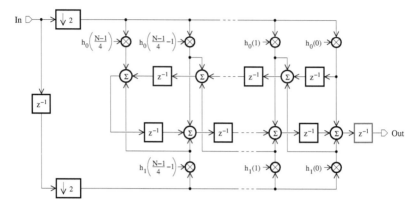

FIGURE 5.64
Block diagram of a linear-phase FIR filter with a decimation factor of 2 (N odd and $(N-1)/2$ even).

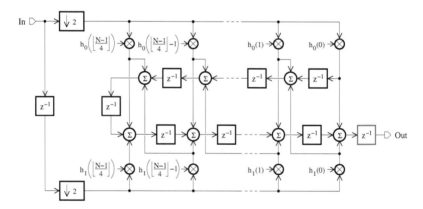

FIGURE 5.65
Block diagram of a linear-phase FIR filter with a decimation factor of 2 (N odd and $(N-1)/2$ odd).

then zero. As a result, the z-domain transfer function can be expressed as

$$H(z) + H(-z) = 2\alpha \tag{5.237}$$

Assuming that $z = e^{j\omega T}$, we have $H(z) \leftrightarrow H(\omega)$ and $H(-z) \leftrightarrow H(\omega - \pi)$, so that

$$H(\omega) + H(\omega - \pi) = 2\alpha \tag{5.238}$$

Hence, the frequency response is symmetric with respect to one-quarter of the sampling frequency, or $f_s/4$. That is,

$$F_p + F_s = f_s/2 \tag{5.239}$$

where F_p and F_s are the edge frequencies of the passband and stopband, respectively, and the passband and stopband ripples should be identical, namely, $\delta_p = \delta_s$.

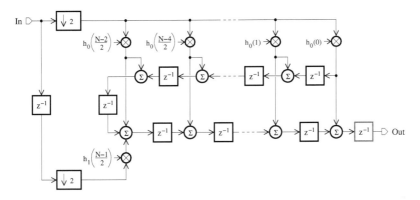

FIGURE 5.66
Block diagram of a half-band FIR filter with a decimation factor of 2.

Because a half-band filter is required to have a magnitude response with the value of $1/2$ at $f_s/4$, it is only suitable for a decimation by a factor of 2. Figure 5.66 shows the block diagram of a half-band FIR filter with a two-fold decimation. The realization of a higher decimation ratio can then be addressed by cascading a series of such filters.

In practice, the decimation filter should have a flat passband response and narrow transition band in order to avoid the signal distortion. These features are not generally offered by the CIC and half-band filters. A solution can then consist of using a compensation FIR filter to cancel the passband droop introduced by the CIC filter and to narrow the overall passband. This FIR filter is designed to have the inverted version of the CIC filter passband response and can achieve a decimation by a factor 2. To prevent the amplification of signal components near the stopband edge frequency of the CIC filter, a good rule of thumb is to limit the upper passband frequency of the compensation FIR filter to about $1/4$ the frequency of the first null in the frequency response of the CIC filter.

An approach for overcoming the threat of runtime overflow in FIR filters is to estimate the maximum value of the signal at the output of the k-th stage, $y_{max,k}$, and to find the worst-case word length of each adder and the size of the corresponding output register using to the following equation,

$$y_{max,k} = x_{max} \sum_{n=0}^{k-1} |h(n)| \le 2^{B_k} \tag{5.240}$$

where x_{max} is the maximum value of the input signal, $h(n)$ denotes the filter coefficients, and B_k is the required number of bits. To minimize the hardware complexity, rounding to lower word length or bit truncation can be performed at each filter stage if the additional noise generated will not affect the target accuracy.

By representing filter coefficients in the canonic signed digit (CSD) form [24, 25], multiplications by a constant value can be transformed into a sequence of shift operations, additions and subtractions. An important reduction in the power consumption, area, and latency in FIR circuits can then be achieved, especially when shift operations are simply carried out at the wiring level, and dedicated shifters are not required. The radix-2 CSD expression of a fractional filter coefficient h is given by

$$h = \sum_{k=P}^{L-1} a_k 2^{p_k} \qquad (5.241)$$

where $a_k \in \{-1, 0, 1\}$, $P \le p_k \le M - 1$ and $M - P$ denotes the number of ternary digits. Note that a CSD code contains no adjacent nonzero digits, and the -1 value is represented by $\bar{1}$.

Starting from the LSB and proceeding toward the MSB, the bits b_k of a two's complement number can be converted into CSD digits a_k for the indexes $k = 0, 1, 2, \cdots, N - 1$. It is assumed that $c_0 = 0$ and $b_N = b_{N-1}$, and the conversion is realized according to the following equation,

$$c_{k+1} = \lfloor (c_k + b_k + b_{k+1})/2 \rfloor \qquad (5.242)$$
$$a_k = b_k + c_k - 2c_{k+1} \qquad (5.243)$$

where c_k is the input auxiliary carry, c_{k+1} represents the output auxiliary carry, and $\lfloor x \rfloor$ denotes the largest integer less than or equal to x.

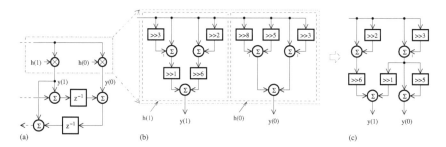

FIGURE 5.67
(a) Section of a transpose filter; filter coefficient implementations (b) without and (c) with sub-expression sharing.

The CSD representation of a number is unique and it has the advantage of containing the fewest number of nonzero digits, which represent additions or subtractions. A substantial hardware saving can then be realized in the implementation of the multiplication with a CSD coefficient.

A section of a linear-phase filter with the coefficients $h(0) = 0.90234$ and $h(1) = 0.45703$ is shown in Figure 5.67(a). The coefficient values in the 10-bit two's complement representation are given by

$$h(0) = 2^{-1} + 2^{-2} + 2^{-3} + 2^{-6} + 2^{-7} + 2^{-8} = 0.111001110 \qquad (5.244)$$

$$h(1) = 2^{-2} + 2^{-3} + 2^{-4} + 2^{-6} = 0.011101001 \qquad (5.245)$$

whereas in the 10-digit radix-2 CSD representation, we have

$$h(0) = 2^0 - 2^{-3} + 2^{-5} - 2^{-8} = 1.00\bar{1}0100\bar{1}0 \qquad (5.246)$$

$$h(1) = 2^{-1} - 2^{-4} + 2^{-6} + 2^{-8} = 0.100\bar{1}01010 \qquad (5.247)$$

By exhibiting a reduced number of nonzero ternary digits, the CSD code appears to be suitable for the implementation of the filter coefficients, as shown in Figure 5.67(b). Note that the symbol $>> i$ indicates the right shift by i bit positions due to the term 2^{-i} and the symbol $<< i$ represents the left shift by i bit positions related to the term 2^i. Provided that the transposed direct-form structure is used for the FIR filter realization, common sub-expressions can be shared between the coefficients to reduce the overall number of operations. The filter coefficient implementation shown in Figure 5.67(b) is based on the identities

$$h(0) = (-2^{-3} + 1)(2^{-5} + 1) \quad \text{and} \quad h(1) = 2^{-1}(-2^{-3} + 1) + 2^{-6}(2^{-2} + 1) \qquad (5.248)$$

For high-order filters, the design of hardware efficient structures is performed using algorithms [26–28] to combine sub-expressions occurring often in coefficients.

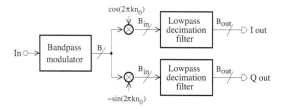

FIGURE 5.68
Architecture of a bandpass modulator with a decimation filter.

The decimation stage for a bandpass modulator can consist of bandpass filters and sample rate down-converters. Because bandpass modulators are generally used to digitize signals at an intermediate-frequency stage of a wireless receiver based on the direct-conversion scheme, where in-phase (I) and quadrature-phase (Q) signals are required to track any changes in magnitude and phase of the incoming message, the decimation filter can be realized, as shown in Figure 5.68. The two mixers, respectively, perform digital sine and cosine multiplications at the same sample rate as the one of the bandpass modulator. The center frequency of the input signal is then shifted to baseband or dc, and I/Q versions of the modulator output are generated. This allows the use of the lowpass decimation filter, which is easier to implement and more hardware efficient than the bandpass decimation filter. However, the word length of the input signal can still be long enough to significantly increase the filter hardware resource.

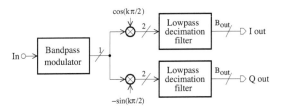

FIGURE 5.69
Architecture of a bandpass $f_s/4$-modulator with a decimation filter.

When the sampling frequency, f_s, is four times the center frequency, f_0, of the modulator, the ratio $n_0 = f_0/f_s$ is reduced to $1/4$ and the coefficients in the multiplications are given by

$$\cos(k\pi/2) = 1, 0, -1, 0, 1, 0, -1, 0, \cdots \quad \text{for} \quad k = 0, 1, 2, 3, 4, 5, 6, 7, \cdots$$
$$(5.249)$$

$$-\sin(k\pi/2) = 0, -1, 0, 1, 0, -1, 0, 1, \cdots \quad \text{for} \quad k = 0, 1, 2, 3, 4, 5, 6, 7, \cdots$$
$$(5.250)$$

Because the multiplication operations are reduced to selectively not change, nullify, or invert the input signal, they can be implemented using a few logic gates in the case of a single-bit modulator. Figure 5.69 shows a bandpass $f_s/4$-modulator with the decimation filter.

Remark
In special cases, the CIC decimation filter can be implemented using polyphase structures [18].

Consider a CIC decimation filter consisting of N cascaded sections. Its transfer function is

$$H(z) = \left(\frac{1 - z^{-D}}{1 - z^{-1}} \right)^N$$
$$(5.251)$$

By choosing the decimation factor to be a power of 2, we have

$D = 2^M$, where M is an integer. The transfer function $H(z)$ can be written as

$$H(z) = \left(\sum_{i=0}^{D-1} z^{-i}\right)^N = \left(\sum_{i=0}^{2^M-1} z^{-i}\right)^N = \prod_{i=0}^{M-1} (1 + z^{-2^i})^N \qquad (5.252)$$

The commutative rule for multirate systems can be used to

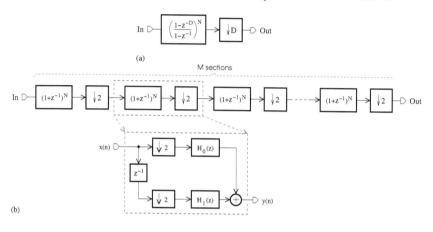

(a)

(b)

FIGURE 5.70
(a) Single-stage CIC decimation filter and (b) its polyphase FIR implementation for $D = 2^M$.

transform the block diagram of the CIC decimation filter shown in Figure 5.70(a) into the polyphase structure of Figure 5.70(b), which is realized by cascading M identical sections with a decimation factor of 2. The transfer function of each section can be decomposed as follows:

$$(1 + z^{-1})^N = \sum_{j=0}^{N} h(j)z^{-j} = H_0(z^2) + z^{-1}H_1(z^2) \qquad (5.253)$$

where H_0 and H_1 denote the polyphase components. This implementation is based on FIR filters. The data path size increases by N bits for each section and the overflow is prevented by setting the minimum word length for the k-th section to $B_{in} + kN$ bits, where B_{in} is the number of bits at the filter input and $k = 1, 2, \cdots, M$. Because the sampling frequency is successively decreased by a factor of 2, the power consumption component, which is proportional to the sampling frequency, can be reduced in comparison with the one of decimation structures where this is not the case.

5.2 Delta-sigma digital-to-analog converter

Delta-sigma ($\Delta\Sigma$) digital-to-analog converters (DACs) are typically used in applications where a high linearity is preferred over a high bandwidth.

FIGURE 5.71
Block diagram of a delta-sigma DAC.

The block diagram of a delta-sigma DAC is depicted in Figure 5.71. It comprises a digital interpolation filter, a digital delta-sigma modulator, a low-resolution DAC, and an analog lowpass filter. The input signal is assumed to be a stream of digital words with N bits. It is processed by a digital interpolation filter, which raises the data rate to $OSR \cdot f_s$, where OSR is the oversampling ratio and f_s is the sampling rate, by inserting $OSR - 1$ equidistant zero-valued samples between two consecutive samples of the input sequence. The oversampled signal is then supplied to a digital modulator or noise shaper, which reduces the word length, generally to 1 bit. An analog version of the modulator output is provided by the reconstruction stage, which includes a low-resolution DAC and a lowpass (smoothing) filter.

5.2.1 Interpolation filter

The interpolation by an integer factor of I, which results in an increase in the output sampling rate, is the process of inserting $I - 1$ zeros between successive samples of the input signal, followed by the filtering of the undesired spectral images. For large values of I, the hardware-efficient implementation of the interpolation filter is generally based on a multistage structure [20].

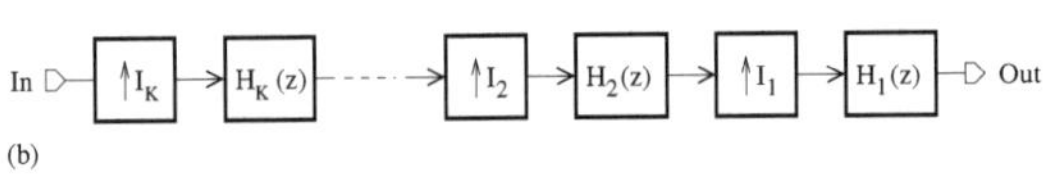

In $\triangleright$— $\uparrow I$ → $H(z)$ —$\triangleright$ Out

(a)

In $\triangleright$— $\uparrow I_K$ → $H_K(z)$ ----- $\uparrow I_2$ → $H_2(z)$ → $\uparrow I_1$ → $H_1(z)$ —$\triangleright$ Out

(b)

FIGURE 5.72
(a) Single-stage and (b) multistage interpolation filters.

Figure 5.72(a) shows the block diagram of a single-stage interpolation filter. The overall interpolation ratio I, which is equal to the oversampling ratio of the modulator, OSR, can also be realized by a cascade of K stages, each achieving a sampling rate increase of I_k (see Figure 5.72(b)). It can then be factored as

$$I = \prod_{k=1}^{K} I_k \tag{5.254}$$

where K is the number of stages. The sampling frequency, F_k, at the output of the k-th stage is given by

$$F_{k-1} = I_k F_k \qquad k = K, K-1, \cdots, 1 \tag{5.255}$$

where it is assumed that the output sampling frequency is $F_0 = I f_s$ with f_s being the sampling frequency of the signal to be interpolated. Here, the stages are numbered backward from K to 1 to show that, for a given rate change and number of stages, an interpolation filter is the dual of a decimation filter, and the input signal is applied to the K-th filter stage. The up-sampling process also creates images of the original spectrum centered at multiples of the original sampling frequency. For the k-th stage, lowpass filters characterized by the transfer functions $H_k(z)$ are used to remove the images of the baseband signal at frequencies above $\omega = \pi(f_s/I_k)$. This requirement can be met by a filter designed to have a passband ripple δ_p/K, a stopband ripple δ_s, a passband specified by

$$0 \leq f \leq F_p \tag{5.256}$$

and a stopband of the form

$$F_k - \frac{F_K}{2} \leq f \leq \frac{F_{k-1}}{2} \tag{5.257}$$

where F_K is the input sampling rate, f_s. It should be noted that the K-th filter stage has the smaller transition band.

Generally, multistage structures can exhibit reduced filter lengths and computational complexity as compared to single-stage designs. This is due to the fact that the specifications of individual interpolation filters are relaxed and low-order filters can provide a sufficient attenuation of unwanted high-frequency signals that can be aliased into the baseband. A hardware-efficient implementation of the interpolation filter will then consist of a compensation FIR filter, followed by half-band FIR filters and a CIC interpolation filter. The role of the compensation filter, which is designed to have the inverse frequency response of the CIC filter, is to pre-equalize the passband droop of the CIC structure.

The block diagram of a twofold interpolation filter based on the direct form structure, as shown in Figure 5.73(a), features the computational complexity of the FIR filter and is not hardware efficient. The number of operations required for the signal processing in interpolation filters can be reduced using a polyphase structure and then swapping the position of the filter and down-sampler as shown in Figure 5.73(b).

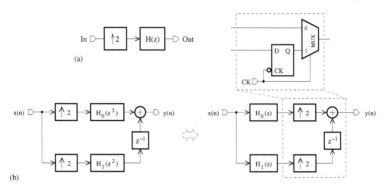

(a)

(b)

FIGURE 5.73
Block diagram of twofold interpolation filters based on (a) direct form and (b) polyphase structures.

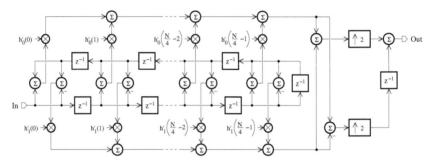

FIGURE 5.74
Block diagram of a linear-phase FIR filter with an interpolation factor of 2 (N even and $N/2$ even).

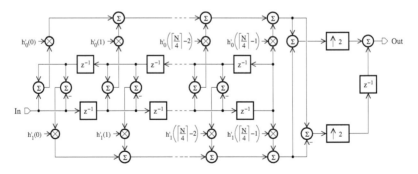

FIGURE 5.75
Block diagram of a linear-phase FIR filter with an interpolation factor of 2 (N even and $N/2$ odd).

By grouping the filter coefficients $h(n)$ into even- and odd-numbered samples, the transfer function of an FIR filter can be expressed as

$$H(z) = H_0(z^2) + z^{-1}H_1(z^2) \tag{5.258}$$

where

$$H_0(z) = \begin{cases} \displaystyle\sum_{n=0}^{\frac{N-2}{2}} h_0(n)z^{-n} & \text{if } N \quad \text{even} \\ \displaystyle\sum_{n=0}^{\frac{N-1}{2}} h_0(n)z^{-n} & \text{if } N \quad \text{odd} \end{cases} \tag{5.259}$$

and

$$H_1(z) = \begin{cases} \displaystyle\sum_{n=0}^{\frac{N-2}{2}} h_1(n)z^{-n} & \text{if } N \quad \text{even} \\ \displaystyle\sum_{n=0}^{\frac{N-3}{2}} h_1(n)z^{-n} & \text{if } N \quad \text{odd.} \end{cases} \tag{5.260}$$

In the case of linear-phase FIR filters, the coefficients are symmetric because $h(n) = h(N-1-n)$.

When N is even, we have $h_0(n) = h_1(N/2-1-n)$. Hence, the coefficients of the polyphase components are not symmetric and exist in time-reversed pairs that can be realized using filter structures with symmetric and anti-symmetric impulse responses [30].

For $N/2$ even, a new set of filter coefficients is defined as

$$h_0'(n) = \frac{1}{2}[h_0(n) + h_0(N/2-1-n)] \tag{5.261}$$

$$h_1'(n) = \frac{1}{2}[h_0(n) - h_0(N/2-1-n)] \tag{5.262}$$

where $n = 0, 1, 2, \cdots, N/4-1$. The transfer functions of the polyphase components can then be given by

$$H_0(z) = \sum_{n=0}^{N/4-1} h_0'(n)(z^{-n} + z^{-(N/2-1-n)}) + \sum_{n=0}^{N/4-1} h_1'(n)(z^{-n} - z^{-(N/2-1-n)}) \tag{5.263}$$

$$H_1(z) = \sum_{n=0}^{N/4-1} h_0'(n)(z^{-n} + z^{-(N/2-1-n)}) - \sum_{n=0}^{N/4-1} h_0'(n)(z^{-n} - z^{-(N/2-1-n)}) \tag{5.264}$$

The block diagram of the resulting linear-phase FIR filter with an interpolation factor of 2 is depicted in Figure 5.74.

For $N/2$ odd, we have

$$h_0'(n) = \frac{1}{2}[h_0(n) + h_0(N/2 - 1 - n)] \tag{5.265}$$

$$h_1'(n) = \frac{1}{2}[h_0(n) - h_0(N/2 - 1 - n)] \tag{5.266}$$

where $n = 0, 1, 2, \cdots, \lceil N/4 \rceil - 2$, and

$$h_0'(\lceil N/4 \rceil - 1) = \frac{1}{2}h_0(\lceil N/4 \rceil - 1) \tag{5.267}$$

$$h_1'(\lceil N/4 \rceil - 1) = -\frac{1}{2}h_0(\lceil N/4 \rceil - 1) \tag{5.268}$$

where $\lceil x \rceil$ denotes the function that returns the smallest integer not less than x. The following expressions can then be derived:

$$
\begin{aligned}
H_0(z) &= \sum_{n=0}^{\lceil N/4 \rceil - 2} h_0'(n)(z^{-n} + z^{-(N/2-1-n)}) + h_0'(\lceil N/4 \rceil - 1)z^{-(\lceil N/4 \rceil - 1)} \\
&+ \sum_{n=0}^{\lceil N/4 \rceil - 2} h_1'(n)(z^{-n} - z^{-(N/2-1-n)}) + h_1'(\lceil N/4 \rceil - 1)z^{-(\lceil N/4 \rceil - 1)}
\end{aligned}
\tag{5.269}
$$

$$
\begin{aligned}
H_1(z) &= \sum_{n=0}^{\lceil N/4 \rceil - 2} h_0'(n)(z^{-n} + z^{-(N/2-1-n)}) + h_0'(\lceil N/4 \rceil - 1)z^{-(\lceil N/4 \rceil - 1)} \\
&- \left[\sum_{n=0}^{\lceil N/4 \rceil - 2} h_0'(n)(z^{-n} - z^{-(N/2-1-n)}) + h_1'(\lceil N/4 \rceil - 1)z^{-(\lceil N/4 \rceil - 1)} \right]
\end{aligned}
\tag{5.270}
$$

The block diagram of the resulting linear-phase FIR filter with an interpolation factor of 2 is shown in Figure 5.75.

In the case of FIR filters with N odd and half-band FIR filters, the coefficients of the polyphase components are also symmetric, and the interpolation structures can be obtained by transposing the polyphase structures of the corresponding decimation filter. The block diagrams of linear-phase two-fold FIR filters with N odd are shown in Figures 5.76 and 5.77, for $(N-1)/2$ even and $(N-1)/2$ odd, respectively. Note that $\lfloor x \rfloor$ is the largest integer less than or equal to the number x. Figure 5.78 shows the block diagram of a half-band FIR filter with the interpolation factor of 2.

The polyphase structure offers increased efficiency in both size and speed. This is due to the fact that the filtering operation occurs at the lower sampling-rate side of the system, and the coefficient symmetry is exploited to derive an optimal form of the filter using resource sharing.

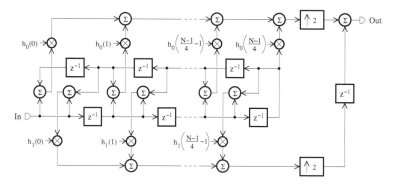

FIGURE 5.76
Block diagram of a linear-phase FIR filter with an interpolation factor of 2
(N odd and $(N-1)/2$ even).

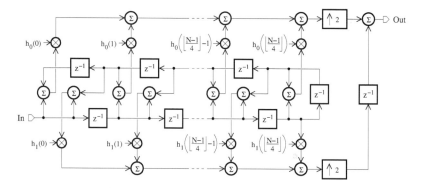

FIGURE 5.77
Block diagram of a linear-phase FIR filter with an interpolation factor of 2
(N odd and $(N-1)/2$ odd).

For large rate changes, a cascaded integrator-comb (CIC) filter [16] has
an advantage over an FIR structure with respect to hardware efficiency in
the context of a high-speed operation. The higher interpolation factor in a
multistage architecture can then be achieved in the CIC filter following the
polyphase FIR systems. Two equivalent block diagrams based on a noble iden-
tity for multi-rate structures are shown in Figure 5.79(a). A CIC interpolation
filter, as illustrated in Figure 5.79(b), includes a differentiator section, an up-
sampler or zero-stuff circuit, and an integrator section. The differentiator
consists of a register and a subtractor. The zero-stuff circuit is realized using
a 2-to-1 multiplexer controlled by a binary counter. The integrator is composed of a register and adder. The differentiator register is clocked at the input
sampling frequency while the counter and integrator register are clocked at
the output sampling frequency.

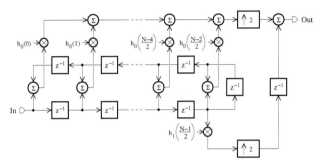

FIGURE 5.78
Block diagram of a half-band FIR filter with an interpolation factor of 2.

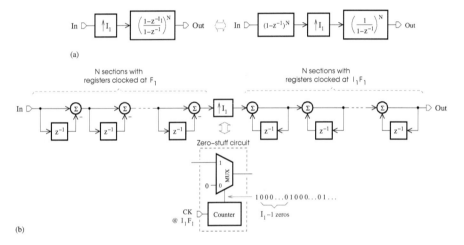

FIGURE 5.79
(a) Single-stage CIC interpolation filter and (b) its implementation.

Because the effect of finite word lengths may be critical in the integrator section, where overflows can cause very large errors and lead to instability due to the unity feedback, rounding is not allowed for integrators. Furthermore, the data widths in the CIC interpolation filter should be set by the worst-case gain at the output of each section to accommodate the maximum value of the signal.

Consider the realization of the interpolation by a factor I_1 using a CIC structure. The transfer function of the interpolation filter can be shown to be

$$H_1(z) = \left(\frac{1 - z^{-I_1}}{1 - z^{-1}} \right)^N \tag{5.271}$$

where N is the number of sections. The CIC interpolation filter [16] is realized by cascading N differentiators, an up-sampler, and N integrators. Each differentiator exhibits the transfer function

$$H_D(z) = 1 - z^{-1} \tag{5.272}$$

With $z = e^{-j\omega T}$, the frequency response is given by

$$H_D(\omega) = 2j \sin(\omega T/2) e^{-j\omega T/2} \tag{5.273}$$

The magnitude, which is then of the form

$$|H_D(\omega)| = 2|\sin(\omega T/2)| \tag{5.274}$$

has a maximum value of 2 because the absolute value of a sine function is bounded by one. Hence, the maximum gain from the input to the output of the k-th section of the interpolation filter can be written as

$$G_j = \begin{cases} 2^j & j = 1, 2, \cdots, N \\ 2^{2N-j} I_1^{j-N-1} & j = N+1, \cdots, 2N \end{cases} \tag{5.275}$$

where, for $j > N$, we have

$$H_1(z) = (1 - z^{-I_1})^{2N-j} \left(\frac{1 - z^{-I_1}}{1 - z^{-1}} \right)^{j-N} \tag{5.276}$$

and the factor $1/I_1$ is introduced to account for the $I_1 - 1$ zeros inserted by the up-sampler between the input samples. For an input data stream with B_{in} bits, the minimum data width at the j-th section is

$$B_j = \lceil B_{in} + \log_2 G_j \rceil \tag{5.277}$$

However, as $G_N > G_{N+1}$, the data width of the last differentiator is larger than the one of the first integrator. Consequently, either the data width of the last differentiator must be reduced to

$$B_N = B_{in} + N - 1 \tag{5.278}$$

when the two's complement arithmetic is employed, or the data width at each integrator should be increased by one to ensure filter stability. To obtain an output data with B_{out} bits, the number of LSBs discarded should be

$$B_T = B_{2N} - B_{out} \tag{5.279}$$

Note that some LSBs will be truncated only if the effect of arithmetic errors at the filter output is maintained at an acceptable level.

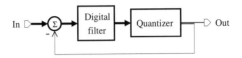

FIGURE 5.80
Block diagram of an output-feedback modulator.

5.2.2   Digital modulator

Generally, the digital modulator is based on a noise-shaping loop, where either the output signal or the quantizer error signal is fed back. This leads to two possible structures, known as the output-feedback and error-feedback modulators.

The block diagram of the output-feedback modulator is shown in Figure 5.80. It consists of a digital filter and quantizer, which has a truncation function. In this case, the quantizer should provide an output data of B-bit length consisting of the input signal MSBs. Let H be the transfer function of the filter. Assuming that the quantizer can be modeled as an additive noise source, the modulator output, V_0, can be related to the input voltage, V_i, in the z-domain by

$$V_0(z) = H_S(z)V_i(z) + H_Q(z)E_Q(z) \qquad (5.280)$$

where

$$H_S(z) = \frac{H(z)}{1 + H(z)} \qquad (5.281)$$

$$H_Q(z) = \frac{1}{1 + H(z)} \qquad (5.282)$$

and E_Q is the quantization error signal. Ideally, the quantization noise has to be suppressed at the frequency band allocated to the signal. The quantizer and the loop filter must be designed to ensure the stability of the modulator.

Another architecture of the digital modulator, referred to as the error-feedback scheme [29], is depicted in Figure 5.81. Here, the feedback signal involves the quantizer error. Assuming that H is the filter transfer function,

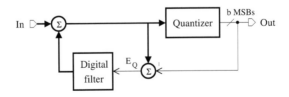

FIGURE 5.81
Block diagram of an error-feedback modulator.

the output of this structure can be expressed in the z-domain as

$$V_0(z) = V_i(z) + H_Q(z)E_Q(z) \qquad (5.283)$$

where the QNTF is given by

$$H_Q(z) = 1 - H(z) \qquad (5.284)$$

and E_Q is the error signal introduced by the quantizer. Note that the STF is reduced to 1. Due to the high sensitivity to component nonidealities, the error feedback scheme is only suitable for digital implementations.

One way to improve the attenuation of the quantization noise in the baseband is to increase the order of the digital modulator, which is similar to the one of the loop filter. There are various structures for the realization of digital modulators based on QNTFs with a highpass frequency response.

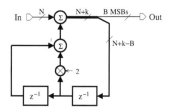

FIGURE 5.82
Block diagram of a second-order error-feedback modulator with a B-bit quantizer.

The block diagram of a second-order error-feedback modulator with a B-bit quantizer, which extracts the MSBs of its input signal, is shown in Figure 5.82. The remaining LSBs are accumulated in the feedback path until they overflow into MSBs and thus contribute to the output. The quantization error is shaped by a second-order transfer function of the form

$$H_Q(z) = (1 - z^{-1})^2 \qquad (5.285)$$

and the transfer function of the filter is given by

$$H(z) = 2z^{-1} - z^{-2} \qquad (5.286)$$

The modulator implementation is greatly simplified because all multiplier coefficients are powers of 2 and the multiplication can be reduced to shift and add operations.

The block diagram of a third-order error-feedback modulator is depicted in Figure 5.83. By placing the zeros of the QNTF on the unit circle, we have

$$H_Q(z) = (1 - z^{-1})^3 \qquad (5.287)$$

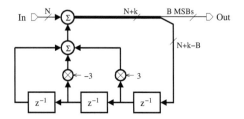

FIGURE 5.83
Block diagram of a third-order error-feedback modulator with a B-bit quantizer.

and the transfer function of the filter is given by

$$H(z) = 3z^{-1} - 3z^{-2} + z^{-3} \tag{5.288}$$

Hence, the filter used in the modulator has an FIR frequency response. The modulator implementation may not be hardware efficient due to the circuit complexity required for multiplications by non-power-of-2 coefficients.

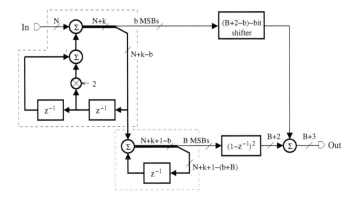

FIGURE 5.84
Block diagram of a third-order, 2-1 cascaded error-feedback modulator.

A third-order error-feedback modulator can also be implemented by the 2-1 cascaded structure shown in Figure 5.84, where it is assumed that $B \geq b$. The first stage is a second-order b-bit modulator, and the inverted version of its quantization noise is applied to the second stage, which is based on a first-order B-bit modulator. Let S be the modulator input signal. The linear analysis yields the equations

$$Y_1(z) = S(z) + (1 - z^{-1})^2 E_{Q1}(z) \tag{5.289}$$

$$Y_2(z) = -E_{Q1}(z) + (1 - z^{-1}) E_{Q2}(z) \tag{5.290}$$

and

$$Y(z) = Y_1(z) + (1 - z^{-1})^2 Y_2(z) \qquad (5.291)$$

where Y_1 and Y_2 denote the outputs of the first and second stage, respectively; E_{Q1} and E_{Q2} represent the quantization noises of the first and second stage, respectively; and the transfer function of the cancelation section is chosen in the form, $(1 - z^{-1})^2$. In order to correctly combine the output signals, Y_1 and Y_2, the output of the first stage must be shifted left to ensure the alignment of both MSBs. The output of the digital modulator can then be written as

$$Y(z) = S(z) + (1 - z^{-1})^3 E_{Q2}(z) \qquad (5.292)$$

The realization of the differentiator at the output of the second stage using digital circuits has the advantage of being more accurate than the implementation in the analog domain. However, the bit growth introduced in the error-cancelation path has the effect of imposing stringent matching requirements on the unit elements used in the implementation of the output DAC.

The single-bit quantization is generally preferred, because the two-level DAC is inherently linear. However, the use of a multi-bit quantizer results in an increase of the modulator resolution due to the reduction of the noise effect both inside and outside the signal band and the elimination of the spectral tones that can be problematic in single-bit structures. Note that a B-bit quantizer can be implemented in modulators based on the two's complement arithmetic by simply truncating the multi-bit input code to its B MSBs. Aside from this advantage, multi-bit converters require calibration schemes for the cancelation of the distortion caused by element mismatches in the DAC.

A high resolution can also be obtained using high-order modulators. This approach is limited by the stability constraint set by the nonoverload requirement of the quantizer. A digital modulator with a QNTF of the form $(1-z^{-1})^L$ will remain stable if the number of bits, B, in the quantizer output is such that $B \geq L + 1$ and the input signal does not exceed the midpoint of the last quantization interval [31]. This is a sufficient but not necessary criterion for the modulator stability, which depends on the input signal. To ensure stability in special cases, simulations can be carried out to determine the allowed maximum magnitude of the input signal.

The representation of the modulator output in the analog domain is achieved by a DAC. The latter produces a signal containing the replica of the digital input and the additive quantization noise. An analog lowpass filter is then designed to eliminate the noisy signal, which lies outside the bandwidth of interest.

5.3   Nyquist DAC design issues

In DAC implementations using an array of unit elements, the output signal is the sum of all contributions due to the unit elements, which are selected based on the input code. Any element mismatch may then affect DAC operation and linearity. Dynamic element-matching (DEM) calibrations can be used to convert mismatch errors into a zero-mean white noise, or to remove the noise caused by mismatches out of the frequency band of interest [1–3, 51]. The objective is to randomize the switching of DAC unit elements such that the mismatch errors are averaged out. This is achieved taking into account the fact that the bit weights are equal, and the unit element switching associated with each digital code conversion can be performed in an arbitrary way.

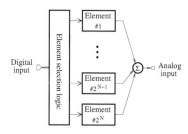

FIGURE 5.85
Block diagram of a DAC with dynamic element-matching calibration.

The general architecture of a DEM DAC is depicted in Figure 5.85. In mismatch-scrambling DEM DACs, mismatch errors are turned into white noise, while for mismatch-shaping DEM DACs, mismatch errors are spectrally shaped by an appropriate filtering function. A straightforward implementation of the digital circuit section, which is needed in DEM DACs, may substantially increase the hardware complexity. Various vector-feedback and data-weighted averaging (tree-structured, butterfly shuffler, barrel shifter) approaches are often adopted to trade the hardware complexity for a lower degree of randomization. In contrast to data-weighted averaging structures, vector-feedback DEM DACs have the advantage of allowing the implementation of any high-order mismatch-shaping transfer functions, but can be limited by hardware complexity.

5.3.1   Vector-feedback DEM

An alternative element selection approach with the advantage of providing a high-order spectral shaping of the DAC noise can be based on the vector feedback structure [3, 33], which uses a sorting mechanism to determine the DAC unit elements to be selected for each conversion.

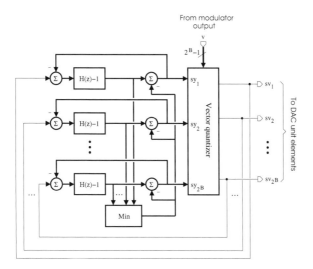

FIGURE 5.86
Vector-feedback ESL.

The block diagram of a vector-feedback ESL is depicted in Figure 5.86. It consists of a vector quantizer, a min block (or a smallest-element sorter), two adders, and a filter. The vector-feedback ESL processes a signal vector, whose length is equal to the number of DAC unit elements, that is, $2^B - 1$ for a resolution of B bits. The signal vector, sy, is sorted and quantized in such a way that the DAC elements associated to the v largest sy components are enabled by the corresponding sv bits, while the remaining elements are deactivated. For each conversion, the number of bits set to 1 in sv must be equal to the number of 1s in the thermometer code of v. To keep the signal values within the range, which is fixed by the finite precision arithmetic, the smallest of all filter output signals is subtracted from the signal available at each filter output.

The noise contribution due to element mismatch is shaped in the z-domain by $H(z)$, where $H(z) - 1$ is the filter transfer function. For applications with lowpass or bandpass signals, the mismatch noise is efficiently removed from the band of interest by highpass and band-reject filters, respectively.

A conventional 4-bit thermometer-coded DAC consists of 16 unit elements. Ideally, all unit elements should be matched. However, in practice, they can be slightly different due to IC process variations. Assuming that the DAC unit elements exhibit random errors with a standard deviation of 1%, a first-order noise-shaping scheme with the transfer function of the form $H(z) = 1 - z^{-1}$ is used to improve the SNR of the DAC. Figure 5.87 shows the unit-element usage for the thermometer coding and first-order noise shaping, respectively. The number of unit elements that are in the on state (boxes filled in gray) corresponds to the decimal equivalent of the input digital code. Figure 5.88

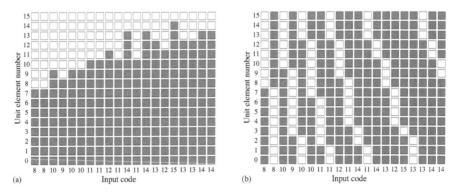

FIGURE 5.87
DAC unit-element usage: (a) Thermometer coding, (b) first-order shaping.

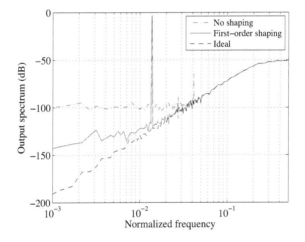

FIGURE 5.88
Output power spectrum of third-order lowpass modulators.

illustrates the output power spectrum of third-order lowpass modulators using thermometer-coded and first-order noise-shaped DACs. The correction scheme improves the SNR for low frequencies by first-order shaping the mismatch-induced noise. In order to whiten the noise generated by the selection algorithm itself, a dither signal can be included in the correction scheme.

Let us consider a vector, $\mathbf{x}$, with four different elements, x_0, x_1, x_2, and x_3. To design the Min block and vector quantizer, six digital multi-bit comparators are necessary. They are configured,

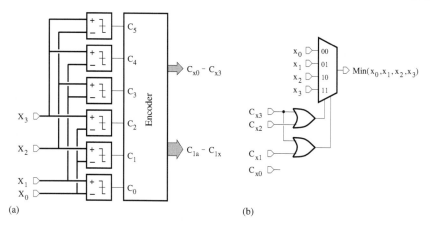

FIGURE 5.89

(a) Generation of the signals C_{x0}-C_{x3} and C_{1a}-C_{1x}; (b) implementation of the Min block for an input vector with four elements.

TABLE 5.3

Boolean Expressions for the Min Block Design

$$\text{Min}(x_0, x_1, x_2, x_3) = x_0$$
$$C_{x0} = C_0 \cdot C_1 \cdot C_2 \cdot \overline{C}_3 \cdot \overline{C}_4 + C_0 \cdot C_1 \cdot C_2 \cdot C_3 \cdot \overline{C}_5 + C_0 \cdot C_1 \cdot C_2 \cdot C_4 \cdot C_5$$

$$\text{Min}(x_0, x_1, x_2, x_3) = x_1$$
$$C_{x1} = \overline{C}_0 \cdot \overline{C}_1 \cdot \overline{C}_2 \cdot C_3 \cdot C_4 + \overline{C}_0 \cdot C_1 \cdot C_3 \cdot C_4 \cdot \overline{C}_5 + \overline{C}_0 \cdot C_2 \cdot C_3 \cdot C_4 \cdot C_5$$

$$\text{Min}(x_0, x_1, x_2, x_3) = x_2$$
$$C_{x2} = \overline{C}_0 \cdot \overline{C}_1 \cdot \overline{C}_2 \cdot \overline{C}_3 \cdot C_5 + C_0 \cdot \overline{C}_1 \cdot \overline{C}_3 \cdot \overline{C}_4 \cdot C_5 + \overline{C}_1 \cdot C_2 \cdot \overline{C}_3 \cdot C_4 \cdot C_5$$

$$\text{Min}(x_0, x_1, x_2, x_3) = x_3$$
$$C_{x3} = \overline{C}_0 \cdot \overline{C}_1 \cdot \overline{C}_2 \cdot \overline{C}_4 \cdot \overline{C}_5 + C_0 \cdot \overline{C}_2 \cdot \overline{C}_3 \cdot \overline{C}_4 \cdot \overline{C}_5 + C_1 \cdot \overline{C}_2 \cdot C_3 \cdot \overline{C}_4 \cdot \overline{C}_5$$

as shown in Figure 5.89(a), to perform the following operations:

$$C_0 : x_1 > x_0 \quad C_1 : x_2 > x_0 \quad C_2 : x_3 > x_0 \quad C_3 : x_2 > x_1$$
$$C_4 : x_3 > x_1 \quad C_5 : x_3 > x_2 \quad \overline{C}_0 : x_0 > x_1 \quad \overline{C}_1 : x_0 > x_2$$
$$\overline{C}_2 : x_0 > x_3 \quad \overline{C}_3 : x_1 > x_2 \quad \overline{C}_4 : x_1 > x_3 \quad \overline{C}_5 : x_2 > x_3$$

The output signal of a comparator is set to either the logic high state or the logic low state, depending on the comparison result. A dichotomy technique can be used to sort a given set of elements. This is achieved by iteratively defining mutually exclusive subsets such that a tree hierarchy can emerge. Here, the sorting process provides six ordered combinations with a

TABLE 5.4

Boolean Functions Useful for the Sorting Procedure Implementation

$x_3 > x_2 > x_1 > x_0$	$x_3 > x_2 > x_0 > x_1$
$C_{1a} = C_0 \cdot C_1 \cdot C_2 \cdot C_3 \cdot C_4 \cdot C_5$	$C_{1b} = \overline{C}_0 \cdot C_1 \cdot C_2 \cdot C_3 \cdot C_4 \cdot C_5$
$x_3 > x_1 > x_2 > x_0$	$x_3 > x_1 > x_0 > x_2$
$C_{1c} = C_0 \cdot C_1 \cdot C_2 \overline{C}_3 \cdot C_4 \cdot C_5$	$C_{1d} = C_0 \cdot \overline{C}_1 \cdot C_2 \overline{C}_3 \cdot C_4 \cdot C_5$
$x_3 > x_0 > x_2 > x_1$	$x_3 > x_0 > x_1 > x_2$
$C_{1e} = \overline{C}_0 \cdot \overline{C}_1 \cdot C_2 \cdot C_3 \cdot C_4 \cdot C_5$	$C_{1f} = \overline{C}_0 \cdot \overline{C}_1 \cdot C_2 \overline{C}_3 \cdot C_4 \cdot C_5$
$x_2 > x_3 > x_1 > x_0$	$x_2 > x_3 > x_0 > x_1$
$C_{1g} = C_0 \cdot C_1 \cdot C_2 \cdot C_3 \cdot C_4 \cdot \overline{C}_5$	$C_{1h} = \overline{C}_0 \cdot C_1 \cdot C_2 \cdot C_3 \cdot C_4 \cdot \overline{C}_5$
$x_2 > x_1 > x_3 > x_0$	$x_2 > x_1 > x_0 > x_3$
$C_{1i} = C_0 \cdot C_1 \cdot C_2 \cdot C_3 \cdot \overline{C}_4 \cdot \overline{C}_5$	$C_{1j} = C_0 \cdot C_1 \cdot \overline{C}_2 \cdot C_3 \cdot \overline{C}_4 \cdot \overline{C}_5$
$x_2 > x_0 > x_3 > x_1$	$x_2 > x_0 > x_1 > x_3$
$C_{1k} = \overline{C}_0 \cdot C_1 \cdot \overline{C}_2 \cdot C_3 \cdot C_4 \cdot \overline{C}_5$	$C_{1l} = \overline{C}_0 \cdot C_1 \cdot \overline{C}_2 \cdot C_3 \cdot \overline{C}_4 \cdot \overline{C}_5$
$x_1 > x_3 > x_2 > x_0$	$x_1 > x_3 > x_0 > x_2$
$C_{1m} = C_0 \cdot C_1 \cdot C_2 \cdot \overline{C}_3 \cdot \overline{C}_4 \cdot C_5$	$C_{1n} = C_0 \cdot \overline{C}_1 \cdot C_2 \cdot \overline{C}_3 \cdot \overline{C}_4 \cdot C_5$
$x_1 > x_2 > x_3 > x_0$	$x_1 > x_2 > x_0 > x_3$
$C_{1o} = C_0 \cdot C_1 \cdot C_2 \cdot \overline{C}_3 \cdot \overline{C}_4 \cdot \overline{C}_5$	$C_{1p} = C_0 \cdot C_1 \cdot \overline{C}_2 \cdot \overline{C}_3 \cdot \overline{C}_4 \cdot \overline{C}_5$
$x_1 > x_0 > x_3 > x_2$	$x_1 > x_0 > x_2 > x_3$
$C_{1q} = C_0 \cdot \overline{C}_1 \cdot \overline{C}_2 \cdot \overline{C}_3 \cdot \overline{C}_4 \cdot C_5$	$C_{1r} = C_0 \cdot \overline{C}_1 \cdot \overline{C}_2 \cdot \overline{C}_3 \cdot \overline{C}_4 \cdot \overline{C}_5$
$x_0 > x_3 > x_2 > x_1$	$x_0 > x_3 > x_1 > x_2$
$C_{1s} = \overline{C}_0 \cdot \overline{C}_1 \cdot \overline{C}_2 \cdot C_3 \cdot C_4 \cdot C_5$	$C_{1t} = \overline{C}_0 \cdot \overline{C}_1 \cdot \overline{C}_2 \cdot \overline{C}_3 \cdot C_4 \cdot C_5$
$x_0 > x_2 > x_3 > x_1$	$x_0 > x_2 > x_1 > x_3$
$C_{1u} = \overline{C}_0 \cdot \overline{C}_1 \cdot \overline{C}_2 \cdot C_3 \cdot C_4 \cdot \overline{C}_5$	$C_{1v} = \overline{C}_0 \cdot \overline{C}_1 \cdot \overline{C}_2 \cdot C_3 \cdot \overline{C}_4 \cdot \overline{C}_5$
$x_0 > x_1 > x_3 > x_2$	$x_0 > x_1 > x_2 > x_3$
$C_{1w} = \overline{C}_0 \cdot \overline{C}_1 \cdot \overline{C}_2 \cdot \overline{C}_3 \cdot \overline{C}_4 \cdot C_5$	$C_{1x} = \overline{C}_0 \cdot \overline{C}_1 \cdot \overline{C}_2 \cdot \overline{C}_3 \cdot \overline{C}_4 \cdot \overline{C}_5$

given element being the minimum. The total number of possible combinations is then 6×4, or 24.

The circuit diagram of the Min block is depicted in Figure 5.89(b). There is no connection between the C_{x0} input and any of the encoder gates because all inputs are in the logic low state when the 00 code is selected. It is assumed that only one of the Boolean functions, C_{x0}, C_{x1}, C_{x2}, or C_{x3}, can be set to the logic high state at a time. Following the design process of combinational logic circuits, each of the Boolean expressions, C_{x0}, C_{x1}, C_{x2}, and C_{x3}, whose logic state can be related to the

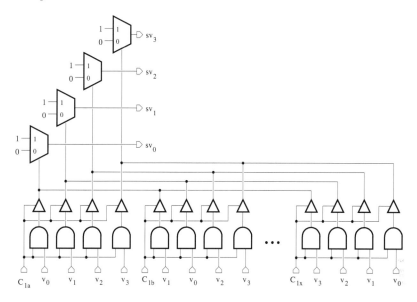

FIGURE 5.90
Implementation of the vector quantizer for a DAC with four unit elements.

fact that the corresponding element is minimum or not, can be derived as shown in Table 5.3.

To design the vector quantizer, the occurrence condition for each of the possible combinations of the vector elements to be sorted is translated into a Boolean expression. The block diagram of the vector quantizer is shown in Figure 5.90. It consists of four 2-to-1 multiplexers and 24 decoders followed by tri-state buffers. The use of tri-state buffers allows all decoders to share the same output line. Table 5.4 summarizes the Boolean functions used as control signals in the vector quantizer.

In practice, the implementation of DACs based on the vector-feedback ESL can be affected by the stability problems of high-order modulator loops and is limited to resolutions less than 4 bits due to the high number of logic gates (e.g., $(B-1)!$ digital comparators and $(B-1)! \times B$ decoders for a resolution of B bits) required by the vector quantizer.

5.3.2   Data-weighted averaging technique

Data-weighted averaging (DWA) algorithms equally select the DAC elements taking part in the data conversion, such that matching errors average to zero over a given time period. The re-selection of a given element is possible only after the choice of all the others. As a consequence, the processing of consecutive input codes should require different DAC units.

Assuming that the deviation from the nominal value of the i-th unit elements of the DAC is denoted by ϵ_i, we have

$$\sum_{i=1}^{I} \epsilon_i = 0 \tag{5.293}$$

where I is the number of elements. A given input code, which can be written as

$$X = qI + r, \quad 0 \le r < I \tag{5.294}$$

where q and r are two integers, is converted with the mismatch error

$$\Delta X = q \sum_{i=1}^{I} \epsilon_i + \sum_{i=1}^{J} b_i \epsilon_i \tag{5.295}$$

$$= \sum_{i=1}^{J} b_i \epsilon_i \tag{5.296}$$

Here, $\sum_{i=1}^{J} b_i = r$ and b_i is either 0 or 1. To reduce the effect of the residual error term on the converter performance, it is advisable to choose b_i randomly.

The result of the conversion is obtained by adding the DAC codes at successive time instants, k $(k = 0, 1, \cdots, K)$. That is, ΔX is written in the z-domain as the product of the error associated with the initial DAC code and the function $1 + z^{-1} + \cdots + z^{-K}$, which can be approximated by $1 - z^{-1}$ for large K. The error caused by mismatch is then first-order shaped.

The block diagram of the N-bit $\Delta\Sigma$ modulator depicted in Figure 5.91 includes a DWA circuit [34], which can reject the tones caused by mismatches of the DAC unit elements out of the baseband. The DWA implementation of Figure 5.92 consists of an adder, a shift register, binary-to-thermometer encoders, AND and exclusive-OR gates, inverters, and multiplexers. The number of unit elements actually required is $N = K + L$, where K and L are the number of elements assuming an ideal DAC and the number of additional elements due to the DWA, respectively. Note that the value of L determines the noise-shaping performance of the DWA technique. For a 4-bit modulator, N can be chosen to be $15 + 1$. The binary-to-thermometer decoder should have 16 outputs and the last one can be maintained at the low state.

5.3.2.1 Element selection logic based on a tree structure and butterfly shuffler

Assuming a 3-bit (or eight-element) DAC, Figures 5.93 and 5.94 show the block diagrams of the element selection logic (ESL) or encoder based on a tree structure and butterfly shuffler, respectively. In both schemes, the number of unit elements is given by 2^N, and the switching section comprises $\log_2(N)$ layers, where N is the number of bits.

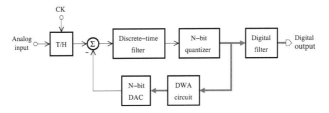

FIGURE 5.91
Block diagram of an N-bit delta-sigma modulator including a DWA circuit.

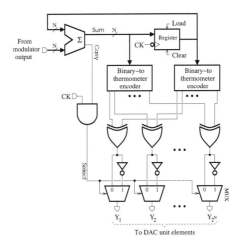

FIGURE 5.92
A digital implementation of the data-weighted averaging technique.

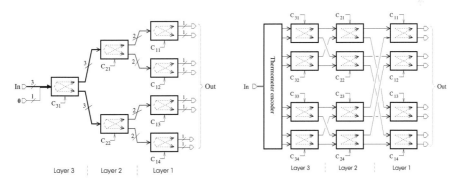

FIGURE 5.93
Tree-structured ESL.

FIGURE 5.94
Butterfly shuffler ESL.

- For the approach based on a tree structure, a combination of the N-bit binary code to be converted and a zero, which represents the first bit (LSB)

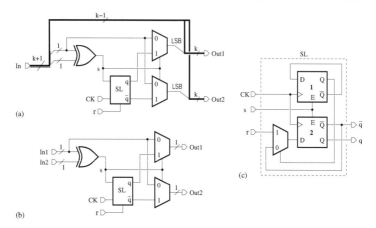

FIGURE 5.95
Circuit diagram of a switching block with (a) $(k+1)$-bit and (b) 2-bit input words; (c) implementation of the selection logic (SL).

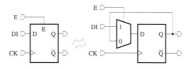

FIGURE 5.96
Circuit diagram of D flip-flops with enable.

of the input data, is applied to the encoder. This latter consists of $2^N - 1$ switching blocks, the operation of which is equivalent to signal processing functions of the form

$$y_1^{ij} = \frac{x^{ij} + C_{ij}}{2} \tag{5.297}$$

$$y_2^{ij} = \frac{x^{ij} - C_{ij}}{2} \tag{5.298}$$

where x^{ij} and y_k^{ij} ($k = 1, 2$) are the input and output signals, respectively, and C_{ij} denotes the difference between the top and bottom outputs of the switching block j on the layer i. To simplify the hardware implementation, other definitions, which satisfy the number conservation rule, may be adopted for switching function. It can be assumed, for instance, that

$$C_{ij} = \begin{cases} 0 & \text{if} \quad x^{ij} \quad \text{is even} \\ \pm 1 & \text{if} \quad x^{ij} \quad \text{is odd.} \end{cases} \tag{5.299}$$

The operation of the encoder is equivalent to an N-to-2^N transformation followed by a scrambling. By processing an input with $(k + 1)$-bits, the switching block provides two k-bit outputs.

- The butterfly shuffler ESL can only operate with an even number of DAC unit elements. It can be composed of a thermometer encoder, which achieves the N-to-2^N conversion, and a selection stage consisting of $N2^{N-1}$ cells called swappers or switching blocks, which can be described as follows:

$$y_1^{ij} = \frac{x_1^{ij} + x_2^{ij}}{2} + C_{ij} \frac{x_1^{ij} - x_2^{ij}}{2} \tag{5.300}$$

$$y_2^{ij} = \frac{x_1^{ij} + x_2^{ij}}{2} - C_{ij} \frac{x_1^{ij} - x_2^{ij}}{2} \tag{5.301}$$

where x_k^{ij} and y_k^{ij} $(k = 1, 2)$ are the input and output signals, respectively; i denotes the layer number; and j represents the position of the swapper in the layer. Depending on the level of the signal C_{ij}, the switching block may either pass the inputs directly to the corresponding outputs or assign the inputs in reverse of the outputs on each clock cycle. The butterfly shuffler ESL (see Figure 5.94) can perform efficiently even if it allows only a selected set of connections. It should be noted that the association of each of the N inputs to all possible N outputs, would require a digital encoder with N factorial paths, or a large die area.

The resulting noise of the DAC is a linear combination of the data, C_{ij}, with weighting factors, which are linearly related to the unit element errors. As long as each selection logic shapes C_{ij} by a specific transfer function, the static errors introduced by the element mismatch in the DAC will also be modeled in the same way.

For a $(k + 1)$-bit input word, the switching block can be implemented, as shown in Figure 5.95(a). The $(k + 1)$-bit input word is split into the upper $k - 1$ bits and the lower 2 bits, which are used to appropriately assign the least significant bit (LSB) of the output data. When the length of the input word becomes equal to 2 bits, the switching block can be realized using the structure of Figure 5.95(b). For a first-order mismatch shaping, the selection logic (SL) can be designed, as illustrated in Figure 5.95(c) [35–37]. The D flip-flops are enabled by the parity signal, s, provided by the XOR gate connected to the 2 LSBs of the input code, and the clock frequency is fixed at the data rate. The dither signal, r, which is delivered by a pseudo-random noise generator (or linear feedback shift register) whose output assumes the values 0 or 1 with the same likelihood and is used to eliminate spurious tones in the DAC output spectrum, should be uncorrelated with the input code. The selection sequence is determined by the output signals (Q and $\overline{Q}$) of the second D flip-flop. The circuit diagram of D flip-flops with enable is shown in Figure 5.96. A logical 1 at the enable node E allows the transfer of the data D to the flip-flop, and the truth table of the overall structure is then similar to the one of a conventional D-flip-flop. The previous state of the flip-flop is maintained in the cases where $E = 0$.

It should be noted that a tree-structured DEM DAC seems to offer a more efficient hardware implementation in the case of high-order mismatch shaping functions than a butterfly shuffler DEM DAC.

5.3.2.2　Generalized DWA structure

In comparison with other DEM approaches, the DWA technique [38, 39] is mainly used because it offers a suitable trade-off between hardware complexity and mismatch noise attenuation performance. The reduction of the hardware complexity, and therefore the delay that can be introduced in the delta-sigma modulator feedback path by the DWA circuit, proves essential in meeting high-frequency specifications.

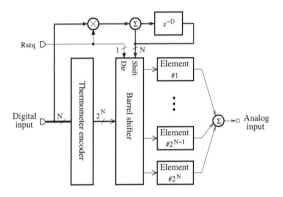

FIGURE 5.97
Generalized structure for DWA implementations.

A generalized structure that can be used for the implementation of various DWA algorithms is shown in Figure 5.97, where N denotes the DAC bit resolution, and the sequence $Rseq$ can follow any arbitrary repeating ± 1 pattern of length D. It must use modulo-N arithmetic in order to satisfy the requirement to limit the selection data range from 0 to $N - 1$.

In particular for a 3-bit DAC, the barrel shifter can be realized as depicted in Figure 5.98. It operates as a reversible rotator and should involve CMOS switches with dual supply voltages (V_{DD} and V_{SS}).

By configuring the generalized DWA structure to realize a mismatch shaping transfer function, $H(z)$, of the form

$$H(z) = 1 + z^{-D} \tag{5.302}$$

where D is a positive integer, the term of the required $Rseq$ sequence can, for instance, be defined by the formula, $Rseq(k) = (-1)^{\lceil k/D \rceil}$, where $\lceil x \rceil$ is the smallest integer not less than x. For the implementation of

$$H(z) = 1 - z^{-D} \tag{5.303}$$

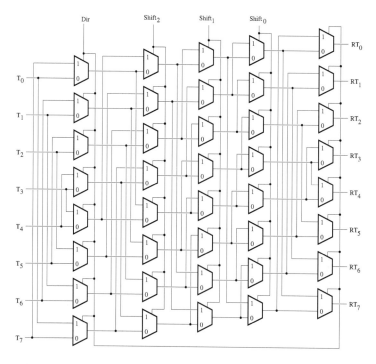

FIGURE 5.98
Circuit diagram of an 8-input barrel shifter.

the term of the *Rseq* sequence can be chosen as $Rseq(k) = 1$, for $k > 0$.

Known implementations of DWA algorithms use various values of D. For lowpass delta-sigma modulators, a conventional DWA circuit based on the transfer function given by, $H(z) = 1 - z^{-1}$, provides a high-pass shaping of the DAC mismatch noise. However, its performance may be limited by spurious tones. A method to reduce the effect of these spurious tones consists of using DWA algorithms with bi-directional rotation of unit elements.

On the other hand, DWA algorithms can also be used to improve the linearity of bandpass delta-sigma modulators with a multi-bit DAC. In a bandpass modulator that has center frequency at $f_s/4$, the shaping transfer function can be chosen as, $H(z) = 1 - z^{-4}$, and has notch frequencies at dc, $\pm f_s/4$, and $\pm f_s/2$, where f_s is the sampling frequency. The extra notch frequencies (at dc and $f_s/2$) due to the term $1 - z^{-2}$ in the shaping transfer function help de-correlate the mismatch error from the input code, thereby leading to an attenuation of spurious tones, but at a price of an increased mismatch noise floor.

By selecting the shaping transfer function as $H(z) = 1 + z^{-2}$, the notch frequencies are only located at $\pm f_s/4$, and the resulting DWA circuit for the bandpass delta-sigma modulator can suppress the mismatch noise in the band of interest. A DAC with the $1+z^{-2}$ shaping transfer function does not substantially contribute to spurious tones in the output frequency spectrum, especially in the implementation of quadrature bandpass modulators.

5.4 Data converter testing and characterization

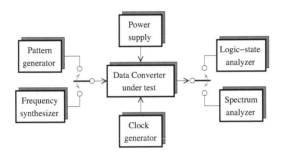

FIGURE 5.99
Block diagram of a test setup.

The performance of data converters can be characterized by static and dynamic parameters. Static linearity, which can be obtained by comparing the ideal and real transfer characteristics of the converter, is generally specified through DNL and INL errors. The analysis of the converter output samples can also provide dynamic measures such as signal-to-noise ratio (SNR), signal-to-noise and distortion (SINAD), effective number of bits (ENOB), spurious-free dynamic range (SFDR), and harmonic distortions. The data converter testing [40–42] is achieved by means of microprocessor-based instrumentation due to the complexity of the required signal processing algorithms. A typical test setup for a data converter is shown in Figure 5.99. The following analysis methods can be used for the converter characterization.

5.4.1 Histogram-based testing

As shown in Figure 5.100, the histogram of an ideal ADC processing a dc signal consists of equal-sized bins for all output codes. When the converter transfer characteristic exhibits a nonlinearity, the bins will not have the same size due to the fact that some output codes occur more frequently than others. Generally, a periodic input sequence is applied to the ADC. The output data are collected as a series of records, each of which contains a given number of samples. They are represented in the form of a normalized histogram or code density showing the occurrence frequency of each converter code. The converter characteristics can then be determined by comparing the code density data to the ideal distribution density function. In the absence of offset errors, for instance, the histogram should be symmetrical. Gain deviations affect the histogram width. A zero occurrence is the result of a missing code. Note that a large spike generally corresponds to a high DNL.

Let Λ_i be the number of occurrences of code i, Λ_T be the total number of samples, and $P(i)$ be the occurrence probability of code i or the bin width of the ideal converter. The next definitions can be adopted for the DNL and INL expressed in LSBs:

$$DNL(i) = \frac{\Lambda_i}{\Lambda_T P(i)} - 1 \tag{5.304}$$

$$DNL = \max|DNL(i)| \tag{5.305}$$

$$INL(i) = \sum_{j=1}^{i} DNL(j) \tag{5.306}$$

$$INL = \max|INL(i)| \tag{5.307}$$

where $i = 0, 1, \cdots, 2^{B+1} - 2$. Note that the definition of the DNL cannot be applied to the last code and Λ_i/Λ_T is the histogram bin width of the converter under test (Λ_i is the number of occurrences of each code i and Λ_T denotes the total number of acquired codes). If the signal applied to the converter is assumed to be a sinusoid of the form

$$x = A\sin(2\pi ft) \tag{5.308}$$

the probability density function is given by[4]

$$p_X(x) = \frac{1}{\pi\sqrt{A^2 - x^2}} \tag{5.309}$$

where A denotes the amplitude. The parameter P for a given input sample is defined as

$$P(i) = \int_{V_i}^{V_{i+1}} p_X(x)\mathrm{d}x \tag{5.310}$$

---

[4]Let $X = \Phi(\Theta) = A\sin(\Theta)$. With $\theta = \Phi^{-1}(x)$, the probability density function (pdf) is derived from the following expression,

$$p_X(x) = p_\Theta(\theta)\left|\frac{\mathrm{d}\theta}{\mathrm{d}x}\right|$$

where the phase, Θ, of the sine wave is assumed to be a random variable uniformly distributed between $-\pi/2$ and $+\pi/2$ with the pdf given by

$$p_\Theta(\theta) = \begin{cases} \frac{1}{\pi} & \text{if } \theta \in (-\frac{\pi}{2}, +\frac{\pi}{2}) \\ 0 & \text{otherwise.} \end{cases}$$

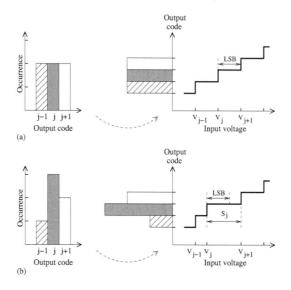

FIGURE 5.100
Correspondence between the histogram and ADC transfer characteristic: (a)
Ideal case, (b) nonideal case.

where V_i and V_{i+1} are the lower and higher transition levels, respectively. That
is,

$$P(i) = \frac{1}{\pi}\left[\arcsin\left(\frac{V_{i+1}}{A}\right) - \arcsin\left(\frac{V_i}{A}\right)\right] \tag{5.311}$$

Taking the cosine of both sides of the above relation and using trigonometric
relations,[5] we can obtain

$$V_{i+1}^2 - [2V_i\cos(\pi P(i))]\,V_{i+1} + V_i^2 - A^2\left[1 - \cos^2(\pi P(i))\right] = 0 \tag{5.312}$$

This quadratic equation can be solved for V_{i+1}. As a result,

$$V_{i+1} = V_i\cos(\pi P(i)) + \sin(\pi P(i))\sqrt{A^2 - V_i^2} \tag{5.313}$$

where only the positive square root was retained so that V_{i+1} can be greater
than V_i. With the assumption that the first decision level is fixed at $-A$, the
other decision levels can be computed as

$$V_{i+1} = -A\cos\left(\frac{\pi\Sigma\Lambda_i}{\Lambda_T}\right) \tag{5.314}$$

---

[5]Given two numbers x and y,

$$\cos(x - y) = \cos(x)\cos(y) + \sin(x)\sin(y)$$

and

$$\cos(\arcsin(x)) = \sqrt{1 - x^2}$$

where P is replaced by the measured frequency of occurrence, $\Sigma\Lambda_i/\Lambda_T$, and $\Sigma\Lambda_i$ denotes the total number of codes included in the bins 1 through i. A missing code corresponds to a DNL equal to -1. The record length and ratio of the sampling rate to the signal frequency are chosen so that dynamic errors (in-phase distortion, information redundancy, etc.) are negligible.

5.4.2 Spectral analysis method

Fast Fourier transform data are used to characterize linearity and noise properties of the ADC in the frequency domain. The output provided by an ADC, which processes a sine-wave signal, comprises a tone at the input frequency, harmonics, spurious components, *dc* offset, and a broadband term characterizing the different kinds of noise. The power estimation of each narrowband component can be affected by the energy leaking from neighboring tones. A solution can then consist of using a suitable window function prior to the Fourier transform. The next parameters can be deduced from the spectrum data.

- The SNR is a measure of the broadband noise introduced by the converting and sampling process into the signal band. It is the ratio of the root-mean square (rms) value or power of the output signal to the one of the sum of all other frequency components below the Nyquist rate, except those representing dc and harmonics of the fundamental.

- The dynamic range (DR) is the ratio of the rms value of a full-scale sinusoidal input signal to the rms noise delivered by the converter with inputs shorted together. It is limited by the Nyquist frequency.

- The total harmonic distortion (THD) is the ratio of the fundamental to the sum of the harmonics, which can be identified from the noise floor. It can also be expressed as a percentage.

- The SINAD[6] is the ratio of the power in the fundamental frequency bin to that in all other bins, including harmonics. It can also be computed as $(\text{SNR}^2 + \text{THD}^2)^{1/2}$.

- The SFDR is the difference in rms magnitudes of the fundamental and the highest spur, which is not due to dc offset.

In another approach to performing a spectral analysis of the converter, the input signal is assumed to be the sum of two sine waves with the same amplitude, and frequencies equal to f_1 and f_2, respectively. The intermodulation distortion (IMD) provides the ratio of the rms sum of intermodulation components at frequencies $if_1 \pm jf_2$ in the spectrum to the rms value of the input signal, where i and j are integers different from zero. The intermodulation

---

[6]The SINAD is also known as signal-to-noise and distortion ratio (SNDR).

order is given by $i + j$. In a practical implementation, the spectral leakage is eliminated either by assuming a coherent relationship between the sampling frequency, f_s, and input frequencies, that is, $m/f_s = m_1/f_1 = m_2/f_2$, where the integers m_1 and m, and m_2 and m are, respectively, prime of each other, or by applying a filtering window such as the Blackmann Harris function to the data.

The noise power test, as shown in Figure 5.101, consists of analyzing the output samples delivered by a converter, which processes a limited band of white noise provided by a generator. The Fourier transform is used to evaluate the noise power ratio (NPR), which is the measure of all contributed errors in the frequency domain. However, the fundamental frequency and dc offset are discarded in the computation. The ENOB can be written as

$$\text{ENOB} = N - \frac{1}{2} \log_2 \frac{\text{NPR}}{\sigma_Q^2} \qquad (5.315)$$

where N is the number of bits of the converter and σ_Q^2 denotes the theoretical quantization noise.

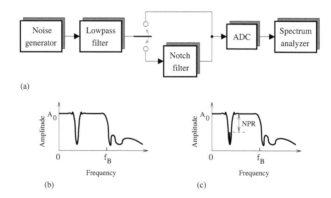

(a)

(b)

(c)

FIGURE 5.101
(a) NPR test setup; (b) signal spectrum at the output of the notch filter; (c) output signal spectrum of the ADC.

Dynamic specifications are generally expressed in decibels (dB). However, they can be referenced to the converter FSR, which is constant, before being transformed into decibels. This results in parameters, whose unit is dBFS, or decibels relative to full-scale.

5.4.3   Walsh transform-based transfer function estimation

The transfer function of an N-bit converter can be represented as the sum of a given number of Walsh functions adequately weighted by the Walsh coefficients. These latter can be obtained by reconstructing the ADC output data using the Walsh transform. To achieve a good resolution, the number of points considered for the computation should be a power of 2 multiple of 2^N. The comparison of the ideal and real transfer functions can then provide the ADC error parameters.

5.4.4 Testing using sine-fit algorithms

The ENOB characterization of an ADC, which processes a sampled sine wave, is carried out by reconstructing the input signal based on the four parameters (amplitude, frequency, phase, and dc offset) computed from the output data. The signal samples of the original input are then subtracted from the ones of the synthesized sine wave to estimate the average noise power. The ENOB at a given input frequency can then be computed as

$$\text{ENOB} = \log_2 \left(\frac{FSR}{\sqrt{12} \cdot RMSE} \right) \tag{5.316}$$

where FSR is the full-scale range of the ADC and $RMSE$ is the root-mean square of the digitized signal or the noise power provided by the test procedure. The achievable accuracy is limited by the convergence performance of the estimation algorithm, and the validity of the stochastic model is guaranteed only for a restricted range of the ratio between the number of ADC quantization levels and the one of the acquired samples.

Note that a pattern generator instead of a frequency synthesizer is required for the DAC testing. The ADC, which can be used to deliver a digital version of the analog output necessary for the different computations, can limit the speed and precision of the evaluation. To test DAC in the frequency-domain, the solution can consist of using analog spectrum analysis techniques. The level of harmonic distortions can then be related to the transfer characteristic deviations.

Generally, the power consumption of data converters increases with performance characteristics such as the dynamic range and bandwidth. The figure-of-merit (FOM) measures the efficiency with respect to the dissipated power. It is defined as

$$\text{FOM} = \frac{\text{DR} \times \text{BW}}{P} \tag{5.317}$$

where DR and P are the dynamic range and the total power dissipation of the converter, respectively, and BW is the signal bandwidth.

5.5 Delta-sigma modulator-based oscillator

Generally, the on-chip generation of high-quality signals is required in built-in self-test structures for mixed-signal circuits. An approach to resolve this problem is to use $\Delta\Sigma$ modulator-based oscillators, which can deliver signals with a spurious-free dynamic range of about 90 dB.

A $\Delta\Sigma$ modulator-based oscillator consists of a loop including a digital resonator with poles on the unit circle and an $N \times N$-bit multiplier, which is implemented by the combination of a $\Delta\Sigma$ modulator with a multiplexer to reduce the silicon area and timing delay. The 1-bit pattern used to control the multiplexer switching is available at the output of the $\Delta\Sigma$ modulator, which should have a unity signal transfer function. It contains the sinusoidal signal generated by the resonator and the out-of-band quantization noise, which can be suppressed by a filter. An analog signal can be obtained by cascading a 1-bit DAC with the oscillator.

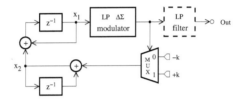

FIGURE 5.102
Lowpass $\Delta\Sigma$ modulator-based oscillator.

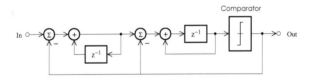

FIGURE 5.103
Block diagram of a second-order lowpass $\Delta\Sigma$ modulator.

The block diagram of a lowpass $\Delta\Sigma$ oscillator [43] is shown in Figure 5.102. It includes two integrators, a lowpass $\Delta\Sigma$ modulator, and a 2-to-1 multiplexer. The delay of one clock period introduced on the signal path by the second-order lowpass $\Delta\Sigma$ modulator shown in Figure 5.103 is compensated for by using one nondelayed integrator. Let x_1 and x_2 be the state variables associated with the output of the first and second integrators, respectively. We can write

$$x_1(n) = x_1(n-1) + x_2(n-1) \tag{5.318}$$
$$x_2(n) = -kx_1(n) + x_2(n-1) \tag{5.319}$$

Using the z-transform, X_1 and X_2 can be eliminated and the next characteristic equation is derived,

$$z^{-2} - (2-k)z^{-1} + 1 = 0 \tag{5.320}$$

To ensure the oscillation, the roots of the above equation should be conjugate complex and located on the unit circle. This is the case for $0 < k < 4$, and

$$z_{1,2}^{-1} = \frac{2 - k \pm j\sqrt{k(4-k)}}{2} \qquad (5.321)$$

The angular frequency of oscillation, ω_0, can then be related to the coefficient, k, and the period of the clock signal, T, according to

$$\tan(\omega_0 T) = \pm\frac{\sqrt{k(4-k)}}{2-k} \qquad (5.322)$$

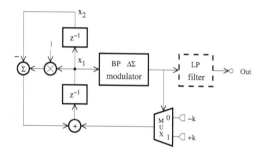

FIGURE 5.104
Bandpass $\Delta\Sigma$ modulator-based oscillator.

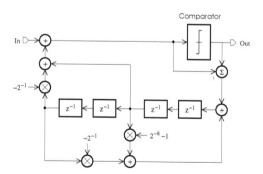

FIGURE 5.105
Block diagram of a fourth-order bandpass $\Delta\Sigma$ modulator.

A bandpass $\Delta\Sigma$ oscillator [44], as shown in Figure 5.104, offers the advantage of possessing a greater usable bandwidth while operating at a sample rate comparable to that of a lowpass structure. It uses two registers (blocks denoted by z^{-1}) included in a loop with a multiplier (coefficient l), a bandpass $\Delta\Sigma$ modulator, and a 2-to-1 multiplexer. Figure 5.105 shows the block diagram of a fourth-order bandpass $\Delta\Sigma$ modulator with a signal transfer function equal to 1, such that the signal level is not modified. By inspection of the oscillator, we can derive

$$x_1(n) = -x_2(n-1) + lx_1(n-1) - kx_1(n-1) \tag{5.323}$$
$$x_2(n) = x_1(n-1) \tag{5.324}$$

where x_1 and x_2 denote the state variables of the register outputs. These last equations can be transformed to the z-domain as

$$z^{-2} + (k-l)z^{-1} + 1 = 0 \tag{5.325}$$

Solving for z^{-1}, the roots of the characteristic equation (5.325) are given by

$$z_{1,2}^{-1} = \frac{l - k \pm j\sqrt{4 - (k-l)^2}}{2} \tag{5.326}$$

where $|k - l| < 2$. Hence, the oscillation frequency can be obtained from the next expression,

$$\tan(\omega_0 T) = \pm\frac{\sqrt{4 - (k-l)^2}}{l - k} \tag{5.327}$$

The multiplication coefficient l can be chosen to be a power of 2 to reduce the hardware complexity. Further reduction is achieved for $l = 0$.

For both oscillator structures, a discrete-time sinusoidal signal of the form

$$x(n) = A\sin(\omega_0 T n + \phi) \tag{5.328}$$

can be obtained at the node labeled x_1. The amplitude A and the phase ϕ are dependent on the coefficient k (and l), and the initial conditions, I_1 and I_2, of the first and second registers.

Principles of time division multiplexing [45] can be exploited for the generation of two-tone signals. This is realized by replacing the 2-to-1 multiplexer with a 4-to-1 multiplexer and each register with a pair of registers. As a result, the effective clock frequency is divided by a factor of two.

5.6 Digital signal processor interfacing with data converters

Due to the difference of processing speed and electrical characteristics existing between input/output (I/O) devices, such as data converters, and the computer processing unit (CPU) of a microprocessor (digital signal processor (DSP), micro-controller), interface chips are required to synchronize data transfer between the CPU and I/O devices. Generally, an interface chip is composed of control registers, data registers, status registers, data direction registers, and control circuit [46]. Control registers include data bits, whose states determine the parameters of the I/O operation. The data transfer direction for each I/O pin is set by the corresponding bit of the data direction registers. The data register is used as a buffer to temporarily store the data being transferred to or from the CPU. The status registers store bits providing information on the progression of the I/O operation. Because access to the data bus is allowed to only one I/O device at a time, an address decoder is used to generate chip-select or chip-enable signals for each device at the request of the microprocessor.

The data transfer between the microprocessor and I/O devices can be either parallel or serial. Parallel communications are based on the use of several wires to simultaneously transmit data. Serial communications involve transmitting digital data, sequentially, over only one wire. To achieve a high transfer speed, parallel data transmissions are preferred, while serial links are the better option when the interconnection hardware overhead should be kept minimal.

Due to the typical speed difference between the microprocessor and I/O devices, a synchronization mechanism is required for proper data transmission. Various types of synchronization can be used to interface I/O devices.

A simple synchronization technique is to design the software such that it can initiate the communication and then wait a fixed amount of time for the I/O operation to complete. This method is known as blind cycle counting because it provides no information about the outcome of the I/O operation back to the microprocessor.

In the gadfly or busy waiting approach, the software routine includes loops that can check the I/O status until the completion of data transfer. This approach is suitable only for I/O operations with a small wait time.

The periodic polling is based on the principle of continually checking the status of the I/O operation to detect whether it is complete. By continuously monitoring the status register, the microprocessor can notice the end of the data transfer. It can then retrieve data and proceed further according to the programmed instructions.

The interrupt technique requires more complex hardware and software, but has the advantage of efficiently using the microprocessor CPU. An interrupt request is generated either when the I/O device is ready or to acknowledge a successful data transfer. As a result, the CPU forces a branch-out of the current program sequence to the appropriate interrupt service routine (ISR). Prior to the transfer of the control to the ISR, the CPU state must be saved on the stack. This is necessary because the program execution should resume after returning from the ISR. However, the achievable response time may be limited due to the microprocessor latency time (the time elapsed between the generation of an interrupt request and the servicing of the corresponding I/O device).

Another I/O synchronization technique is based on direct memory access (DMA). DMA controllers can transfer data from I/O devices directly to the main memory, and vice versa, without the intervention of the CPU. They can generate an address sequence to access blocks of data and manage access priorities. Here, the load of the CPU is reduced and a higher data throughput can be achieved by manipulating data blocks.

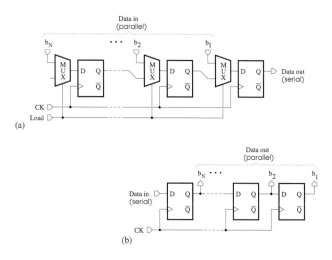

(a)

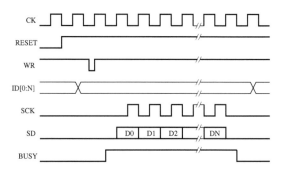

(b)

FIGURE 5.106

(a) Parallel-to-serial converter; (b) serial-to-parallel converter.

FIGURE 5.107

Timing waveforms for the parallel-to-serial converter.

Interfacing a DSP with data converters involves both physical connections and software routines that steer the transmission of data. Data converters used in the interface implementation should exhibit more flexibility. This is achieved using a set of on-chip registers to achieve programmability and control the data flow. The write (WR) and read (RD) operation of each register is determined by a given signal. The communication between a digital signal processor (DSP) and data converters can be done either in parallel or serially [47]. A DSP with only one type (parallel or serial) of port can still communicate with any I/O device, provided data can be converted from parallel to serial form, and vice versa. Figures 5.106(a) and (b) show the circuit diagrams, which realize the parallel-to-serial and serial-to-parallel transformation, respectively. The first structure uses time-division multiplexing for the placement of N-bit input data in a single channel, while the second structure relies on the delay, which can be introduced on a data stream by shift registers. Note that various architectures are available for the same interface type, which is efficiently implemented as a combination of hardware and software.

It should be noted that the above converters include additional input and output nodes in a data acquisition environment. The timing waveforms are shown in Figure 5.107 for the specific case of the parallel-to-serial converter. After the initialization step steered by the RESET signal, the input data $ID[0:N]$ are applied to the circuit and the write (WR) pulse is enabled. The signal BUSY changes to the high level and data are transferred one bit after the other to the serial output (SD) under control of the clock signal (CK). The initiation and end of the transmission are detected from the information in SCK, which is an inverted version of CK.

5.6.1 Parallel interfacing

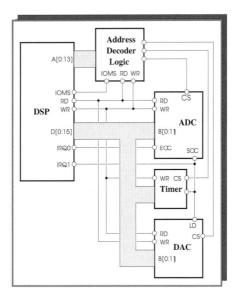

FIGURE 5.108
Parallel interfacing of a DSP.

The block diagram of a parallel interface implementation for a fixed-point DSP is depicted in Figure 5.108. Data are transferred between the DSP memory and ADC outputs (WR operation) or DAC inputs (RD operation) in one clock period. Due to the short memory access time of high-speed processors, the data transfer flow must be regulated by programming the DSP to insert wait-states in the converter access cycle. Alternatively, the DSP can include a different external input/output memory space (IOMS) for converters or other nonmemory peripherals. Each data bit requires a pin, as well as the control signals (WR, RD, and chip select (CS)). The timer must generate an interrupt request (IRQ), which determines the start of the conversion (SOC) of the ADC or the load of data (LD) into the DAC. The end-of-conversion (EOC) goes high to indicate that the conversion is complete and ADC output data are ready to be read.

The converter resolution can be lower than the one of the DSP data bus. The appropriate connection is then determined by the number representation system. For instance, the right justification of buses is needed for binary coding, while the left justification provides an adequate transfer in the case of two's complement representations. This latter situation can be applied to the interface structure of Figure 5.108, where a 12-bit ADC and 14-bit DAC are used. The MSB (B11) of the ADC should be joined to D15 down to the LSB (B0) wired to D4. The DAC inputs must be connected to the data bus starting from B13 to D15 through B0 to D2.

Parallel interfacing has the advantage of higher transfer speed, but it results in a chip package with a high number of pins.

5.6.2 Serial interfacing

By serially interfacing a DSP, the number of pins can be reduced. This approach is constrained by the requirement that the transfer rate must be greater than the required data bandwidth. Various serial protocols based on different bit encoding and basic packet structure are available for the communication between the DSP and data converters. Serial ports (SPORTs) can be used to transmit or receive data words of length 4 to 16 bits. A DSP is able to communicate in both directions simultaneously, that is, in full duplex mode. In contrast to microcontrollers, DSPs use a frame sync (FS) signal to indicate the beginning of the data stream and can operate with a continuous serial clock (SCK) signal together with FS pulses. For a microcontroller, the data transfer takes place with respect to the SCK signal, which must be active. The data synchronization can be conducted either by the DSP or data converters, but it is often convenient to have the sample timing being determined by the ADC and DAC.

The block diagram of a serial interface is shown in Figure 5.109. The DSP features pins corresponding to the data receive (DR), data transmit (DT), receive frame sync (RFS), and transmit frame sync (RFS) operations. Before the start of the transfer, synchronization pulses must be generated on the

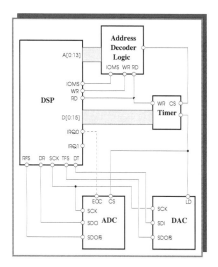

FIGURE 5.109
Serial interfacing of a DSP.

corresponding pin. When the SPORT is enabled, the digital data from the ADC are sent out on the serial data output (SDO) pin, and the ones from the serial data input (SDI) pin are transmitted to the DAC. The EOC flag is raised at the end of the analog-to-digital conversion and can be reset to account for the DSP interrupt signal. The timer is used to generate the chip select (CS) and DAC load (LD) inputs.

5.7 Built-in self-test structures for data converters

Due to the increase in circuit complexity, testing is becoming an integral part of the integrated circuit design. Built-in self-test (BIST) structures provide the advantage of reducing the test cost and improving the testing accuracy in high-density circuits.

BIST structures based on code density test principles can be used to determine low-frequency spectral characteristics of data converters [48, 49]. The generation of the test signal can rely on the use of pattern memory and a DAC. To reduce the required chip area, the digital version of the signal, which is available at the memory output, is transformed into an analog waveform by a 1-bit DAC, whose linearity is generally excellent. Thus, the quality of the signal is primarily determined by the number of samples, which is bounded by the memory size. By using a linear ramp as a test stimulus, the code width associated with the converter output signal can be computed. The number of occurrences in each bin should be equal in the ideal case, and any deviation can be related to the imperfection of a practical converter. The DNL for a given input sample is derived by subtracting the ideal code width from the measured code width. The sum of the DNL from the first up to the current code is equal to the INL. The accuracy of the DNL and INL determination is limited by the noise and quantization errors to about 0.05 LSB. For data converters embedded in a mixed-signal circuit, including a digital signal processor (DSP), the self-test program and test data can be stored in the read-only memory (ROM) and random access memory (RAM). However, BIST structures using logic gates can feature a low area and a high speed.

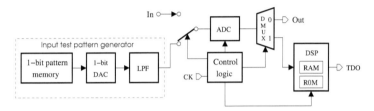

FIGURE 5.110
BIST structure for ADCs.

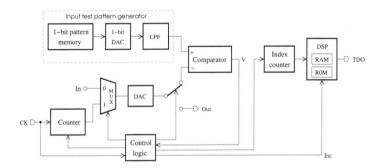

FIGURE 5.111
BIST structure for DACs.

The block diagram of the BIST structure for the ADC is shown in Figure 5.110, where CK is the clock signal and TDO is the test data output. It consists of an input test generator including a pattern memory, a 1-bit DAC and a lowpass filter (LPF), a control logic, a DSP, and the ADC to be tested. Let the test pattern be a linear ramp, the magnitude of which is greater than the full-scale range of the ADC. The output code of the converter can be written as $b_{2^N-1}, b_{2^N-2}, \cdots, b_0$, where N denotes the number of bits, and b_{2^N-1} and b_0 are the MSB and LSB, respectively. The DNL and INL can be estimated from the array of $2^N - 2$ elements obtained by excluding the MSB and LSB, which correspond to non-doubly bounded input ranges, and the occurrence number, Λ_i, of each code b_i. That is,

$$\text{DNL}_i = \frac{\Lambda_i}{\Lambda} - 1 \tag{5.329}$$

and

$$\text{INL}_i = \begin{cases} 0 & \text{if} \quad i = 0 \\ \text{INL}_{i-1} + \dfrac{\text{DNL}_i + \text{DNL}_{i-1}}{2} & \text{otherwise,} \end{cases} \tag{5.330}$$

where

$$\Lambda = \frac{\displaystyle\sum_{i=1}^{2^N-2} \Lambda_i}{2^N - 2} \tag{5.331}$$

The above static parameters can be expressed as a fraction of the LSB.

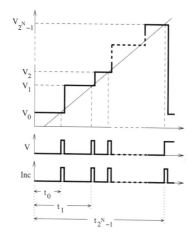

FIGURE 5.112
Operation principle of a DAC BIST structure.

The BIST architecture for the DAC, as shown in Figure 5.111, includes a test pattern generator, a counter for digital code generation, an analog comparator, an index counter, a DSP, a control logic, and the circuit under test, which is a DAC. An analog version of the encoded sawtooth signal stored in the memory of the pattern generator and the DAC output signal are applied, respectively, to the positive and negative input nodes of the analog comparator. During the test, the output signal, V, delivered by the analog comparator, gives an estimation of the magnitude levels associated with the different input codes of the DAC. It will assume the high or low state if the signal level at V^+ is greater or lower than the one at V^-. By detecting the rising edge of V, the control logic can increment the index counter, the content of which represents the different indexes, t_i, to be stored in the DSP memory. The DNL and INL (in fraction of the LSB) of the DAC can be derived as

$$\text{DNL}_i = \frac{t_i - t_{i-1}}{\Lambda} - 1 \tag{5.332}$$

and

$$\text{INL}_i = \begin{cases} 0 & \text{if } i = 0 \\ \text{INL}_{i-1} + \text{DNL}_i & \text{otherwise,} \end{cases} \tag{5.333}$$

where

$$\Lambda = \frac{t_{2^N - 1} - t_0}{2^N - 1} \tag{5.334}$$

and N is the number of bits of the DAC. Note that $i = 1, \cdots, 2^N - 2$ and $\text{INL}_{2^N - 1} = 0$ because the determination of the DAC parameters relies on the use of a linearized output line, whose support points are located in the first and last levels of the transfer characteristic.

The BIST performance depends on the quality of the test signal generated on-chip. Figure 5.112 illustrates the testing principle when the DAC transfer characteristic is a monotonically increasing function and the levels of the adjacent codes are sufficiently separated to be detected by the analog comparator. In cases where these last requirements are not fulfilled for the codes i and $i + 1$, after the estimation of t_i, the determination process of the next index t_{i+1} should be restarted with the code $i + 1$ held constant at the DAC input.

5.8   Circuit design assessment

1.   Truncation quantizer model

Delta-sigma digital modulators rely on a truncation quantizer to reduce the number of bits of digital code. The conversion of digital code x into a truncated version $\hat{x}$ incurs a quantization error defined as $e_Q = \hat{x} - x$.

– Considering the characteristic and quantization error of the truncation quantizer shown in Figure 5.113(a) in the case of the two's complement representation, the probability density function is given by

$$p(e_Q) = \begin{cases} \dfrac{1}{\triangle} & \text{if } -\triangle < e_Q \leq 0 \\ 0 & \text{otherwise.} \end{cases} \tag{5.335}$$

Show that

$$E(e_Q) = \int_{-\infty}^{+\infty} e_Q p(e_Q) de_Q = \int_{-\triangle}^{0} e_Q p(e_Q) de_Q = -\frac{\triangle}{2} \tag{5.336}$$

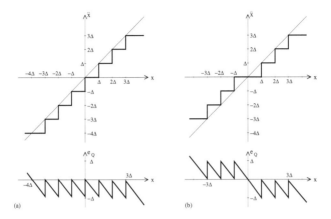

FIGURE 5.113
Characteristics and errors of a truncation quantizer: (a) Two's complement and (b) sign-magnitude representations.

and

$$E(e_Q^2) = \int_{-\infty}^{+\infty} e_Q^2 p(e_Q) de_Q = \int_{-\Delta}^{0} e_Q p(e_Q) de_Q = \frac{\Delta^2}{3} \qquad (5.337)$$

Deduce that the variance or power of the quantization noise is of the form

$$\sigma_Q^2 = E(e_Q^2) - [E(e_Q)]^2 = \frac{\Delta^2}{12} \qquad (5.338)$$

where Δ is the quantizer step size.

– Suppose now that a sign-magnitude representation is adopted. The characteristic and quantization error of the truncation quantizer are depicted in Figure 5.113(b) and the probability density function can be obtained as

$$p(e_Q) = \begin{cases} \dfrac{1}{2\Delta} & 0 \le e_Q < \Delta \quad \text{if} \quad x < 0 \\[2mm] \dfrac{1}{2\Delta} & -\Delta < e_Q \le 0 \quad \text{if} \quad x \ge 0 \\[2mm] 0 & \text{otherwise.} \end{cases} \qquad (5.339)$$

Show that

$$E(e_Q) = E(e_Q)|_{x<0} + E(e_Q)|_{x\ge0} = 0 \qquad (5.340)$$

and

$$E(e_Q^2) = E(e_Q^2)|_{x<0} + E(e_Q^2)|_{x\ge0} = \frac{\Delta^2}{3} \qquad (5.341)$$

Deduce that the variance of the quantization noise is of the form

$$\sigma_Q^2 = \{E(e_Q^2) - [E(e_Q)]^2\}|_{x<0} + \{E(e_Q^2) - [E(e_Q)]^2\}|_{x\geq0} = \frac{5\Delta^2}{2\;12}$$
$$(5.342)$$

where Δ is the quantizer step size.

2. **Analysis of a second-order DT $\Delta\Sigma$ modulator**
   Consider the block diagram of the second-order $\Delta\Sigma$ modulator depicted in Figure 5.114. The integrator can be implemented such that $I(z) = z^{-1}/(1 - z^{-1})$.
   Assuming a linear model for the comparator, or in other words,

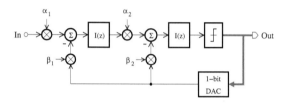

FIGURE 5.114
Block diagram of a second-order $\Delta\Sigma$ modulator.

$Y(z) = qX(z) + E_Q(z)$, where Y is the comparator output, X is the comparator input, E_Q is the quantization noise, and q is the comparator gain, show that

$$Y(z) = H_S(z)S(z) + H_Q(z)E_Q(z) \qquad (5.343)$$

where

$$H_S(z) = \frac{q\alpha_1\alpha_1 z^{-2}}{1 + (q\beta_2 - 2)z^{-1} + [1 + q(\alpha_2\beta_1 - \beta_2)]z^{-2}} \qquad (5.344)$$

and

$$H_Q(z) = \frac{(1 - z^{-1})^2}{1 + (q\beta_2 - 2)z^{-1} + [1 + q(\alpha_2\beta_1 - \beta_2)]z^{-2}} \qquad (5.345)$$

Let $z = e^{j\omega T}$, where T is the clock signal period. Evaluate the coefficients α_1, α_2, β_1, and β_2 in terms of q so that the stability criterion, $|H_Q(\omega)| < 1.5$ for $0 \leq \omega T \leq \pi$, is fulfilled.

3. **State-space representation of a second-order DT modulator**
   A second-order DT modulator, as shown in Figure 5.115, can be implemented using two switched-capacitor integrators, a comparator, and a 1-bit DAC. To proceed further, let $\mathbf{X}(z) = [X_1(z) \; X_2(z)]^T$ be the state vector and use a linear quantizer model, that is, $Y(z) = X_2(z) + E_Q(z)$, where $E_Q(z)$ is the quantization error.

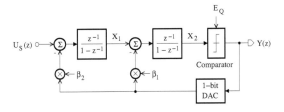

FIGURE 5.115
Second-order DT modulator.

Assuming that $E_Q(z) = 0$, show that the second-order DT modulator can be described by a state-space representation of the form:

$$z\mathbf{X}(z) = \mathbf{A}\mathbf{X}(z) + \mathbf{B}U_S(z) \tag{5.346}$$
$$Y(z) = \mathbf{C}\mathbf{X}(z) + DU_S(z) \tag{5.347}$$

where

$$\mathbf{A} = \begin{bmatrix} 1 & -\beta_2 \\ 1 & 1 - \beta_1 \end{bmatrix} \qquad \mathbf{B} = \begin{bmatrix} 1 \\ 0 \end{bmatrix}$$
$$\mathbf{C} = \begin{bmatrix} 0 & 1 \end{bmatrix} \qquad D = 0 \tag{5.348}$$

Verify that tne STF can written as follows:

$$STT(z) = \frac{Y(z)}{U_S(z)} = \mathbf{C}(z\mathbf{I} - \mathbf{A})^{-1}\mathbf{B} + D \tag{5.349}$$

$$= \frac{z^{-2}}{1 - (2 - \beta_1)z^{-1} + (1 + \beta_2 - \beta_1)z^{-2}} \tag{5.350}$$

Assuming that $U_S(z) = 0$, show that the state-space representation of the second-order DT modulator is given by:

$$z\mathbf{X}(z) = \mathbf{A}\mathbf{X}(z) + \mathbf{B}E_Q(z) \tag{5.351}$$
$$Y(z) = \mathbf{C}\mathbf{X}(z) + DE_Q(z) \tag{5.352}$$

where

$$\mathbf{A} = \begin{bmatrix} 1 & -\beta_2 \\ 1 & 1 - \beta_1 \end{bmatrix} \qquad \mathbf{B} = \begin{bmatrix} -\beta_2 \\ -\beta_1 \end{bmatrix}$$
$$\mathbf{C} = \begin{bmatrix} 0 & 1 \end{bmatrix} \qquad D = 1 \tag{5.353}$$

Verify that the QNTF can be obtained as follows:

$$QNTF(z) = \frac{Y(z)}{E_Q(z)} = \mathbf{C}(z\mathbf{I} - \mathbf{A})^{-1}\mathbf{B} + D \tag{5.354}$$

$$= \frac{(1 - z^{-1})^2}{1 - (2 - \beta_1)z^{-1} + (1 + \beta_2 - \beta_1)z^{-2}} \tag{5.355}$$

Show that the modulator poles are given by

$$p_1, p_2 = \left(1 - \frac{\beta_1}{2}\right) \pm j\sqrt{\beta_2 - \frac{\beta_1^2}{4}} \tag{5.356}$$

Deduce that the modulator poles are exactly located on the unit circle if the coefficients are chosen as $\beta_2 = \beta_1 = 1$ or $\beta_2 = \beta_1 = 1/2$.

An all-pole Butterworth STF with cutoff frequency at $1/4$ of the sampling frequency exhibits the poles: $p_1, p_2 = 0.375 \pm j0.320$. Verify that the modulator coefficients should be chosen as: $\beta_2 = 0.493$ and $\beta_1 = 1.250$.

4. **Second-order modulator with complex poles**
   Consider the second-order modulators described by the block diagrams shown in Figures 5.116 and 5.117.

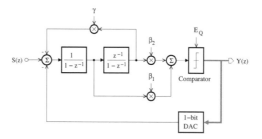

FIGURE 5.116
Second-order modulator with a feedforward summation.

Assuming that the comparator can be modeled with an additive quantization noise source and the feedback DAC exhibits a propagation delay of one clock period, show that

$$Y(z) = H_S(z)S(z) + H_Q(z)E_Q(z) \tag{5.357}$$

where

$$H_S(z) = \frac{N_i(z)}{1 - (2 - \beta_1)z^{-1} + (1 + \gamma + \beta_2 - \beta_1)z^{-2}} \tag{5.358}$$

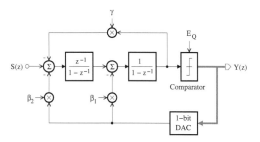

FIGURE 5.117
Second-order modulator with a two feedback paths.

and

$$H_Q(z) = \frac{(1 - z^{-1})^2 + \gamma z^{-2}}{1 - (2 - \beta_1)z^{-1} + (1 + \gamma + \beta_2 - \beta_1)z^{-2}} \qquad (5.359)$$

where $H_S(z)$ and $H_Q(z)$ are the signal and quantization noise transfer functions, respectively. For the first modulator $(i = 1)$ and we have $N_1(z) = (\beta_2 - \beta_1)z^{-1} + \beta_1$, while for the second modulator $(i = 2)$ and $N_1(z) = z^{-1}$.

A second-order transfer function, $N(z)/D(z)$, with a denominator of the form, $D(z) = 1 + a_1 z^{-1} + a_2 z^{-2}$, is stable provided that $|a_2| < 1$ and $-(1 + a_2) < a_1 < 1 + a_2$.

Deduce the constraints to be set on the modulator coefficients to satisfy the stability conditions based on the linear model.

For less in-band noise, the quantization noise transfer function should have complex zeros.

Determine the range of γ values for which the zeros of the quantization noise transfer function are complex.

5. **Second-order modulator with one-bit and multi-bit quantizers**

   The second-order delta-sigma modulator of Figure 5.118 is composed of two stages and requires a single-bit and multi-bit quantizers (comparator and B-bit ADC). Ideally, it can help attenuate the level of the quantization noise at the output by a factor 2^{B-1}, where B is the number of bits, because the output contribution associated to the quantization error of the comparator is replaced by that of the B-bit ADC.

   The combined latency of the quantizer and the feedback DAC is modeled as a unit delay in the feedback path. By modeling the quantization effect due to the comparator and the B-bit ADC as

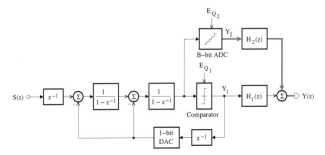

FIGURE 5.118
Second-order modulator with one-bit and multi-bit quantizers.

additive noises, E_{Q_1} and E_{Q_2}, that are uniformly distributed and uncorrelated with the input signal, show that

$$Y_1(z) = z^{-1}S(z) + (1 - z^{-1})^2 E_{Q_1}(z) \tag{5.360}$$

and

$$\begin{aligned} Y_2(z) &= Y_1(z) - E_{Q_1}(z) + E_{Q_2}(z) \\ &= z^{-1}S(z) + E_{Q_2}(z) - (2z^{-1} - z^{-2})E_{Q_1}(z) \end{aligned} \tag{5.361}$$

Verify that the output signal, Y, can be written as

$$\begin{aligned} Y(z) &= z^{-1}(H_1(z) + H_2(z))S(z) + H_2(z)E_{Q_2}(z) \\ &\quad + [(1 - z^{-1})^2 H_1(z) - (2z^{-1} - z^{-2})H_2(z)]E_{Q_1}(z) \end{aligned} \tag{5.362}$$

For $H_1(z) = 2z^{-1} - z^{-2}$ and $H_2(z) = (1 - z^{-1})^2$ show that

$$Y(z) = z^{-1}S(z) + (1 - z^{-1})^2 E_{Q_2}(z) \tag{5.363}$$

For a 1-kHz sinewave input sampled at 1024 kHz with an oversampling ratio of 64, use simulations to verify that a 1% error in the pole location of each integrator of the single-bit loop can degrade the SNR by more than 30 dB if the amplifier dc gain is not greater than 75 dB and the number of bits, B, exceeds 4.

6. **Low-distortion third-order DT modulator**
   For the feedforward third-order modulator shown in Figure 5.119, the loop filter transfer function can be expressed as,

$$H(z) = \sum_{i=1}^{3} \frac{a_i z^{-i}}{(1 - z^{-1})^i} = \sum_{i=1}^{3} a_i I^i(z) \tag{5.364}$$

where

$$I(z) = \frac{z^{-1}}{1 - z^{-1}} \qquad (5.365)$$

Verify that the signal and quantization noise transfer functions are, respectively, given by

$$H_S(z) = \frac{Y(z)}{S(z)} = \frac{1 + H(z)}{1 + H(z)} = 1 \qquad (5.366)$$

and

$$H_Q(z) = \frac{Y(z)}{E_Q(z)} = \frac{1}{1 + H(z)} \qquad (5.367)$$

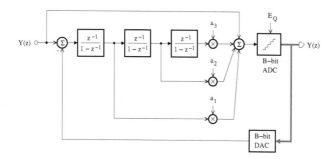

FIGURE 5.119
Feedforward third-order modulator.

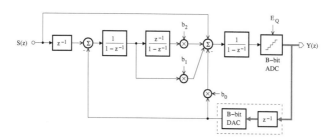

FIGURE 5.120
Low-distortion third-order modulator.

Consider the low-distortion third-order modulator of Figure 5.119. Verify that the signal and quantization noise transfer functions can, respectively, be expressed as

$$H_S(z) = \frac{Y(z)}{S(z)} = \frac{1}{1 + G(z)}\left[1 + \frac{1}{1 - z^{-1}}\sum_{i=1}^{2} b_i I^i(z)\right] \qquad (5.368)$$

and

$$H_Q(z) = \frac{Y(z)}{E_Q(z)} = \frac{1}{1 + G(z)} \tag{5.369}$$

where

$$G(z) = \frac{1}{1 - z^{-1}}\left[b_0 z^{-1} + \sum_{i=1}^{2} b_i I^i(z)\right] \tag{5.370}$$

Using $H(z) = G(z)$, show that

$$\begin{bmatrix} a_1 \\ a_2 \\ a_3 \end{bmatrix} = \begin{bmatrix} 1 & 1 & 0 \\ 0 & 1 & 1 \\ 0 & 0 & 1 \end{bmatrix} \begin{bmatrix} b_0 \\ b_1 \\ b_2 \end{bmatrix} \tag{5.371}$$

or, equivalently,

$$b_i = a_i - b_{i-1} \quad \text{for} \quad i = 1, 2 \tag{5.372}$$

$$b_2 = a_3 \tag{5.373}$$

where

$$b_0 = \sum_{i=1}^{3} a_i(-1)^{i+1} = -H(0) = 1 - \frac{1}{H_Q(0)} = 1 \tag{5.374}$$

7. **Discrete-time to continuous-time transformation**

A $\Delta\Sigma$ modulator can be designed to process either a continuous-time or discrete-time signal, as shown in Figures 5.121(a) and (b), respectively. The feedback digital-to-analog converter (DAC) is characterized by the transfer function $H_{DAC}(s) = (1 - e^{sT})/s$, where $T = 1/f_s$ is the sampling period, and f_s is the sampling frequency.

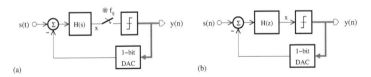

FIGURE 5.121
Block diagram of first-order $\Delta\Sigma$ modulators.

Suppose that

$$H(s) = \frac{1}{sT} \tag{5.375}$$

and use s-transform tables to show that

$$h(t) = \mathcal{L}^{-1}\big[H(s)H_{DAC}(s)\big] = \frac{t}{T}u(t) - \frac{t-T}{T}u(t-T) \quad (5.376)$$

With reference to z-transform tables, show that

$$H(z) = \mathcal{Z}\big[h(t)|_{t=nT}\big] = \frac{z^{-1}}{1-z^{-1}} \quad (5.377)$$

Now consider that

$$H(s) = \frac{s_k}{sT + s_k} \quad (5.378)$$

where $\omega_c = s_k/T$ is the 3-dB bandwidth frequency of the analog lowpass filter, and use s-transform tables to show that

$$h(t) = \mathcal{L}^{-1}\big[H(s)H_{DAC}(s)\big] = \big(1-e^{-s_k t}\big)u(t) - \big[1-e^{-s_k(t-T)}\big]u(t-T) \quad (5.379)$$

Show that the equivalent transfer function in the z-domain can be put into the form

$$H(z) = \mathcal{Z}\big[h(t)|_{t=nT}\big] = \frac{(1-z_k)z^{-1}}{1-z_k z^{-1}} \quad (5.380)$$

where $z_k = e^{-s_k T}$.

8. **First-order CT lowpass modulator**
   Consider the circuit diagram of a first-order modulator shown in Figure 5.122(a). The input stage is an active RC integrator and

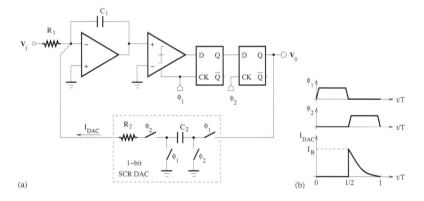

(a)
(b)

FIGURE 5.122
(a) Circuit diagram of a first-order CT modulator; (b) waveforms of a 1-bit SCR DAC ($p_2 = 1$ and $p_1 = 1/2$).

the quantizer output is assumed to be held for a full clock period. Figure 5.122(b) shows waveforms of the switched-capacitor with a series resistor (SCR) DAC, with ϕ_1 and ϕ_2 being non-overlapping clock phases.

Assuming that the clock signal period is T and the desired DT transfer function of the quantization noise is $H_Q(z) = 1 - z^{-1}$, use the following DT-to-CT transform

$$\frac{1}{z-1} \to \frac{r_0}{s} \tag{5.381}$$

where

$$r_0 = \frac{1}{\tau_d(1 - e^{-(p_2 - p_1)T/\tau_d})} \tag{5.382}$$

and $\tau_d = R_2C_2$, to show that $R_1C_1 = 1/r_0$.

9. **Analysis of a second-order CT $\Delta\Sigma$ modulator**
   Consider the second-order CT $\Delta\Sigma$ modulator shown in Figure 5.21.

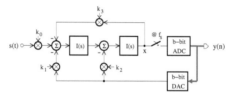

FIGURE 5.123
Block diagram of a second-order CT $\Delta\Sigma$ modulator.

Assuming that $I(s) = 1/(sT)$, where T denotes the clock signal period, show that

$$H(s) = \frac{X(s)}{Y(s)} = -\frac{k_2 sT + k_1}{s^2 T^2 + k_3} \tag{5.383}$$

In the case where the digital-to-analog converter (DAC) is of the half-return-to-zero (HRZ) type, or $H_{DAC}(s) = (e^{-sT/2} - e^{-sT})/s$, find the equivalent discrete-time (DT) transfer function $H(z)$.

Generate the pole-zero plot of the DT modulator noise transfer function based on the following coefficient values: $k_0 = 1$, $k_1 = 2$, $k_2 = 3.5$, and $k_3 = 0.015$.

10. **Second-order bandpass CT modulator**
    A second-order bandpass CT modulator [32] can be implemented as shown in Figure 5.124. It consists of transconductors, a LC resonator, a 1-bit return-to-zero (RZ) DAC, and a 1-bit half-clock-period delayed return-to-zero (HZ) DAC. The output voltage of the

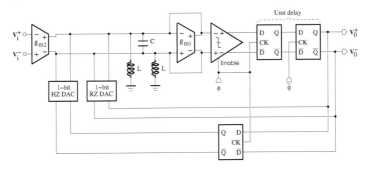

FIGURE 5.124
Circuit diagram of a second-order LC bandpass CT modulator.

CT filter is given by

$$V_0(s) = I_i Z_{LC}(s) \tag{5.384}$$

where $I_i = g_m V_i$ and $Z_{LC}(s) = 1/[Cs + (1/Ls)]$.

Verify that the filter transfer function can be put into the form,

$$\hat{H}(s) = \frac{V_0(s)}{V_i(s)} = \frac{k\omega_0 s}{s^2 + \omega_0^2} \tag{5.385}$$

where $\omega_0 = 1/\sqrt{LC}$ and $k = g_m\sqrt{L/C}$.

Assuming that the desired loop transfer function of the DT modulator prototype is $z^{-2}/(1 + z^{-2})$, and the transfer functions of the DAC pulses are given by,

$$P_{rz}(s) = \frac{1 - e^{-sT/2}}{s} \tag{5.386}$$

$$P_{hz}(s) = \frac{e^{-sT/2} - e^{-sT}}{s} = \frac{e^{-sT/2}(1 - e^{-sT/2})}{s} \tag{5.387}$$

the feedback coefficients implemented by each of the DACs can be determined by solving the next equation

$$z^{-1}H(z) = k_{rz}\mathcal{Z}\left\{\mathcal{L}^{-1}\left[P_{rz}(s)\frac{\omega_0 s}{s^2 + \omega_0^2}\right]\Big|_{t=nT}\right\} +$$
$$k_{hz}\mathcal{Z}\left\{\mathcal{L}^{-1}\left[P_{hz}(s)\frac{\omega_0 s}{s^2 + \omega_0^2}\right]\Big|_{t=nT}\right\}$$

(5.388)

$$= k_{rz}\frac{[\sin(\omega_0 T) - \sin(\omega_0 T/2)]z^{-1} - \sin(\omega_0 T/2)]z^{-2}}{1 - 2\cos(\omega_0 T)z^{-1} + z^{-2}} +$$
$$k_{hz}\frac{\sin(\omega_0 T/2)z^{-1} + [\sin(\omega_0 T/2) - \sin(\omega_0 T)]z^{-2}}{1 - 2\cos(\omega_0 T)z^{-1} + z^{-2}}$$

(5.389)

$$= k_{rz}\frac{(1 - \frac{\sqrt{2}}{2})z^{-1} - \frac{\sqrt{2}}{2}z^{-2}}{1 + z^{-2}} + k_{hz}\frac{\frac{\sqrt{2}}{2}z^{-1} - (1 - \frac{\sqrt{2}}{2})z^{-2}}{1 + z^{-2}}$$

(5.390)

where $\omega_0 T = \pi/2$ and $H(z) = z^{-1}/(1 + z^{-2})$.

Deduce that

$$k_{rz} = -\frac{\sqrt{2}}{2} \quad \text{and} \quad k_{hz} = 1 + \frac{\sqrt{2}}{2}$$

(5.391)

A practical CT modulator can be affected by the excess loop delay caused by the imperfection of the feedback components. Taking into account the excess loop delay, the sampling period can be modeled as $(1+\delta)T$, where $0 < \delta \leq 1$, and the loop transfer function becomes

$$z^{-1}H(z, \delta) = \frac{z^{-2}}{1 + z^{-2}}\left(\sin\frac{(1-\delta)\pi}{2} + z^{-1}\sin\frac{\delta\pi}{2}\right)$$

(5.392)

To study the modulator stability, plot the pole-zero locus of the resulting noise transfer in the z domain and verify that both poles can move out of the unit circle for an excess loop delay of about 58% of the clock signal period, T.

11. **DT and CT 2-1 cascaded $\Delta\Sigma$ modulators**

For the 2-1 discrete-time (DT) cascaded $\Delta\Sigma$ modulator depicted in Figure 5.125(a), verify that

$$H_1(z) = \frac{X_2(z)}{Y_1(z)} = -\frac{\alpha_1\alpha_2 z^{-2}}{(1 - z^{-1})^2} - \frac{2\alpha_1\alpha_2 z^{-1}}{1 - z^{-1}}$$

(5.393)

$$H_2(z) = \frac{X_3(z)}{Y_2(z)} = -\frac{\alpha_3 z^{-1}}{1 - z^{-1}}$$

(5.394)

and

$$H_3(z) = \frac{X_3(z)}{Y_1(z)} = \left[\frac{\kappa_2}{\alpha_1\alpha_2} H_1(z) - \kappa_1\kappa_2\right]\frac{\alpha_3 z^{-1}}{1 - z^{-1}} \qquad (5.395)$$

$$= -\frac{\kappa_2\alpha_3 z^{-3}}{(1 - z^{-1})^3} - \frac{2\kappa_2\alpha_3 z^{-2}}{(1 - z^{-1})^2} - \frac{\kappa_1\kappa_2\alpha_3 z^{-1}}{1 - z^{-1}} \qquad (5.396)$$

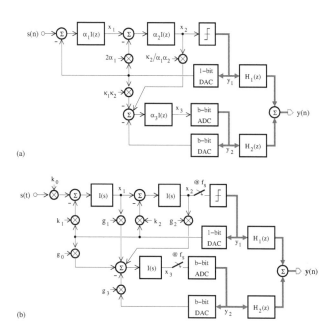

FIGURE 5.125
Block diagrams of 2-1 cascaded $\Delta\Sigma$ modulators.

TABLE 5.5
Equivalent CT Transfer Functions of Some DT Transfer Functions

$\dfrac{z^{-1}}{1 - z^{-1}} \rightarrow \dfrac{1}{sT}$	$\dfrac{z^{-2}}{(1 - z^{-1})^2} \rightarrow \dfrac{1}{s^2 T^2} - \dfrac{1}{2}\dfrac{1}{sT}$
$\dfrac{z^{-3}}{(1 - z^{-1})^3} \rightarrow \dfrac{1}{s^3 T^3} - \dfrac{1}{s^2 T^2} + \dfrac{1}{3}\dfrac{1}{sT}$	

Use the impulse invariant transform relations of Table 5.5, where it was assumed that the digital-to-analog converter (DAC) is modeled by a transfer function of the form $H_{DAC}(s) = (1 - e^{-sT})/s$, where

T is the period of the clock signal, to show that the equivalent s-domain transfer functions of the z-domain transfer functions, H_p, $(p = 1, 2, 3)$, are given by

$$H_1(s) = -\frac{\alpha_1 \alpha_2}{s^2 T^2} - \frac{3\alpha_1 \alpha_2/2}{sT} \tag{5.397}$$

$$H_2(s) = -\frac{\alpha_3}{sT} \tag{5.398}$$

and

$$H_3(s) = -\frac{\kappa_2 \alpha_3}{s^3 T^3} - \frac{\kappa_2 \alpha_3}{s^2 T^2} + \frac{(2\kappa_2 \alpha_3/3)(1 - 3\kappa_1/2)}{sT} \tag{5.399}$$

With reference to the continuous-time (CT) 2-1 cascaded $\Delta\Sigma$ modulator shown in Figure 5.125(b), where $k_0 = k_1$, determine the transfer functions $H_1(s) = X_2(s)/Y_1(s)$, $H_2(s) = X_3(s)/Y_2(s)$, and $H_3(s) = X_3(s)/Y_1(s)$.

Find k_i, $(i = 0, 1, 2)$ and g_j, $(j = 0, 1, 2, 3)$ in terms of α_p $(p = 1, 2, 3)$ and κ_q $(q = 1, 2)$.

With the assumption that $\alpha_1 = 1/2$, $\alpha_2 = 1/2$, $\alpha_3 = 1$, $\kappa_1 \kappa_2 = 1$, and $\kappa_2 = 1/2$, generate the pole-zero plot of each of the modulator noise transfer functions.

12. **Third-order CT modulator with a PDWA DAC**
    MATLAB® (Delta Sigma or Control) toolboxes can be used to design a third-order CT $\Delta\Sigma$ lowpass modulator with an SQNR of 89 dB at an oversampling rate of 16.

    Assuming a 4-bit quantizer and choosing the inverse Chebyshev filter response, the following quantization noise transfer function can be derived:

    $$H_Q(z) = \frac{(z - 1)(z^2 - 1.977z + 1)}{(z - 0.3474)(z^2 - 0.6295z + 0.2935)} \tag{5.400}$$

    Next, the transfer function of the corresponding DT loop filter is obtained as,

    $$H(z) = \frac{H_Q(z) - 1}{H_Q(z)} = \frac{2(z^2 - 1.232z + 0.449)}{(z - 1)(z^2 - 1.977z + 1)} \tag{5.401}$$

    Assuming an NRZ DAC waveform and a total excess-loop delay of $\tau = T/2$, where T denotes the period of the clock signal, the DT transfer function $H(z)$ is transformed into a CT loop-filter transfer function of the form:

    $$\hat{F}(s) = -\frac{0.8374s^3 + 1.984s^2 + 1.321s + 0.4342}{s(s^2 + 0.0231)} \tag{5.402}$$

In the special case of an NRZ DAC pulse, this can performed using the *d2c* function of MATLAB Control toolbox.

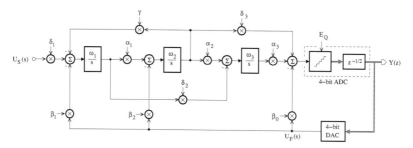

FIGURE 5.126
Block diagram of a third-order CT lowpass modulator.

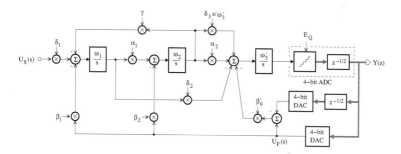

FIGURE 5.127
Block diagram of the third-order CT lowpass modulator after scaling.

Consider the block diagram of a third-order CT lowpass modulator shown in Figure 5.126, where the feedback coefficient β_0 is employed to compensate the excess-loop delay that is generally associated to CT modulators with an NRZ DAC pulse. However, DACs using NRZ pulses are less sensitive to clock jitter. Assuming that $\omega_i = 1$ for $i = 1, 2, 3$, it can be shown that

$$s\mathbf{X}(s) = \mathbf{A}\mathbf{X}(s) + \mathbf{B}\mathbf{U}(s) \tag{5.403}$$
$$Y(s) = \mathbf{C}\mathbf{X}(s) + \mathbf{D}\mathbf{U}(s) \tag{5.404}$$

where

$$\mathbf{A} = \begin{bmatrix} 0 & -\gamma & 0 \\ \alpha_1 & 0 & 0 \\ -\delta_2 & \alpha_2 & 0 \end{bmatrix} \quad \mathbf{B} = \begin{bmatrix} -\beta_1 & \delta_1 \\ -\beta_2 & 0 \\ 0 & 0 \end{bmatrix}$$
$$\mathbf{C} = \begin{bmatrix} 0 & \delta_3 & \alpha_3 \end{bmatrix} \quad \mathbf{D} = \begin{bmatrix} -\beta_0 & 0 \end{bmatrix} \tag{5.405}$$

and $\mathbf{U}(s) = [U_F(s)\ U_S(s)]^T$, with $U_S(s)$ being the modulator input

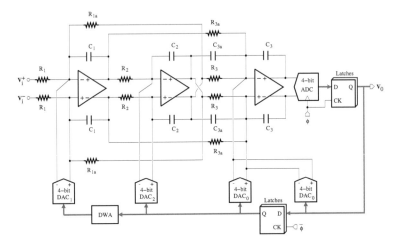

FIGURE 5.128
Circuit diagram of the third-order CT lowpass modulator.

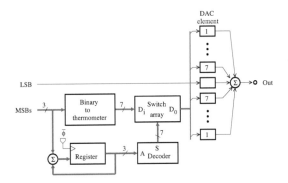

FIGURE 5.129
DAC with PDWA circuit.

and $U_F(s)$ the DAC output. The transfer functions can be obtained as:

$$T(s) = \mathbf{C}(s\mathbf{I} - \mathbf{A})^{-1}\mathbf{B} + \mathbf{D} \qquad (5.406)$$

where

$$T(s) = \left[\frac{Y(s)}{U_F(s)} \ \frac{Y(s)}{U_S(s)} \right]^T \qquad (5.407)$$

The feedback and feedforward transfer functions are, respectively, given by

$$F(s) = \frac{Y(s)}{U_S(s)} = -\frac{a_3 s^3 + a_2 s^2 + a_1 s + a_0}{s(s^2 + \alpha_2 \gamma)} \qquad (5.408)$$

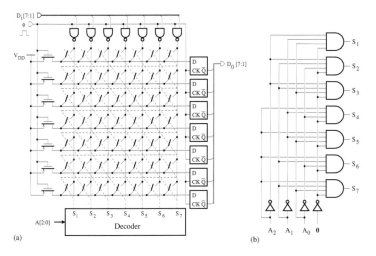

FIGURE 5.130
Circuit diagram of the switch array (a) and the decoder (b).

4–bit code						DAC elements									
MSBs	LSB	1	2	3	4	5	6	7 LSB	7	6	5	4	3	2	1
1	0	■	□	□	□	□	□	□ □	□	□	□	□	□	□	■
3	0	□	■	■	■	□	□	□ □	□	□	□	■	■	■	□
4	0	■	□	□	□	■	■	■ □	□	■	■	■	□	□	■
7	1	■	■	■	■	■	■	■ ■	■	■	■	■	■	■	■
4	1	□	■	■	■	□	□	■ ■	□	□	■	■	■	■	□
2	0	□	□	□	□	□	■	■ ■	■	■	□	□	□	□	□
0	1	□	□	□	□	□	□	■ ■	□	□	□	□	□	□	□
1	0	■	□	□	□	□	□	□ □	□	□	□	□	□	□	■
Usage		4	3	3	3	3	3	3 3	3	3	3	3	3	3	4

FIGURE 5.131
Illustration of the PDWA DAC operation.

and

$$G(s) = \frac{Y(s)}{U_F(s)} = \frac{d_1 s + d_0}{s(s^2 + \alpha_2 \gamma)} \tag{5.409}$$

where

$$a_3 = \beta_0 \tag{5.410}$$

$$a_2 = \beta_2 \delta_3 \tag{5.411}$$

$$a_1 = (\alpha_1 \delta_3 - \alpha_3 \delta_2)\beta_1 + \alpha_2 \alpha_3 \beta_2 + \alpha_1 \beta_0 \gamma \tag{5.412}$$

$$a_0 = \alpha_1 \alpha_2 \alpha_3 \beta_1 + \alpha_3 \beta_2 \gamma \delta_2 \tag{5.413}$$

$$d_1 = (\alpha_1 \delta_3 - \alpha_3 \delta_2)\delta_1 \tag{5.414}$$

and

$$d_0 = \alpha_1 \alpha_2 \alpha_3 \delta_1 \tag{5.415}$$

By matching $F(s)$ and $\hat{F}(s)$, modulator coefficients are obtained as follows:

$$
\begin{aligned}
&\alpha_1 = 0.5 \quad \alpha_2 = 1 \quad \alpha_3 = 2 \quad \beta_1 = 0.39 \quad \beta_2 = 0.65 \\
&\beta_0 = 0.84 \quad \gamma = 0.05 \quad \delta_1 = 0.43 \quad \delta_2 = 0.76 \quad \delta_3 = 3.04
\end{aligned}
\tag{5.416}
$$

For a power-efficient design, the block diagram of the third-order CT lowpass modulator is scaled to eliminate the output summer, as illustrated in Figure 5.127, where $\omega_3' = \omega_3 \alpha_3$. To differentiate the feedforward and feedback signals that are now applied at the input of the last integrator, the gains δ_3' and β_0' of the corresponding paths are scaled by s. The feedforward gain is of the form, $\delta_3' = \delta_3/\omega_3'$. For the feedback path, the differentiator is implemented in the digital domain, by mapping s as $(1 - z^{-1/2})/\tau$. Hence, $\beta_0' = \beta_0/(\omega_3'\tau)$, where $\tau = T/2$.

Assuming that $T = 1$, modulator coefficients can be obtained as follows:

$$
\begin{aligned}
&\alpha_1 = 0.5 \quad \alpha_2 = 1 \quad \alpha_3 = 2 \quad \beta_1 = 0.39 \quad \beta_2 = 0.65 \\
&\beta_0' = 0.84 \quad \gamma = 0.05 \quad \delta_1 = 0.43 \quad \delta_2 = 0.76 \quad \delta_3' = 1.5
\end{aligned}
\tag{5.417}
$$

Due to the fact the coefficient α_3 is equal to a power of 2, it can be implemented by decreasing the voltage reference level of the quantizer (4-bit flash ADC) by a factor of 2.

The circuit diagram of the third-order CT lowpass modulator is shown in Figure 5.128. It is based on RC circuits that are known to feature a high linearity and signal swing.

A data-weighted averaging (DWA) circuit can be associated only to the first DAC, that has the most stringent noise and linearity requirements. It can provide a first-order shaping of errors due to DAC element mismatches.

To reduce the hardware size by almost a factor of 4, the first DAC is implemented with a partial DWA (PDWA) circuit, as shown in Figure 5.129. Elements of the current steering DAC are laid out symmetrically around the central unit element that is directly controlled by the LSB bit. The other three MSBs of the codeword are applied at the input of the accumulator, whose output is decoded and then used as the selection code for the switch array. This ensures that the DAC elements are chosen equally and cyclically so that the random mismatch error is averaged out.

Figure 5.130 shows the circuit diagrams of the switch array and the decoder. The operation of the PDWA DAC circuit is illustrated in Figure 5.131. The use of the PDWA DAC leads to an SNDR degradation of less than 3 dB assuming a standard deviation of $\sigma = 1.5\%$ for the DAC element mismatch.

Use simulation tools to verify that the designed third-order CT modulator exhibit an SQNR of 87.7 dB at the OSR of 16, and a DR of 76 dB over of the signal bandwidth of 5 MHz.

Bibliography

[1] L. R. Carey, "A noise shaping coder topology for 15+ bit converters," *IEEE J. of Solid-State Circuits*, vol. 24, pp. 267–273, April 1989.

[2] A. Keady and C. Lyden, "Tree structure for mismatch noise-shaping multibit DAC," *Electronics Letters*, vol. 33, pp. 1431–1432, Aug. 1997.

[3] T. Shui, R. Schreier, and F. Hudson, "Mismatch shaping for a current-mode multibit delta-sigma DAC," *IEEE J. of Solid-State Circuits*, vol. 34, pp. 331–338, March 1999.

[4] D. B. Ribner, "A comparison of modulator networks for high-order over-sampled $\Sigma\Delta$ analog-to-digital converters," *IEEE Trans. on Circuits and Systems*, vol. 38, pp. 145–159, Feb. 1991.

[5] K. C.-H. Chao, S. Nadeem, W. L. Lee, and C. G. Sodini, "A higher order topology for interpolative modulators for oversampling A/D converters," *IEEE Trans. on Circuits and Systems*, vol. 37, pp. 309–318, March 1990.

[6] R. Schreier, "An empirical study of high-order single-bit delta-sigma modulators," *IEEE Trans. on Circuits and Systems–II*, vol. 40, pp. 461–466, Aug. 1993.

[7] J. A. Cherry, and W. M. Snelgrove, "Excess loop delay in continuous-time delta-sigma modulators," *IEEE Trans. on Circuits and Systems–II*, vol. 46, pp. 376–389, April 1999.

[8] C.-L. Lin, "Dithering noise cancellation for a delta-sigma modulator," U.S. Patent 7,301,489, filed Dec. 6, 2005; issued Nov. 27, 2007.

[9] T. Ndjountche and R. Unbehauen, "Adaptive calibration techniques for time-interleaved ADCs," *Electronics Letters*, vol. 37, pp. 412–414, March 2001.

[10] M. Ortmanns, F. Gerfers, and Y. Manoli, "A continuous-time $\Sigma\Delta$ modulator with reduced sensitivity to clock jitter through SCR feedback," *IEEE Trans. on Circuits and Systems, Part I*, vol. 52, no. 5, pp. 875–884, May 2005.

[11] M. Anderson and L. Sundström, "Design and measurement of a CT $\Delta\Sigma$ ADC with switched-capacitor switched-resistor feedback," *IEEE J. of Solid-State Circuits*, vol. 44, no. 2, pp. 473–483, Feb. 2009.

[12] M. Sarhang-Nejad and G. C. Temes, "A high-resolution multibit sigma-delta ADC with digital correction and relaxed amplifier requirements," *IEEE J. of Solid-State Circuits*, vol. 28, No 6, pp. 648–660, June 1993.

[13] B. P. Brandt and B. A. Woley, "A 50-MHz multibit sigma-delta modulator for 12-b 2-MHz A/D conversion," *IEEE J. of Solid-State Circuits*, vol. 26, pp. 1746–1756, Dec. 1991.

[14] G. Mitteregger, C. Ebner, S. Mechnig, T. Blon, C. Holuigue, and E. Romani, "A 20-mW 640-MHz CMOS continuous-time $\Sigma\Delta$ ADC with 20-MHz signal bandwidth, 80-dB dynamic range and 12-bit ENOB," *IEEE J. Solid-State Circuits*, vol. 41, no. 12, pp. 2641–2649, Dec. 2006.

[15] M. Ranjbar and O. Oliaei, "A multibit dual-feedback CT $\Delta\Sigma$ modulator with lowpass signal transfer function," *IEEE Trans. Circuits Syst. I*, vol. 58, no. 9, pp. 2083–2095, Sep. 2011.

[16] E. B. Hogenauer, "An economical class of digital filters for decimation and interpolation," *IEEE Trans. on Acoustics. Speech, and Signal Processing*, vol. 29, pp. 155–162, April 1981.

[17] J. C. Candy, "Decimation for sigma delta modulation," *IEEE Trans. on Communications*, vol. 34, pp. 72–76, Jan. 1986.

[18] S. Chu and C. S. Burrus, "Multirate filter designs using comb filters," *IEEE Trans. on Circuits Systems*, vol. 31, pp. 913–924, Nov. 1984.

[19] B. P. Brandt and B. A. Wooley, "A low-power, area-efficient digital filter for decimation and interpolation," *IEEE J. of Solid-State Circuits*, vol. 29, pp. 679–687, June 1994.

[20] R. E. Crochiere and L. R. Rabiner, "Interpolation and decimation of digital signals—A tutorial review," *Proc. of the IEEE*, vol. 69, pp. 300–331, March 1981.

[21] S. K. Mitra, A. Mahalanobis, and T. Saramäki, "Generalized structural subband decomposition of FIR filters and its application in efficient FIR filter design and implementation," *IEEE Trans. on Circuits Systems–II*, vol. 40, pp. 363–374, June 1993.

[22] P. P. Vaidyanathan and T. Q. Nguyen, "A trick for the design of FIR halfband filters," *IEEE Trans. on Circuits and Systems*, vol. 34, pp. 297–300, March 1987.

[23] F. Mintzer, "On half-band, third-band and Nth band FIR filters and their design," *IEEE Trans. on Acoustics, Speech, Signal Processing*, vol. 30, pp. 734–738, Oct. 1982.

[24] G. W. Reitwiesner, "Binary arithmetics," *Advances in Computers*, vol. 1, pp. 231–308, 1960.

[25] A. Avizienis, "Signed-digit number representations for fast parallel arithmetic," *IRE Trans. on Electronic Computers*, vol. 10, pp. 389–400, Sept. 1961.

[26] A. G. Dempster and M. D. Macleod, "Use of minimum-adder multiplier blocks in FIR digital filters," *IEEE Trans. on Circuits and Systems–II*, vol. 42, pp. 569–577, Sept. 1995.

[27] R. I. Hartley, "Subexpression sharing in filters using canonic signed digit multipliers," *IEEE Trans. on Circuits and Systems–II*, vol. 43, pp. 677–688, Oct. 1996.

[28] M. Martínez-Peiró, E. I. Boemo, and L. Wanhammar, "Design of high-speed multiplierless filters using a nonrecursive signed common subexpression algorithm," *IEEE Trans. on Circuits and Systems–II*, vol. 49, pp. 196–203, March 2002.

[29] S. R. Norsworthy, "Optimal nonrecursive noise shaping filters for oversampling data converters, part 1: Theory," *Proc. of 1993 IEEE Int. Symp. on Circuits and Systems*, vol. 2, pp. 1353–1356, May 1993.

[30] Z.-J. (Alex) Mou, "Symmetry exploitation in digital interpolators/decimators," *IEEE Trans. on Signal Processing*, vol. 44, pp. 2611–2615, Oct. 1996.

[31] I. Løkken, A. Vinje, T. Sæther, and B. Hernes, "Quantizer nonoverload criteria in sigma-delta modulators," *IEEE Trans. on Circuits and Systems–II*, vol. 53, pp. 1383–1387, Dec. 2006.

[32] W. Gao and W. M. Snelgrove, "A 950-MHz IF second-order integrated LC bandpass delta-sigma modulator," *IEEE J. of Solid-State Circuits*, vol. 33, no. 5, pp. 723–732, May 1998.

[33] A. Yasuda, H. Tanimoto, and T. Iida, "A third-order Δ-Σ modulator using second-order noise-shaping dynamic element matching," *IEEE J. of Solid-State Circuits*, vol. 33, pp. 1879–1886, Dec. 1998.

[34] T.-H. Kuo, K.-D. Chen, and H.-R. Yeng, "A wideband CMOS sigma-delta modulator with incremental data weighted averaging," *IEEE J. of Solid-State Circuits*, vol. 37, pp. 11–17, Jan. 2002.

[35] H. T. Jensen and I. Galton, "A reduced-complexity mismatch shaping DAC for delta-sigma data converter," *Proc. of 1998 IEEE Int. Symp. on Circuits and Systems*, vol. I, pp. 504–507, May 31–June 3, 1998.

[36] E. Fogleman, I. Galton, W. Huff, and H. Jensen, "A 3.3-V single-poly CMOS audio ADC delta-sigma modulator with 98-dB peak SINAD and 105-dB peak SFDR," *IEEE J. of Solid-State Circuits*, vol. 35, pp. 297–307, March 2000.

[37] J. Welz, I. Galton, and E. Fogleman, "Simplified logic for first-order and second-order mismatch-shaping digital-to-analog converters," *IEEE Trans. on Circuits and Systems–II*, vol. 48, pp. 1014–1027, Nov. 2001.

[38] C. Jabbour, H. Fakhoury, V. T. Nguyen, and P. Loumeau, "Delay-reduction technique for DWA algorithms," *IEEE Trans. Circuits Systems–II*, Exp. Briefs, vol. 61, no. 10, pp. 733–737, Oct. 2014.

[39] M. Neitola and T. Rahkonen, "Generalized quadrature data weighted averaging," *IEEE Trans. Circuits Systems–II*, Exp. Briefs, vol. 62, no. 3, pp. 261–265, Mar. 2015.

[40] J. Doernberg, H.-S. Lee, and D. A. Hodges, "Full-speed testing of A/D converters," *IEEE J. of Solid-State Circuits*, vol. 19, pp. 820–827, Dec. 1984.

[41] A. Brandolini and A. Gandelli, "Testing methodologies for analog-to-digital converters," *IEEE Trans. on Instrumentation and Measurement*, vol. 41, pp. 595–603, Oct. 1992.

[42] Y.-C. Jenq, "Discrete-time method for signal-to-noise power ratio measurement," *IEEE Trans. on Instrumentation and Measurement*, vol. 45, pp. 431–434, April 1996.

[43] A. K. Lu, G. W. Roberts, and D. A. Johns, "A high-quality analog oscillator using oversampling D/A conversion techniques," *IEEE Trans. on Circuits and Systems–II*, vol. 41, pp. 437–444, July 1994.

[44] B. R. Veillette and G. W. Roberts, "High frequency sinusoidal generation using delta-sigma modulation," *Proc. of 1995 IEEE Int. Symp. on Circuits and Systems*, pp. 637–640, April 30–May 3, 1995.

[45] A. K. Lu and G. W. Roberts, "An analog multi-tone signal generator for built-in self-test applications," *Proc. of the IEEE International Test Conference*, pp. 650–659, Washington, Oct. 1994.

[46] H.-W. Huang, *MC68HC12 An Introduction: Software and Hardware Interfacing*, Clifton Park, NY: Thompson Delmar Learning, 2003.

[47] Jim Ryan, "Interfacing DSPs with high performance analog converters," *DSP Workshops, ICSPAT*, Orlando, FL, 1999.

[48] J.-L. Huang, C.-K. Ong, and K.-T. Cheng, "A BIST scheme for on-chip ADC and DAC testing," *Proc. of the Design, Automation and Test in Europe (DATE) Conference*, pp. 216–220, Paris, 2000.

[49] Y.-C. Wen and K.-J. Lee, "An on chip ADC test structure," *Proc. of the Design, Automation and Test in Europe (DATE) Conference*, pp. 221–225, Paris, 2000.

[50] S. Rabii and B. A. Wooley, "A 1.8-V digital-audio sigma-delta modulator in 0.8 μm CMOS," *IEEE J. of Solid-State Circuits*, vol. 32, pp. 783–796, June 1997.

[51] A. Yasuda and H. Tanimoto, "Noise shaping dynamic element matching method using tree structure," *Electronics Letters*, vol. 33, pp. 130–131, Jan. 1997.

6

Circuits for Signal Generation and Synchronization

CONTENTS

A circuit for the clock signal generation or recovery is often required to achieve accurate data transfer between the different building blocks (see Figure 6.1) of very large-scale ICs operating at high speed. In the case of transmission systems, as shown in Figure 6.2, the multiplexer converts the input data into a serial stream of non-return-to-zero data, which then drives a high-speed buffer. At the receiver, the signal level is determined by an amplifier and the clock signal is recovered from the transmitted data and used to control the demultiplexer. The resulting data synchronization determines the accuracy of the information regeneration. The rising edges of the clock signal, whose frequency is set equal to the data rate, should coincide with the midpoint of each data bit, such that the sampling occurs farthest from the preceding and following transitions, yielding a maximum tolerance margin for the jitter and other timing uncertainties.

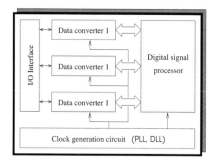

FIGURE 6.1
A typical integrated-circuit floorplan.

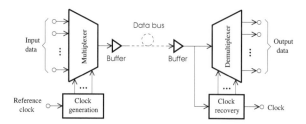

FIGURE 6.2
Transmission system.

Precision timing circuits are generally based on a phase-locked loop (PLL) or delay-locked loop (DLL). They should typically feature a low sensitivity to process and temperature variations, and generate a clock signal with very low skew and jitter. However, achieving these requirements can be difficult due to a number of design trade-offs to be made between the circuit characteristics.

Furthermore, a high level of integration will only be achieved if the objectives of reducing the area and power consumption are met.

6.1 Generation of clock signals with nonoverlapping phases

To keep negligible the charge leakage, the duration of the sampling phase and hold phase required in the operation of switched-capacitor circuits is controlled by nonoverlapping clock signals.

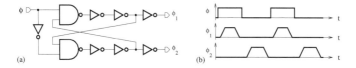

FIGURE 6.3
(a) Circuit diagram of a two-phase nonoverlapping clock signal generator; (b) clock signals.

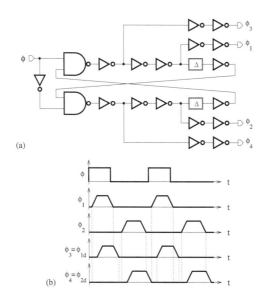

FIGURE 6.4
(a) Circuit diagram of a four-phase nonoverlapping clock signal generator; (b) plot of output signals.

The circuit diagram of a two-phase nonoverlapping clock signal generator is shown in Figure 6.3(a). It consists of NAND gates and inverters. Here, the generator is designed to provide two outputs, which are not allowed to be in the high state during the same time. The nonoverlapping time can be increased by augmenting the number of inverters included between each output and the NAND gate. Each output node is buffered with an inverter sized to drive the on-chip clock bus. Figure 6.3(b) shows the plot of input and output signals. By applying a 50% duty-cycle reference clock signal at the input, the rising edge occurs at one output after the falling edge is produced at the other output. The main advantage of the aforementioned signal generator is its simplicity. However, the propagation delay introduced by the input inverter and changes in clock waveforms due to variations in load conditions can become critical in high-speed applications.

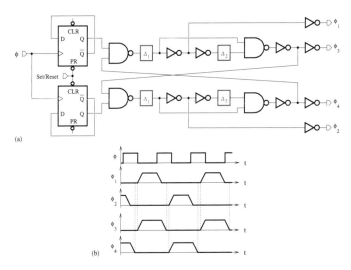

FIGURE 6.5
(a) Circuit diagram of a four-phase nonoverlapping clock signal generator with equal pulse-width complementary signals; (b) plot of output signals.

Charge injection errors can be minimized using a four-phase clock signal generator [4]. In the case of the structure shown in Figure 6.4(a), the delay Δ blocks, which can be implemented by an even number of inverters connected in series, are purposely introduced to increase the nonoverlap time between the clock phases ϕ_1 and ϕ_2. The plot of the input and output signals is depicted in Figure 6.4(b), where ϕ_3 and ϕ_4 represent the delayed versions of ϕ_1 and ϕ_2, respectively. The propagation delay of the NAND gate and inverters, which determines the nonoverlap time, is a critical design parameter. If it is chosen with a very small value, clock skew may affect the accuracy of the clock timing. Conversely, if it is sized to be excessively large, the effective time period of the clock signals will be considerably reduced. Under these conditions, the clock speed may be reduced. Furthermore, the inverter required at the input of the clock generator delivering nonoverlapping clock signals with the same periodicity as the reference clock introduces a difference in the duty cycle of complementary clock signals.

A solution to improve the performance of the clock generator can rely on forcing the rising edges of a clock phase and its delayed version to occur simultaneously. The circuit diagram of the four-phase nonoverlapping clock signal generator [5] with equal pulse-width complementary signals is shown in Figure 6.5(a). To make the pulse widths of complementary clock signals equal, the input reference clock is applied to two divide-by-2 circuits based on D flip-flops initialized to the high state and low state, respectively. The resulting signals are then applied to a cross-coupled section including a NAND gate, a delay block Δ_1, and two signal-edge synchronization structures, each of which is composed of a series connection of two inverters, a delay block Δ_2, a NAND gate, and an inverter. The synchronization is achieved within a specified maximum amount of time equal to the propagation delay introduced by the series of two inverters and the delay block Δ_2. Each of the delay blocks, Δ_1 and Δ_2, should be realized by an even number of inverters connected in series. Figure 6.4(b) shows the plot of the input and output signals. The rising edges of ϕ_1 and ϕ_2 are aligned to the ones of ϕ_3 and ϕ_4, respectively, while the falling edges of ϕ_3 and ϕ_4 occur after the ones of ϕ_1 and ϕ_2, respectively. In this approach, the clock generator has the advantage of not introducing a difference of pulse width between both complementary clock phases and is then suitable for the control of double-sampled or time-interleaved switched-capacitor circuits.

6.2 Phase-locked loop

The block diagram of the PLL is shown in Figure 6.6. The PLL is a feedback

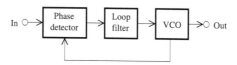

FIGURE 6.6
Phase-locked loop.

system, that operates by generating an oscillation signal, whose frequency has to match the one of the input signal. It consists of a phase detector (PD), loop filter, and voltage-controlled oscillator (VCO). The PD output waveform is proportional to the phase difference between the input and VCO output signals. It is then smoothed by the loop filter and the resulting dc signal is applied to the VCO control node. The VCO can then be driven to minimize the phase difference.

During the initial transient, the PLL operates in the nonlinear region as the VCO tries to find the correct frequency. The PLL linear model is valid when the locked condition is obtained. In this case, the phases of the input and VCO output signals are relatively equal.

Note that the choice of PD architecture has an impact on the overall PLL performance, such as the lock-in range and static phase error. The linear range of a PD detector, which can consist of a simple XOR gate, spreads from $-\pi$ to π. On the other hand, an alternative structure known as a phase and frequency detector (PFD), which generally exhibits the advantage of indicating the lead or lag relation between input waveforms, can handle differences between the clock signals in the range of -2π to 2π.

6.2.1　PLL linear model

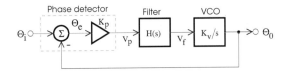

FIGURE 6.7
PLL linear model.

Even if the PLL exhibits a nonlinear behavior, its design often starts with the linear model. Figure 6.7 shows the linear model of the PLL. Let Θ_i and Θ_0 be the phase angles associated with the input and output signals, respectively; the output of the phase detector has the form

$$V_p(s) = K_p(\Theta_i(s) - \Theta_0(s)) \tag{6.1}$$

where K_p is the PD conversion gain in units of volts per radian. The transfer function of the filter is denoted by $H(s)$ and that of the VCO is K_v/s, where K_v is the VCO gain in units of radians per volt·second, because the frequency is related to the time derivative of the phase. The phase and error transfer functions can be, respectively, computed as

$$T(s) = \frac{\Theta_0(s)}{\Theta_i(s)} = \frac{K_v K_p H(s)}{s + K_v K_p H(s)} \tag{6.2}$$

and

$$T_e(s) = \frac{\Theta_e(s)}{\Theta_i(s)} = \frac{s}{s + K_v K_p H(s)} \tag{6.3}$$

The frequency and transient responses of the loop appear to be affected by the choice of filter characteristic. Due to the contribution of the VCO first-order transfer function, the order of the PLL is equal to that of the filter plus 1. In the special case of a second-order loop, the PLL transfer function contains two poles at the origin due to the VCO and the loop filter implemented by an integrator. To counteract the effect of these poles, the loop transfer function must include a stabilizing zero, which is implemented by connecting a resistor in series with the integrating capacitor.

A simple second-order PLL, as illustrated in Figure 6.8, consists of a phase detector (PD), a loop filter, and a voltage-controlled oscillator (VCO).

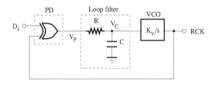

FIGURE 6.8
Second-order PLL with XOR PD.

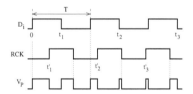

FIGURE 6.9
XOR PD waveforms.

An XOR logic gate is used to implement the PD. The XOR PD output takes the high logic level when the inputs are set to different logic levels, otherwise, it is at the low logic level. Figure 6.9 shows a representation of XOR PD waveforms.

The PD characteristic can be obtained by plotting the average output voltage, $\bar{V}_P$, as a function of the phase difference, θ. For a given integer k, the phase difference between the PD inputs can be defined as,

$$\theta = \pm 2\pi \frac{t'_k - t_k}{T} \tag{6.4}$$

where T is the signal period.

The XOR PD with a single-ended output has a sawtooth characteristic, as shown in Figure 6.10(a), that is periodic and non-monotonic, while the XOR PD with differential outputs possess

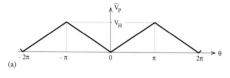

(a)

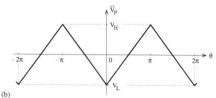

(b)

FIGURE 6.10
XOR PD characteristics: (a) single-ended output, (b) differential outputs.

a triangular characteristic, as illustrated in Figure 6.10(b). In general, the PD gain is $K_p = (V_H - V_L)/\pi$, where $V_L \leq 0$. The PD characteristic is linear over a range of π radians and is sensitive to the duty cycle of the input signals.

Let θ_0 and θ_i be the output and input phase angles, respectively. The phase transfer function is of the form,

$$T(s) = \frac{\theta_0(s)}{\theta_i(s)} = \frac{K_v K_p H(s)}{s + K_v K_p H(s)} \tag{6.5}$$

where the filter transfer function is given by

$$H(s) = \frac{V_c(s)}{V_p(s)} = \frac{1}{1 + s/\omega_c} \tag{6.6}$$

and $\omega_c = 1/RC$ represents the -3 dB cutoff angular frequency. Hence,

$$T(s) = \frac{\theta_0(s)}{\theta_i(s)} = \frac{\omega_n^2}{s^2 + 2\xi\omega_n s + \omega_n^2} \tag{6.7}$$

where

$$\omega_n = \sqrt{K_p K_v \omega_c} \tag{6.8}$$

and

$$\xi = \frac{1}{2}\sqrt{\frac{\omega_c}{K_p K_v}} = \frac{1}{2}\frac{\omega_c}{\omega_n} \tag{6.9}$$

with ω_n and ξ being the natural frequency and damping ratio, respectively. The -3 dB bandwidth of the PLL can be obtained as

$$BW = \omega_n(1 - 2\xi^2 + \sqrt{2 - 4\xi^2 + 4\xi^4})^{1/2} \tag{6.10}$$

The phase error transfer function is given by

$$T_e(s) = \frac{\theta_e(s)}{\theta_i(s)} = 1 - T(s) \tag{6.11}$$

$$= \frac{s(s + 2\xi\omega_n)}{s^2 + 2\xi\omega_n s + \omega_n^2} = \frac{s(s + \omega_c)}{s^2 + 2\xi\omega_n s + \omega_n^2} \tag{6.12}$$

In the time domain, the PLL phase error in response to a unit phase step at the input can be obtained by taking the inverse Laplace transform of $(1/s)T_e(s)$. It depends on the damping ratio ξ and can exhibit damped oscillations or overshoots before reaching the steady state. The settling time, t_s, is minimum when $\xi = \sqrt{2}/2$, and the steady-state value of the phase error is then reached to within a tolerance of 10% for $\omega_n t_s \simeq 2.5$.

Given a settling time t_s of 10 μs, $V_H = V_{DD} = 2.5$ V, and choosing $\xi = \sqrt{2}/2$, the loop characteristics can be obtained as follows:

$$\omega_n \simeq 2.5/t_s = 2.5 \times 10^5 \text{ rad} \tag{6.13}$$
$$\omega_c = 2\xi\omega_n = 3.5 \times 10^5 \text{ rad} \tag{6.14}$$

and

$$K_v = \omega_n^2/(\omega_c K_p) = \omega_n/(\sqrt{2}K_p) = 2.2 \times 10^5 \text{ Hz/V} \tag{6.15}$$

where $K_p = V_{DD}/\pi$ in V/rad. By selecting $C = 1$ nF, we can obtain $R = 1/(\omega_c C) = 2.8$ kΩ.

The hold-in range, that is the range over which the PLL will remain in the phase-locked state, but not necessarily acquire that state, is given by $\triangle\omega_H = \pm K_p K_v \pi/2$ for the PD with a triangular characteristic and $\triangle\omega_H = \pm K_p K_v \pi$ for the PD with a sawtooth characteristic. Thus, the VCO can be tuned by $\triangle\omega_H$ without loss of lock.

A PLL with XOR PD can be limited by the fact that it can lock to fractional harmonics of the input signal frequency, f_i, or Mf_i/N, with M and N being odd integers.

6.2.2 Charge-pump PLL

To achieve an extended tracking range, the PLL can be implemented as shown in Figure 6.11. This structure includes a PFD, a charge-pump circuit, a loop filter, and a VCO. The PFD has the advantage of also being sensitive to

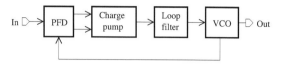

FIGURE 6.11
Charge-pump phase-locked loop.

frequency error. It acts as an extended-range phase detector and generates a signal, which is indicative of the difference between its input signals. The purpose of the charge-pump circuit is to convert the logic states of the PFD into analog signals, which are appropriate for the VCO control.

6.3   Charge-pump PLL building blocks

A charge-pump PLL system generally consists of a phase and frequency detector (PFD) or phase detector (PD), a charge-pump circuit, a lowpass filter, and a VCO. The PFD or PD compares the phase of the input data and that of the recovered clock signal generated by the VCO, and produces an error signal, typically consisting of Up and Dn signals used to drive a charge-pump circuit differentially. The output signal delivered by the charge-pump circuit is dependent on the phase difference between the data and clock signals. The resulting average signal, which is provided by the loop filter operating as an integrator, is applied to the VCO control input in order to appropriately change the frequency of the clock signal.

6.3.1   Phase and frequency detector

The ideal characteristic of a three-state PFD, as shown in Figure 6.12, is linear for the range of input phase differences from -2π to 2π. The gain of the PFD can be defined as $K_P = (V_H - V_L)/2\pi$, where V_H and V_L denote the highest and lowest output levels, respectively.

When the PFD is followed by a charge-pump circuit, the threshold levels V_H and V_L are used to control the value of the charge-pump current, I_P, and the combination of the PFD and charge-pump circuit then exhibits a gain of the form, $K_P = I_P/2\pi$. For the absolute value of the phase difference not exceeding 2π, the PFD is said to be in the lock state.

In the case of a comparison of two input signals with the same period and amplitude, the PFD operation principle can be illustrated by the timing diagrams shown in Figure 6.13.

If both input signals are of the same frequency and in phase, both outputs will be set at zero. If CK leads RCK, pulses will be generated at the output

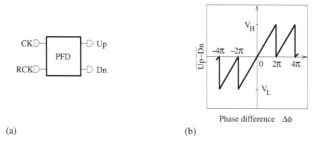

(a)   (b)

FIGURE 6.12

(a) Phase and frequency detector symbol; (b) ideal characteristic of a three-state PFD.

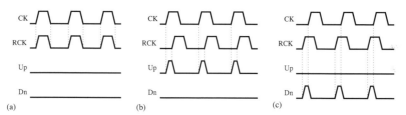

FIGURE 6.13

(a)–(c) Timing diagrams illustrating the three states of a PFD.

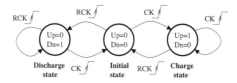

FIGURE 6.14

State graph of the three-state PFD.

Up while the Dn signal will remain at zero. On the other hand, if CK lags RCK, the Dn signal will be pulsed while the Up signal level will remain low. By applying input signals with different frequencies to the PFD, one output signal is more often set to the high level than the other. As a result, the average value of the output will be positive or negative, depending on the sign of the frequency and phase difference.

The PFD operation can be described using finite-state machine, as shown in Figure 6.14, where CK and RCK represent the two periodic input signals. There are only three allowed combinations for the outputs Up and Dn. From the initial state, Up = 0 and Dn = 0, a rising edge of CK causes a transition to the charge state, Up = 1 and Dn = 0, while a rising edge of RCK causes a transition to the discharge state, Up = 0 and Dn = 1. If the PFD is in the charge state, a rising edge of CK will cause no state change, but a rising edge of RCK will cause a change to the initial state. From the discharge state, the PFD can change state only in response to a rising edge of CK. The charge, initial, and discharge states can be, respectively, associated with the three different values, $I_P, 0$, and $-I_P$, of the current to be generated by a charge-pump circuit.

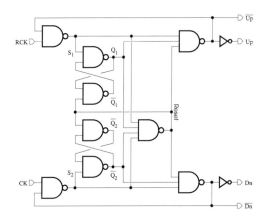

FIGURE 6.15
Gate implementation of the phase and frequency detector.

At the gate level, a PFD can be realized as shown in Figure 6.15 [6]. Each of the logic states Q_1 and Q_2 are generated by an $\overline{S}\,\overline{R}$ latch, including a pair of cross-coupled two-input NAND gates. From the initial state where both outputs are in the low state, the PFD can move to either the state Up (high at the output Up and low at the output Dn) or state Dn (high at the output Dn and low at the output Up) after the detection of the rising edge of one of the input signals. It remains in this last state until the second input goes high, causing the generation of a reset signal, which enables the return to the initial state.

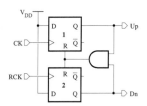

FIGURE 6.16
Circuit diagram of a D flip-flop-based PFD.

An implementation of the PFD based on two resettable D flip-flops, which are clocked by the input signals, is depicted in Figure 6.16. The input terminal D is connected to the positive supply voltage, V_{DD}, and the CK terminals serve, respectively, as inputs for the signals to be compared. The Dn and Up output pulses are applied to the AND gate, which generates the flip-flop reset signal. When both output signals are high, the AND gate output is set to the high level and this in turn resets both D flip-flops. The reset time is about several gate delays and determines the circuit speed.

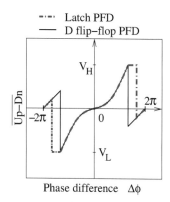

FIGURE 6.17
Characteristics of PFD circuits based, respectively, on the $\overline{R}\,\overline{S}$ latch and D flip-flop.

In practice, due to the nonzero gate delays in the reset path and violations of the setup and hold time, the PFD characteristic, as shown in Figure 6.17 [7], may exhibit a reduced linear range and a dead zone, where small changes in the input signals are not detected. The dead zone effect can be significantly reduced by introducing an extra delay in the reset path to increase the over-lapping time between the Up and Dn output signals. However, the efficiency of this approach appears to be limited by the fact that the delay size is generally dependent on variations in the IC process. For instance, the power consumption will increase in the lock state if the predicted value of the delay time is somewhat high. For a given PFD architecture, the dead zone, the operation frequency range, the power dissipation, and the phase noise are dependent on the design technique (standard logic gates, true single-phase clock logic circuits, differential logic circuits). To extend the operating frequency range, it may be necessary to use a PFD with more than three logic states.

6.3.2 Phase detector

A PD circuit, which can exhibit either a linear or binary transfer character-istic [3], is required for the generation of the phase error signal. Linear PDs, such as the Hogge phase detector, deliver a continuous error signal that is responsible for a linear behavior in the tracking characteristic of the acquisition loop, while binary PDs, also known as Alexander (or bang-bang) phase detectors, generate a quantized phase error signal that contributes to a non-linear tracking characteristic. The input dynamic range of a phase detector is smaller than that of a PFD.

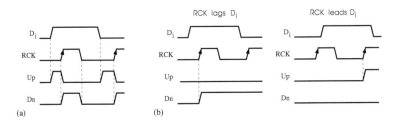

FIGURE 6.18
Timing diagram of a (a) linear and (b) binary phase detectors.

Figures 6.18(a) and (b) show the timing diagram of a linear and binary phase detectors, respectively. A linear PD generates an Up pulse whose width is proportional to the phase difference between the input data D_i and the reference clock RCK and a Dn pulse that can be used to eliminate the dependency on data pattern. A binary PD delivers Up and Dn pulses with two possible levels that can be combined into a tri-state binary sequence indicating whether the input data D_i lags or leads the reference clock RCK.

6.3.2.1 Linear phase detector

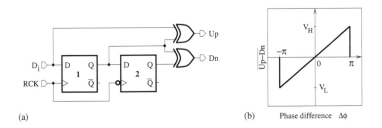

FIGURE 6.19
(a) Circuit diagram and (b) characteristic of a linear (Hogge) PD.

The circuit diagram of a Hogge PD [8] and its transfer characteristic are depicted in Figure 6.19. The Hogge PD consists of two flip-flops, which are, respectively, enabled at the rising and falling edge of the clock signal, and two XOR gates. The input data, D_i, and the output signal of the first flip-flop are processed by the first XOR gate to generate an Up phase error signal, while the output signal of the first flip-flop and the output signal of the second flip-flop are applied to the second XOR gate to produce a Dn phase error signal.

For each data transition, a pulse, whose width varies with the phase difference between the reference clock signal, RCK, and the input data, D_i, is first generated at the Up terminal, and a pulse with a fixed width of half of the clock signal period, or $T_{CK}/2$, is then produced at the Dn terminal. The ideal sampling points of the data correspond to instants where the rising edge of the clock occurs near the center of the data sample, thus yielding the maximum noise margin. The width of the Up pulse will be $T_{CK}/2$ if the rising edge of the signal RCK is nearly aligned with the data center; otherwise, the Up pulse will become smaller or larger than the Dn pulse for early or late clock signals. The average value of the difference between the Up and Dn signals is a linear function of the phase error.

In practical implementations of the Hogge PD, two of the three signals applied to the XOR gates are affected by the RCK-to-Q delay of flip-flops, resulting in an increased width of the Up pulses. These delay variations, which are particularly critical at high frequencies, can cause an increase in the static phase error, and a reduction of the clock phase margin and jitter tolerance. A design solution for the equalization of the RCK-to-Q delay can consist of placing extra delay elements on the path of the data signal to the XOR gate input.

6.3.2.2   Binary phase detector

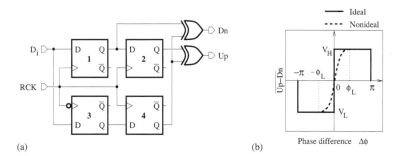

(a)   (b)

FIGURE 6.20
(a) Circuit diagram and (b) characteristic of a binary PD.

The binary PD, which is also known as the Alexander or bang-bang PD [9], is shown in Figure 6.20 and is generally used in high-speed clock and data recovery circuits. It consists of four flip-flops and a pair of XOR gates. The input signal is received at the D terminals of the first and third flip-flops. The first, second, and fourth flip-flops are enabled by the rising edge of the clock pulse, while the sampling instants of the third flip-flop correspond to the falling edges of the clock pulse. In addition to the retimed data that can be obtained at the output of the second or fourth flip-flop, the binary PD generates the Up and Dn pulses indicating whether the clock signal is leading or lagging the input data signal.

By sampling two adjacent input data bits and the in-between data transition, the binary PD can deliver the early, late, or no transition information. If the logical states of the first data bit and data transition are identical but differ from that of the second data bit, the clock is early; and if the logical states of the signal transition and the second data bit are equal but differ from that of the first data bit, the clock is late. On the other hand, if all logical states are similar or if the logical states of the first and second data bits are identical but differ from that of the data transition, there is no data transition and a zero dc output is generated.

With Q_1, Q_2, and Q_4 denoting the output of the first, second, and fourth flip-flops, it can be deduced that

- CK is early (E),   if $Q_2 = Q_4 \neq Q_1$
- CK is late (L),   if $Q_2 \neq Q_4 = Q_1$
- There is no transition (X),   if $Q_2 = Q_4 = Q_1$ or $Q_2 = Q_1 \neq Q_4$

While the binary PD can effectively align the clock with the data signal, the average value of its Up and Dn output signals is not proportional to the magnitude of the phase difference. Instead, the output characteristics is a discrete function, which can assume only one of two logical states depending on the result of each phase comparison. By detecting the phase information only at the zero-crossings, the performance of the binary PD can be sensitive to the data transition density.

In practice, due to non-ideal effects such as jitter and metastability, the characteristic of the phase detector, that is ideally binary, may exhibit a finite slope for a small range, $|\triangle\phi| < \phi_L$, of input phase around zero (see Figure 6.20).

6.3.2.3 Half-rate phase detector

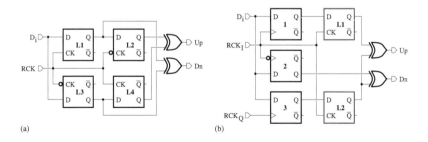

(a)   (b)

FIGURE 6.21
Circuit diagram of half-rate PDs with the (a) linear and (b) binary characteristics.

The PDs described previously are assumed to operate at the full rate or to use a clock frequency equal to the baud rate of the input data. Generally, half-rate PDs are employed to enable operation at a higher speed with a clock frequency equal to half the input data rate, as they can relax the requirements set for the acquisition loop components. This is in contrast to full-rate architectures, which require more system components to work at higher frequencies and then quickly reach the operating limit of the IC manufacturing process.

- Half-rate linear PD

The circuit diagram of a half-rate linear PD is depicted in Figure 6.21(a). This PD circuit, which is the half-rate version of Hogge's detector, consists of four latches and two XOR gates [10].

The outputs of the first and third latches follow the data at the D inputs whenever the clock signal is in the high and low state, respectively. They are processed by the XOR gate to generate the signal Dn, whose width is dependent on the phase difference between the half-rate clock and the input data. The retimed data at half rate can be obtained at the outputs of the second and fourth latches, which are, respectively, enabled on the low and high levels of the clock signal. A full-rate output can be obtained by interleaving the two retimed data streams using a multiplexer controlled by the half-rate clock signal.

The two latches included in each PD path operate as a master-slave flip-flop. Hence, the outputs of the second and fourth latches change on both edges of the clock signal, and the Up signal provided by the XOR gate is a pulse with a constant width of half the clock period, or $T_{CK}/2$, in the case where a data transition is detected. In the locked state, the width of the Dn becomes $T_{CK}/4$, while that of the Up pulse is $T_{CK}/2$. To equalize the effect of both PD outputs, the Up signal can be scaled down by a factor of 2. This is realized in practice by sizing the charge pump circuit such that the current source controlled by the Up signal is two times smaller than the one steered by the Dn signal.

- Half-rate binary PD

The circuit diagram of a half-rate binary PD is shown in Figure 6.21(b) [11]. The incoming data are applied to flip-flops enabled by the in-phase clock signal, the complement of the in-phase clock signal, and the quadrature clock signal, respectively. The output signals of the first and second flip-flops are then compared to that of the third flip-flop using XOR gates to generate the Up and Dn signals. Note that the latches L1 and L2 are introduced to align in time the flip-flop output signals. If a data transition occurs between the rising edge of the in-phase and quadrature clock signals, a pulse will be generated at the Up or Dn terminal indicating whether the clock signal is leading or lagging the input data; otherwise, both the Up and Dn signals remain at the low level.

The aforementioned PD architecture samples the input data at the rising and falling edges of the in-phase clock, while the quadrature clock is used

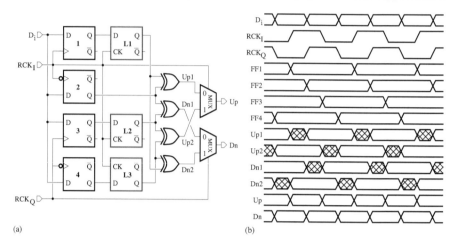

(a)   (b)

FIGURE 6.22
(a) Circuit diagram of a half-rate binary interleaved PD; (b) timing diagram.

to track the data transition. Its jitter tolerance is degraded as the operating frequency is increased. A solution can consist of increasing the number of sampling instants [12]. This approach is exploited in the half-rate PD implementation shown in Figure 6.22(a) [13,14]. The symmetric architecture of this PD circuit helps ensure the matching of delays in the signal paths.

The input data are sampled at $0°$, $90°$, $180°$, and $270°$ of the clock phases using four flip-flops in parallel. The synchronization of data samples, which is particularly useful at high frequencies, is performed using three latches controlled by the in-phase clock signal. The decoding logic for the Up and Dn signals is based on XOR gates and 2-to-1 multiplexers. Each XOR gate produces either an Up1/Up2 or Dn1/Dn2 signal by comparing data samples of two adjacent signal paths. The two multiplexers with the select line controlled by the in-phase and quadrature clock signals, respectively, keep the transfer of the input signals to the Up and Dn terminals within a phase angle of $180°$. The timing diagram of the half-rate binary interleaved PD is depicted in Figure 6.22(b), where the shaded sections represent the invalid time intervals of the Up1/Up2 and Dn1/Dn2 waveforms.

In the locked state, the quadrature clock edges are aligned with the data transitions and the retimed data can be obtained at the output of the first or third flip-flop. The multiplexer selection signals can be delayed by the total amount of signal propagation delay up to the multiplexer input and the multiplexer setup time to improve the timing margin.

6.3.3 Charge-pump circuit

In practice, the PFD or PD incorporated in PLLs does not provide sufficient drive currents to achieve an adequate loop bandwidth. A charge-pump circuit is therefore required to convert the logic pulses generated by the PFD or PD into current signals that are used to drive the loop filter providing the VCO control voltage. The associated current amplification contributes to an increase in the loop bandwidth. The transfer characteristic of the charge pump is generally determined in accordance with the operation principle of the PFD or PD.

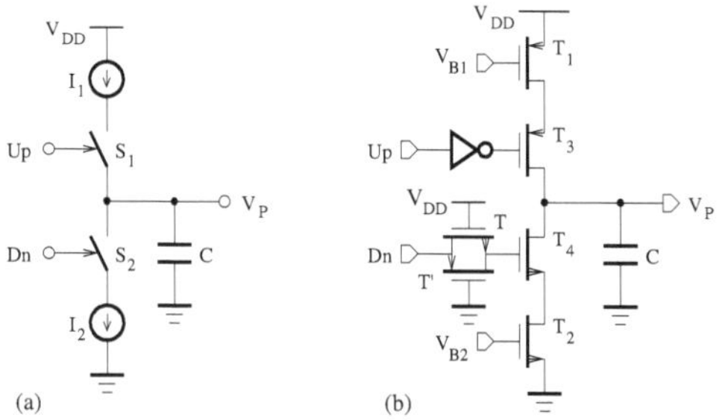

FIGURE 6.23
(a) Conceptual diagram and (b) implementation of a charge-pump circuit.

The conceptual diagram of a charge-pump circuit is shown in Figure 6.23(a). The current sources I_1 and I_2 should be identical. Their connections to the output node are controlled by the Up and Dn signals. The current obtained at the output node will be ideally zero, if the Up and Dn signals are identical. Otherwise, the average output current over a cycle is $I_p \triangle\phi/2\pi$, where I_p is the maximum output current and $\triangle\phi$ is the phase difference between the input signals.

A charge-pump circuit can be implemented as shown in Figure 6.23(b). The complementary transistor pair, $T - T'$, is used to equalize the delay of the inverter. Due to the difference in the electron mobility of nMOS and pMOS transistors and the asymmetry in rise and fall times of the Up and Dn signals, a dynamic mismatch between the output currents corresponding, respectively, to the Up and Dn control voltages can be observed during the operation of this charge-pump circuit. The design of the charge-pump circuit can be improved by minimizing the charge injection errors due to switches and the charge sharing from parasitic capacitances. These charge errors are known to result in a phase offset when a PLL is in the lock state.

For the charge-pump circuit of Figure 6.24(a) [15], two transistors, T_5 and T_6, are used to remove the residual charge from the nodes, x and y, during the inactive phases of the Up and Dn signals, thereby mitigating the charge-sharing problem caused by parasitic capacitances at these nodes. The charge-pump current flowing through the transistors, T_3 and T_4, is defined

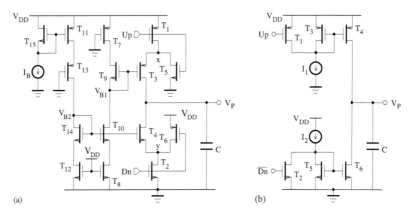

(a)   (b)

FIGURE 6.24
Circuit diagrams of charge-pump structures with reduced charge sharing due
to the use of (a) compensation switches and (b) nonoverlapping switching
pulses.

by the voltages, V_{B1} and V_{B2}, set by the biasing circuit section, $T_7 - T_{15}$. A
drawback to this charge-pump circuit is the limited dynamic range available
at the output node, due to the overdrive voltages required to maintain the
transistors, T_3 and T_4, in the saturation region.

An alternative charge-pump structure is shown in Figure 6.24(b) [16, 17].
The current I_1 applied to the current mirror $T_3 - T_4$ will be directed to the
output if the signal Up is high, while the input current of the current mirror
$T_5 - T_6$ will be transferred to the output if the signal Dn is low. Due to the
fact that this charge-pump circuit operates without switching pulse overlap,
the charge redistribution associated with overlap capacitances of switches and
parasitic capacitances of current sources is reduced.

For the aforementioned charge-pump circuits with switched currents I_1
and I_2, the switches controlled by the Up and Dn pulses can be assumed to
be, respectively, closed for the time periods T_{Up} and T_{Dn}. The output charge
can then be expressed as

$$\triangle Q = I_1 T_{Up} - I_2 T_{Dn} \qquad (6.16)$$

Ideally, no charge should be transferred to the output in the lock state because
$I_1 = I_2$ and $T_{Up} = T_{Dn}$. In practice, the mismatch between the current levels
and the difference in arrival times of the Up and Dn pulses can introduce
a steady-state phase offset and increase frequency spurs in the acquisition
loop. Furthermore, the transient glitch caused by parasitic capacitors can also
contribute to the increase of the spur and jitter levels.

In the case of the charge-pump circuit structure shown in Figure 6.25, the
effect of the charge sharing between the parasitic and output capacitors is attenuated using an operational amplifier configured as a unity-gain buffer [18].

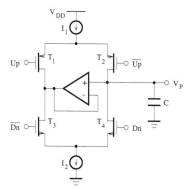

FIGURE 6.25
Charge-pump circuit with a reduced charge-sharing error.

The charge pump is implemented as switched current source and sink I_1 and I_2, which can be connected to the capacitor C defining the output voltage, V_P. The upper switches consist of p-channel transistors, while the lower switches are realized with n-channel transistors. For Up and Dn pulses associated with a phase difference, the capacitor C should ideally be charged by either the current source or the current sink. In the lock state, no current should be supplied to the capacitor C by the charge-pump circuit so that the output voltage will remain unchanged.

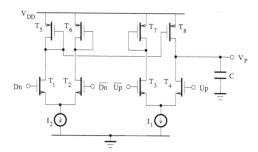

FIGURE 6.26
Charge-pump circuit with an improved current matching.

The charge-sharing errors that can cause mismatch between the current source and sink are reduced by maintaining the output node of all switches connected to the output voltage. In this way, the nonideal voltage variation applied to parasitic capacitors is reduced to the small level of the amplifier offset voltage. However, the accuracy of the nonideal charge cancelation, especially in the lock state, can be limited by the inherent mismatching between the n-channel and p-channel transistors.

The inherent mismatch between the nMOS and pMOS transistors can be eliminated using the charge-pump circuit of Figure 6.26 [19]. Here, only nMOS switches are required. The currents I_1 and I_2 are transferred to the output when either the Up signal or the Dn signal assumes the high level. However, the overall circuit performance can be limited by the dynamic range of the current mirror $T_5 - T_6$. In general, it should also be noted that performance improvement is achieved at the price of a power consumption increase.

6.3.4 Loop filter

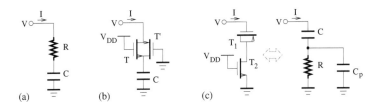

FIGURE 6.27
(a) Circuit diagram and (b) CMOS implementation of a first loop filter; (c) alternative MOS implementation and its equivalent circuit.

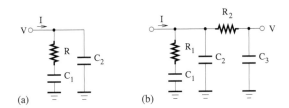

FIGURE 6.28
Circuit diagrams of (a) second-order and (b) third-order loop filters.

The loop filter plays an important role in the determination of PLL characteristics. Generally, it is designed based on the trade-off to be achieved between the lock time, the phase noise, and the residual level of reference spurs. A large loop filter bandwidth helps reduce the lock time, but is also associated with a low attenuation of the phase noise and reference spurs, whereas a narrow bandwidth leads to an improved suppression of the phase noise and reference spurs while increasing the lock time. In practice, the order of the loop filter is increased to satisfy the high attenuation requirement of all unwanted frequencies beyond the cutoff frequency. However, the calculation of the loop filter components then becomes cumbersome, and computer methods must be employed.

The circuit diagram of a first-order filter is shown in Figure 6.27(a). It consists of a resistor connected in series with a capacitor, and can be characterized by a transfer function of the form

$$H(s) = \frac{V(s)}{I(s)} = \frac{R\left(s + \frac{1}{RC}\right)}{s} = \frac{1 + \frac{s}{\omega_z}}{sC} \qquad (6.17)$$

where $\omega_z = 1/RC$. The closed loop of a PLL using a first-order filter is characterized by a second-order transfer function due to the extra single pole introduced by the VCO. Because the damping factor of the closed-loop transfer function is inversely proportional to the zero at ω_z, the loop bandwidth is enlarged by decreasing the zero frequency. In the s-domain, the stability of a second-order linear system is not affected by the loop gain.

Due to the fact that the loop filter generally requires a resistor with a high resistance, the resistor R can be implemented using a CMOS transistor pair, as illustrated in Figure 6.27(b). However, this simple approach may be limited at low supply voltages, as the effective resistance of the CMOS transistor pair becomes a function of the input node voltage (or VCO control voltage). An alternative implementation of a first-order filter is shown in Figure 6.27(c). To reduce the resistance dependence on the voltage, the resistor R, which is realized by an nMOS transistor, is permutated with the pMOS transistor-based capacitor. But, the equivalent model of the filter should take into account the well-substrate parasitic capacitance, C_p, associated here with the MOS capacitor structure.

For a second-order filter, as depicted in Figure 6.28(a), the transfer function is given by

$$H(s) = \frac{V(s)}{I(s)} = \frac{s + \frac{1}{RC_1}}{sC_2\left(s + \frac{C_1 + C_2}{RC_1C_2}\right)} = \frac{1 + \frac{s}{\omega_z}}{sC_T\left(1 + \frac{s}{\omega_p}\right)} \qquad (6.18)$$

where $\omega_z = 1/RC_1$, $\omega_p = (C_1 + C_2)/RC_1C_2$, and $C_T = C_1 + C_2$. In practice, the filter components are designed to satisfy the constraints of robust stability, and noise and spur rejection while keeping the component sizes as small as possible.

To improve the suppression of the reference spurs while keeping the bandwidth sufficiently large to meet the requirement of a high lock speed, a third-order filter, as shown in Figure 6.28(b), can be used. This is achieved by setting the pole frequency due to R_2 and C_3 to be lower than the reference frequency, but higher than the loop bandwidth. Using the voltage division principle, the voltage across the R_1C_1 or C_2 branch is of the form $(sR_2C_3 + 1)V(s)$. The input current I is then given by

$$I(s) = \frac{(sR_2C_3 + 1)V(s)}{Z(s)} + \frac{(sR_2C_3 + 1)V(s) - V(s)}{R_2} \qquad (6.19)$$

where

$$Z(s) = \left(R_1 + \frac{1}{sC_1}\right) \middle\| \frac{1}{sC_2} = \frac{R_1C_1s+1}{R_1C_1C_2s^2 + (C_1 + C_2)s} \qquad (6.20)$$

Hence,

$$H(s) = \frac{V(s)}{I(s)} = \frac{R_1C_1s+1}{D(s)} \qquad (6.21)$$

where

$$D(s) = R_1R_2C_1C_2C_3s^3 + [R_1C_1(C_2+C_3)+R_2C_3(C_1+C_2)]s^2 + (C_1+C_2+C_3)s \qquad (6.22)$$

Finally, the transfer function, $H(s)$, can be put into the form

$$H(s) = \frac{V(s)}{I(s)} = \frac{1 + \dfrac{s}{\omega_z}}{sC_T\left(1 + \dfrac{s}{\omega_p Q_p} + \dfrac{s^2}{\omega_p^2}\right)} \qquad (6.23)$$

where

$$\omega_z = \frac{1}{R_1C_1} \qquad (6.24)$$

$$\omega_p^2 = \frac{C_1 + C_2 + C_3}{R_1R_2C_1C_2C_3} \qquad (6.25)$$

$$\omega_p Q_p = \frac{C_1 + C_2 + C_3}{R_1C_1(C_2 + C_3) + R_2C_3(C_1 + C_2)} \qquad (6.26)$$

and

$$C_T = C_1 + C_2 + C_3 \qquad (6.27)$$

Note that the use of a third-order loop filter has a considerable impact on the design requirements, as the PLL can be prone to instability.

6.3.5   Voltage-controlled oscillator

Generally, an oscillator can be considered a feedback system similar to the one in Figure 6.29(a).

The oscillator closed-loop transfer function can obtained as

$$\left.\frac{V_0(s)}{V_i(s)}\right|_{s=j\omega} = \frac{A(j\omega)}{1 + A(j\omega)} \qquad (6.28)$$

The system will oscillate at the frequency ω_0 if $A(j\omega_0) = -1$. The oscillation

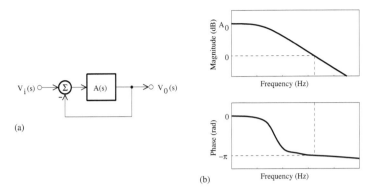

FIGURE 6.29
(a) Linear model of an oscillator; (b) open-loop frequency response.

condition or Barkhausen criteria (see Figure 6.29(b)) can then be summarized as

$$|A(j\omega_0)| = 1 \qquad (6.29)$$
$$\arg[A(j\omega_0)] = -\pi \qquad (6.30)$$

The oscillatory feature of the system relies on the fact that the feedback signal is added in phase to the forward one. A phase shift of π is introduced by the negative feedback and the overall phase shift is 2π. In practice, the use of the oscillation condition can be limited due to the component imperfections. Then, an open-loop gain larger than unity at the desired oscillation may be required in order to ensure the normal circuit operation.

• **Fully differential VCO**
A voltage-controlled oscillator (VCO) structure with a differential architecture is shown in Figure 6.30 [20]. It consists of a loop of N delay stages with a wire inversion. The ring will oscillate with a period of 2N times the stage delay. The differential delay stage using a replica biasing circuit is depicted in Fig 6.31(a) [20]. Fig 6.31(b) shows the circuit diagram of a differential delay stage with symmetric loads. The tail current, which is applied to the differential transistor pair $T_1 - T_2$, is driven by T_7. The voltages V_c and V_B are generated by the replica biasing circuit, which should adjust the bias currents of the delay stage to provide a wide tuning range over the temperature and process variations.

An alternative structure of the delay cell, which also achieves a good power-supply noise rejection, is shown in Figure 6.32 [15]. Transistors $T_5 - T_6$ should fix the output voltage at the minimum value of $V_{DD} - V_T$, where V_T is the transistor threshold voltage, and set an output swing and a common-mode level without the requirement for a replica biasing circuit. That is, the bias current I_B must be greater than the current, I_L, flowing through the pMOS

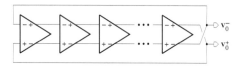

FIGURE 6.30
Oscillator using differential delay stages.

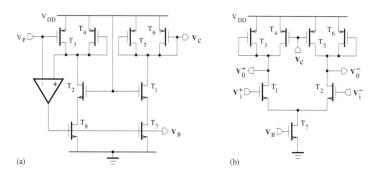

(a)   (b)

FIGURE 6.31
(a) Replica biasing circuit; (b) delay buffer.

transistor loads. A common practice is to have $I_B = 2I_L$. It should be noted that the parasitic capacitance introduced at the output nodes by $T_5 - T_6$ can limit the operating frequency range.

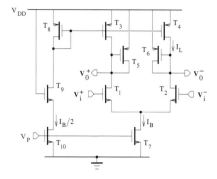

FIGURE 6.32
Delay buffer using pMOS transistor diodes.

Let us consider a delay buffer consisting of a differential pair loaded by transistors in the triode region. The related equivalent model is shown in Figure 6.33.

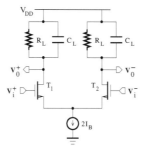

FIGURE 6.33
Equivalent model of the delay buffer.

The differential output voltage can be written as

$$\frac{dV_0(t)}{dt} = -\frac{V_0(t)}{R_L C_L} + \frac{I_B}{C_L} \tag{6.31}$$

where R_L is the output load capacitor and C_L denotes the output load resistor. An expression of V_0 is then given by

$$V_0(t) = R_L I_B (1 - 2e^{-t/\tau}) \tag{6.32}$$

where $\tau = R_L C_L$ is the time constant. It was assumed that the output voltage initially takes the value $-I_B R_L$ and can increase up to $I_B R_L$. The delay of the buffer can be defined as the time required for the change of V_0 from the initial value to zero, that is,

$$t_d = \ln(2)\tau \tag{6.33}$$

In practical circuits, the parameter T_d involves a variable contribution due to transistor noise. This delay uncertainty gives rise to clock jitter. Using the relation

$$\left.\frac{dV_0(t)}{dt}\right|_{t=t_d} = \frac{I_B}{C_L} \tag{6.34}$$

the average jitter component can be obtained as

$$\overline{\Delta t_d^2} = \frac{\overline{v_n^2}}{[(dV_0(t)/dt)|_{t=t_d}]^2} = \overline{v_n^2}\frac{C_L^2}{I_B^2} \tag{6.35}$$

where $\overline{v_n^2}$ denotes the voltage variance of the total noise.

It should be noted that an additional buffer may be required to provide the single-ended version of the output signal [22]. Figure 6.34 shows the circuit diagram of a differential-to-single-ended converter. Two source followers, a differential stage, and two inverters are required in this design. The input impedance of the buffer is increased by the source followers $T_6 - T_7$ and $T_8 - T_9$, while the output drive capability is improved by the two inverters $T_{10} - T_{11}$ and $T_{12} - T_{13}$.

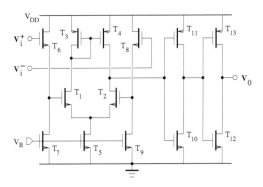

FIGURE 6.34
Differential-to-single-ended buffer.

Using a VCO, the jitter of the output clock is only affected by that of the reference signal, because the loop acts as a lowpass filter. The periodicity of the signal delivered by a VCO is useful in PLL applications such as clock and data recovery. Furthermore, by inserting a frequency divider in the loop, the output clock period can be a fraction of the reference signal to meet the frequency synthesizer specifications.

• **Pseudo-differential VCO**
A delay stage can also be implemented using a pseudo-differential structure, as shown in Figure 6.35. Cross-coupled inverters help align the complimentary signal phases.

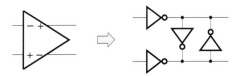

FIGURE 6.35
Pseudo-differential implementation of the delay stage using inverters.

An alternative VCO structure with quadrature outputs is represented in Figure 6.36. It is designed by adding feedforward inverters between the nodes with opposite signal phases of a ring of four inverters. A pair of cross-coupled inverters exhibits a negative transconductance and can operate as a regenerative circuit whose positive feedback causes the switching of output levels.

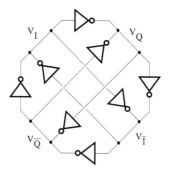

FIGURE 6.36
Quadrature ring oscillator using inverters.

Assuming that each inverter stage exhibits a propagation delay of T_d, the oscillation frequency is approximatively given by, $f_0 \simeq 1/(4T_d)$. To sustain a stable oscillation, feedforward inverters should be sized notably smaller than other inverters. The oscillation frequency can be controlled by tuning the bias current that is related to the inverter propagation delay.

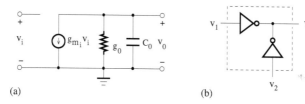

FIGURE 6.37
(a) Equivalent model of an inverter; (b) representation of a node of the ring oscillator.

To analyze the ring oscillator, each inverter can be modeled as the transconductance stage depicted in Figure 6.37(a), where g_{m_i} is the transconductance, g_0 denotes the output conductance, and C_0 is the output capacitance. With reference to Figure 6.37(b), the following node equations can be written

$$h_1 g_m v_1(t) + g_L v_0(t) + C_L \frac{dv_0(t)}{dt} = 0 \qquad (6.36)$$

$$h_2 g_m v_2(t) + g_L v_0(t) + C_L \frac{dv_0(t)}{dt} = 0 \qquad (6.37)$$

where h_i $(i = 1, 2)$ is the scaling factor of the i-th inverter, and g_L and C_L are the output load conductance and capacitance, respectively. Combining (6.36) and (6.37), we obtain a general expression given by

$$G \sum_{i=1}^{P} x_i v_i(t) + v_0(t) + \tau \frac{dv_0(t)}{dt} = 0 \tag{6.38}$$

where $P = 2$, $x_i = h_i/P$, $G = g_m/g_L$ and $\tau = C_L/g_L$ represent the *dc* gain and time constant, respectively.

The total phase shift around the loop should be a multiple of 2π, and the waveforms at each inverter output and input nodes are assumed to be sinusoidal and characterized by

$$v_0(t) = A\cos(\omega_j t - \phi) \tag{6.39}$$

$$v_i(t) = A\cos(\omega_j t - \phi \times i)) \tag{6.40}$$

where $\phi = 2\pi j/N$ and j is an integer that can take any value from 0 to $N-1$, with N being the number of input (or output) nodes (here $N = 4$). Substituting (6.39) and (6.40) into (6.38) yields

$$G \sum_{i=1}^{P} x_i \cos\left(\omega_j t - \frac{2\pi j}{N}i\right) + \cos\left(\omega_j t - \frac{2\pi j}{N}\right)$$
$$- \omega_j \tau \sin\left(\omega_j t - \frac{2\pi j}{N}\right) = 0 \tag{6.41}$$

By expanding (6.41) and then equating the $\cos(\omega_j t)$ and $\sin(\omega_j t)$ terms, the oscillation frequency and the required minimum *dc* gain for the *j*-th mode are obtained as follows

$$\omega_j \tau = \frac{\sum\limits_{i=1}^{P} x_i \sin\left(\frac{2\pi j}{N}(i-1)\right)}{-\sum\limits_{i=1}^{P} x_i \cos\left(\frac{2\pi j}{N}(i-1)\right)} \tag{6.42}$$

$$G = G_j = \frac{1}{-\sum\limits_{i=1}^{P} x_i \cos\left(\frac{2\pi j}{N}(i-1)\right)} \tag{6.43}$$

Initially, each inverter is in the linear region, where it can operate as a transconductor. All oscillation modes characterized by a *dc* gain, G_j, that is lower than the inverter gain, G, start to build up. They are then sustained in the steady state provided the above frequency and gain equations are satisfied.

In practice, however, the ring oscillator generates a signal that differs from a sinusoidal waveform and may look like a triangular or square waveform depending on the value of the inverter transition time. Hence, the oscillation amplitude is determined by the nonlinear switching characteristic of inverters.

In the special case of a ring oscillator consisting of only one loop of single-ended inverters, $x_P = 1$ and all other x_i coefficients are reduced to zero. Because the oscillation condition requires a gain of -1 at dc, the number, N, of inverters should be odd. The oscillation frequency is given by

$$f_0 = \frac{1}{2\pi\tau} \tan\left(\frac{\pi}{N}\right) \tag{6.44}$$

where τ denotes the time constant.

6.4 Applications

PLLs find use in a wide variety of applications, including, but not limited to, frequency synthesizers, clock, and data recovery circuits.

6.4.1 Frequency synthesizer

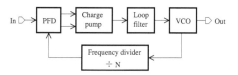

FIGURE 6.38
Block diagram of a frequency synthesizer based on the PLL.

In general, a frequency synthesis consists of generating a desired frequency from one or more reference signals, each at a given frequency and generated by a precise crystal oscillator. The block diagram of a frequency synthesizer based on the PLL is shown in Figure 6.38. It includes an integer-N programmable divider in the feedback path. The output frequency is given by

$$f_0 = N \cdot f_{ref} \tag{6.45}$$

where N is the division ratio and f_{ref} is the frequency of the input reference signal.

Based on the continuous-time linear model of the PLL, the relationship between the phase angles of the error, input, and output signals is of the form

$$\Theta_e = \Theta_i - \Theta_0/N \tag{6.46}$$

Assuming that the transfer functions of the filter and VCO are $H(s)$ and K_v/s, respectively, the closed-loop transfer function can be expressed as

$$T(s) = \frac{\Theta_0(s)}{\Theta_i(s)} = \frac{G(s)}{1 + G(s)} \tag{6.47}$$

and the error transfer function is

$$T_e(s) = \frac{\Theta_e(s)}{\Theta_i(s)} = \frac{1}{1 + G(s)} \tag{6.48}$$

where $G(s)$ denotes the open-loop transfer function given by

$$G(s) = \frac{K_p K_v H(s)}{sN} \tag{6.49}$$

By reducing $H(s)$ to a constant, the frequency synthesizer becomes a first-order system that is unconditionally stable. However, in practice, the transfer function of the loop filter should include a pole/zero pair, which is used to increase the frequency range.

Although the linear model in the s-domain provides a helpful set of design equations, a charge-pump-based PLL should be accurately described and optimized in the z-domain. The direct conversion of the PLL continuous-time equations into the z-domain can be computationally intensive. For this reason, the method based on the impulse invariant transformation[1] is generally adopted. The closed-loop transfer function is written as

$$\widehat{T}(z) = \frac{\widehat{\Theta}_0(z)}{\widehat{\Theta}_i(z)} = \frac{\widehat{G}(z)}{1 + \widehat{G}(z)} \tag{6.50}$$

where

$$\widehat{G}(z) = \frac{K_p K_v \widehat{H}(z)}{N} \tag{6.51}$$

and

$$\widehat{H}(z) = (1 - z^{-1}) \mathcal{Z} \left\{ \mathcal{L}^{-1} \left(\frac{H(s)}{s^2} \right) \Big|_{t=nT/N} \right\} \tag{6.52}$$

Here, T is the sampling frequency, and $\mathcal{Z}$ and $\mathcal{L}^{-1}$ denote the z-transform (or modified z-transform) and inverse Laplace transform, respectively. Note that the correspondence between the models is ensured only at the time nT/N. For the evaluation of $\widehat{H}$, the value of the time-domain function should be zero at the initial time instant.

Let us consider the third-order frequency synthesizer shown in

---

[1] Let $F(s)$ be the transfer function of a system in the s-domain. Using the impulse invariant transformation, its z-domain version can obtained as

$$\widehat{F}(z) = \mathcal{Z} \left\{ \frac{1 - e^{-Ts}}{s} F(s) \right\} = (1 - z^{-1}) \cdot \mathcal{Z} \left\{ \mathcal{L}^{-1} \left(\frac{F(s)}{s} \right) \Big|_{t=kT} \right\}$$

It is assumed that the equivalent discrete representation is provided by the series connection of an ideal sampler with the sampling period T, a zero-order hold stage, and the analog model of the system.

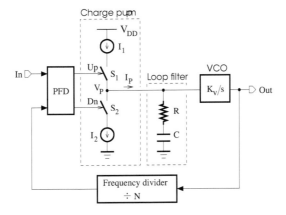

FIGURE 6.39
Block diagram of a third-order frequency synthesizer.

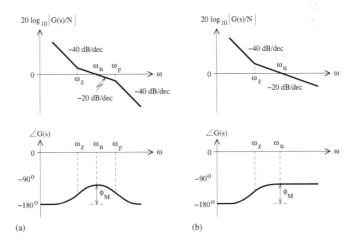

FIGURE 6.40
Magnitude and phase of the open-loop transfer function: (a) $C_2 \neq 0$, (b) $C_2 = 0$.

Figure 6.39. The loop filter has a second-order transfer function of the form

$$H(s) = \frac{V_P(s)}{I_P(s)} = \frac{1 + \dfrac{s}{\omega_z}}{sC_T\left(1 + \dfrac{s}{\omega_p}\right)} \tag{6.53}$$

where $\omega_z = 1/RC_1$, $\omega_p = (C_1 + C_2)/RC_1C_2$, and $C_T = C_1 + C_2$.

The open-loop transfer function is given by

$$G(s) = \frac{K_p K_v H(s)}{sN} \tag{6.54}$$

$$= K_p K_v \frac{1 + \dfrac{s}{\omega_z}}{s^2 N C_T \left(1 + \dfrac{s}{\omega_p}\right)} \tag{6.55}$$

The combination of the PFD and charge-pump circuit can be characterized by the gain factor, $K_p = I_p/2\pi$. The magnitude and phase of the open-loop transfer function are depicted in Figure 6.40. At the frequency, ω_u, the magnitude of the open-loop transfer function is equal to 1 (or 0 dB).

To determine the zero and pole frequencies needed to obtain the desired phase margin, we compute

$$\phi_M = 180° - \angle G(j\omega_u) = \arctan\left(\frac{\omega_u}{\omega_z}\right) - \arctan\left(\frac{\omega_u}{\omega_p}\right) \tag{6.56}$$

where ω_u is the unity-gain frequency. The phase margin is maximum for a value of ω_u, which can be obtained by setting the derivative of ϕ_M with respect to ω_u equal to zero. That is,

$$\frac{d\phi_M}{d\omega_u} = \frac{1}{\omega_z} \cdot \frac{1}{1 + \left(\dfrac{\omega_u}{\omega_z}\right)^2} - \frac{1}{\omega_p} \cdot \frac{1}{1 + \left(\dfrac{\omega_u}{\omega_p}\right)^2} = 0 \tag{6.57}$$

and

$$\omega_u = \sqrt{\omega_z \omega_p} = \omega_z \sqrt{1 + \eta} \tag{6.58}$$

where

$$\eta = C_1/C_2 \tag{6.59}$$

By substituting Equation (6.58) into (6.56), the maximum value of ϕ_M is obtained as

$$\phi_M = \arctan\left(\sqrt{1 + \eta}\right) - \arctan\left(\frac{1}{\sqrt{1 + \eta}}\right) \tag{6.60}$$

Recalling that $\arctan x + \arctan y = \arctan[(x - y)/(1 + xy)]$ and solving Equation (6.60) for η gives

$$\eta = 2\tan^2 \phi_M + 2\tan \phi_M \sqrt{1 + \tan^2 \phi_M} \tag{6.61}$$

The stability of the loop in the s-domain is guaranteed, provided the capacitors are chosen such that Equation (6.61) is satisfied. The phase margin can be selected between 30° and 70°. For a PLL design with a high phase margin, a wide stability range is traded off for a slow loop response speed and reduced attenuation of the reference (or input) frequency. A common rule of thumb is to begin the design with a 43° phase margin.

Given ω_u and ϕ_M, the initial values of the filter components are determined as follows. By definition, we have

$$|G(j\omega_u)| = \frac{K_p K_v}{N C_T \omega_u^2} \cdot \frac{\sqrt{1 + \left(\dfrac{\omega_u}{\omega_z}\right)^2}}{\sqrt{1 + \left(\dfrac{\omega_u}{\omega_p}\right)^2}} = 1 \tag{6.62}$$

By combining Equations (6.57) and (6.62), and recalling that $\omega_p/\omega_z = 1 + \eta$ and $C_T = C_1(1 + 1/\eta)$, it can be shown that

$$\frac{K_p K_v}{N C_T \omega_u^2}\sqrt{\frac{\omega_p}{\omega_z}} = \frac{K_p K_v}{N C_1 \omega_u^2} \cdot \frac{\eta}{\sqrt{1+\eta}} = 1 \tag{6.63}$$

Therefore, according to Equation (6.63), we find that

$$C_1 = \frac{K_p K_v}{N \omega_u^2} \cdot \frac{\eta}{\sqrt{1+\eta}}. \tag{6.64}$$

From Equations (6.58) and (6.59), we respectively obtain

$$R = \frac{\sqrt{1+\eta}}{\omega_u C_1} \tag{6.65}$$

and

$$C_2 = \frac{C_1}{\eta} \tag{6.66}$$

The s-domain analysis is limited by the fact that it does not take into account the sampling nature of the loop. It is then not suitable for the prediction of jitter performance and nonlinear acquisition process. Simulations are necessary to fine-tune the values of the filter components. Note that a fast settling response requires a wide closed-loop bandwidth.

As a rule of thumb, the closed-loop bandwidth of a charge-pump PLL should be chosen to be less than approximately one-tenth of the reference frequency. Otherwise, the stability, speed, and phase noise of a charge-pump PLL will be affected by the sampling process.

Stability analysis in the z-domain

To take into account the sampling effects, it is necessary to perform the z-domain analysis of the loop [23, 24]. The open-loop transfer function, $G(s)$, can be converted from the s-domain to the z-domain using the impulse invariant transformation. By performing the partial expansion of $G(s)/s$, we obtain

$$\frac{G(s)}{s} = \frac{K_p K_v}{N C_T}\left[\frac{1}{s^3} + \frac{\omega_p - \omega_z}{\omega_z \omega_p} \cdot \frac{1}{s^2} - \frac{\omega_p - \omega_z}{\omega_z \omega_p^2} \cdot \frac{1}{s(1 + s/\omega_p)}\right] \tag{6.67}$$

The inverse Laplace transform of $G(s)/s$ can then be written as

$$\mathcal{L}^{-1}\left(\frac{G(s)}{s}\right) = \frac{K_p K_v}{N C_T}\left[\frac{t^2}{2} + \frac{\omega_p - \omega_z}{\omega_z \omega_p}t - \frac{\omega_p - \omega_z}{\omega_z \omega_p^2}(1 - e^{-\omega_p t})\right] \tag{6.68}$$

The equivalent transfer function in the z-domain is given by

$$\widehat{G}(z) = \frac{z - 1}{z} \cdot \mathcal{Z}\left\{\mathcal{L}^{-1}\left(\frac{G(s)}{s}\right)\Big|_{t=kT}\right\} \tag{6.69}$$

Using z transform tables, it can be shown that

$$\widehat{G}(z) = \frac{K_p K_v}{N C_T} \times$$
$$\left[\frac{T^2}{2} \cdot \frac{z + 1}{(z - 1)^2} + \frac{\omega_p - \omega_z}{\omega_z \omega_p} \cdot \frac{T}{z - 1} - \frac{\omega_p - \omega_z}{\omega_z \omega_p^2} \cdot \frac{1 - e^{-\omega_p T}}{z - e^{-\omega_p T}}\right] \tag{6.70}$$

or equivalently,

$$\widehat{G}(z) = \frac{K_p K_v}{N C_T} \frac{p z^2 + q z + r}{z^3 - (2 + \alpha)z^2 + (1 + 2\alpha)z - \alpha} \tag{6.71}$$

where

$$\alpha = e^{-\omega_p T} \tag{6.72}$$

$$p = \frac{T^2}{2} + \left(\frac{T}{\omega_z} - \frac{1 - \alpha}{\omega_z \omega_p}\right)\left(1 - \frac{\omega_z}{\omega_p}\right) \tag{6.73}$$

$$q = \frac{T^2(1 - \alpha)}{2} - \left(\frac{T(1 + \alpha)}{\omega_z} - \frac{2(1 - \alpha)}{\omega_z \omega_p}\right)\left(1 - \frac{\omega_z}{\omega_p}\right) \tag{6.74}$$

$$r = -\frac{T^2 \alpha}{2} + \left(\frac{T\alpha}{\omega_z} - \frac{1 - \alpha}{\omega_z \omega_p}\right)\left(1 - \frac{\omega_z}{\omega_p}\right) \tag{6.75}$$

and T denotes the period of the clock signal. The parameter of the charge-pump PLL should then be chosen such that the roots of the characteristic equation, $1 + \widehat{G}(z)$, remain inside the unit circle defined by $|z| = 1$ in the z-plane. In general, the stability condition in the z-domain is more constraining than that in the s-domain.

Time-domain analysis

The operation or acquisition process of a charge-pump PLL can be accurately modeled using a time-domain analysis based on difference equations and state-space representation. The input phase, θ_i, and the output phase, θ_0, are related by difference equations of the form

$$\theta_i(t) = \theta_i(0) + \omega_i t \tag{6.76}$$

$$\theta_0(t) = \theta_0(0) + \omega_v t + K_v \int_0^t v_p(\tau) d\tau \tag{6.77}$$

$$\theta_e(t) = \theta_i(t) - \theta_0(t) \tag{6.78}$$

where θ_e is the phase error, ω_i represents the input frequency, ω_v is the free-running frequency of the oscillator, K_v is the VCO gain, and v_p denotes the output voltage of the loop filter or the VCO control voltage, which is identical to the voltage across the capacitor C_2. It was assumed that the initial conditions for the input and the output phases are $\omega_i(0)$ and $\omega_0(0)$, respectively. The state-space representation of the filter can be written as

$$\frac{dv_p}{dt} = -\frac{v_p}{RC_2} + \frac{v_{C_1}}{RC_2} + \frac{i_p}{C_2} \tag{6.79}$$

$$\frac{v_{C_1}}{dt} = \frac{v_p}{RC_1} - \frac{v_{C_1}}{RC_1} \tag{6.80}$$

where v_{C_1} represents the voltage across the capacitor C_1 and i_p is the charge-pump output current, which is used to drive the loop filter. Assuming that the output capacitor of the charge-pump circuit is either charged for a positive phase error or discharged for a negative phase error, the current i_p is given by

$$i_p = \begin{cases} K_p \cdot \text{sign}(\theta_e) & \text{if} \quad 0 \le t \le t_p \\ 0 & \text{if} \quad t_p < t < t_r \end{cases} \tag{6.81}$$

where K_p represents the gain associated to the PFD and charge-pump circuit, $\text{sign}(\theta_e)$ denotes the polarity (i.e., 1 or -1) of the phase error, θ_e, t_p represents the turn-on duration of the current I_P, and t_r is the time at which the next rising edge of either the VCO or the reference signal occurs. Tools for symbolic analysis can be used to solve the system of equations characterizing the charge-pump PLL. By using linear models, the analysis is valid only near locked states. However, due to the fact that the PFD, for instance, is actually a nonlinear and time-variant component, the loop may be affected by nonlinear mechanisms such as cycle slip, which is caused by a large frequency difference between the reference signal and the feedback signal. In practice, this difference is minimized by setting the initial control voltage appropriately, and thereby facilitating the acquisition process.

Note that the synthesis of a frequency, which is N/M times the reference frequency, can simply be performed by adding a divider with a ratio of M at the input of the PLL.

The aforementioned topology is commonly used due to its simplicity. However, the output frequency can only change by integer multiples of the reference frequency. Furthermore, the reference spurs, which appear centered on the reference frequency and its harmonics, are related to the amount of jitter in the retimed signals. Hence, the achievable resolution of the output frequency is limited because the loop bandwidth, as set by the loop filter, should be at least ten times smaller than the reference frequency to prevent undesirable signal components caused by the sampling action in the phase detector from reaching the input of the VCO and corrupting the output frequency. This can result in a slow settling (or lock) time for the frequency synthesizer.

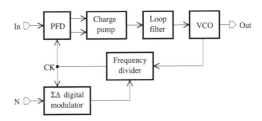

FIGURE 6.41

Block diagram of a $\Delta\Sigma$ fractional-N frequency synthesizer.

To overcome the resolution-bandwidth trade-off of integer-N frequency synthesizers, a $\Delta\Sigma$ fractional-N architecture can be used [27]. This last approach is capable of generating frequencies over wide bandwidths with a very fine frequency resolution to accommodate the narrow channel spacing of wireless telephony applications. The block diagram of a $\Delta\Sigma$ fractional-N frequency synthesizer is depicted in Fig 6.41. It is generally based on the concept of division ratio averaging, which is implemented using a PLL whose feedback path includes a multi-modulus divider controlled by a $\Delta\Sigma$ modulator. The division ratio is dynamically switched between two or more values to realize the fractional frequency division.

An dual-modulus frequency divider can be implemented as shown in Fig-

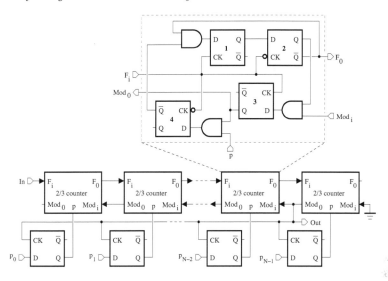

FIGURE 6.42
Block diagram of a dual-modulus frequency divider.

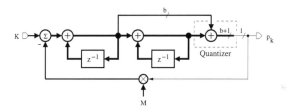

FIGURE 6.43
Block diagram of a second-order $\Delta\Sigma$ modulator.

ure 6.42. This modular structure consists of a plurality of 2/3 counters arranged in a cascade combination [28]. Each 2/3 counter realizes a division by a factor of 2 or 3, depending on whether the control signal p_k is set to a logic state 0 or 1.

In operation, the prescaler logic block of each 2/3 counter first divides the frequency of the incoming signal, F_i, and the resulting signal, F_0, is applied to the following counter cell. The division ratio is determined by the logic state of the control signals applied to the end-of-cycle logic block of each 2/3 counter cell. Upon completion of a division cycle, the end-of-cycle logic block of the last 2/3 counter cell in the divider chain generates a signal, Mod_0, which is transferred successively to the preceding 2/3 counter cells after being retimed by each cell. The division cycle corresponds to the clock period of the F_0 signal available at the output of the last 2/3 counter, whose Mod_0 terminal can serve as the divider output. To avoid the perturbation of the current division operation, the updated control signals for the division ratio should transit through latches, which are synchronized with the respective division cycle, on their way to each 2/3 counter cell. A new division ratio can be set only when a division cycle is completed.

The basic principle of the 2/3 counter cell can be extended to any $P/(P+1)$ cell. Even if the division ratio of each counter cell of the frequency divider is an integer at any instant, its repetitive switching between the integers P and $P+1$ by the modulator will give rise to a fractional division with a ratio comprised between these last two integer values. With the division ratio equal to P during $(1-\eta)T$ and $P+1$ during ηT over a time period of T, the average output frequency is given by

$$F_0 = [\eta(P+1)T + (1-\eta)PT]f_{ref}/T = (P+\eta)Fi, \qquad (6.82)$$

where F_i is the input frequency of the $P/(P+1)$ counter and η is determined by the modulator output.

The overall division ratio of the frequency divider can be put into the form

$$N = 2^n + 2^{n-1}p_{n-1} + 2^{n-2}p_{n-2} + \cdots + 2p_1 + p_0 \qquad (6.83)$$

where $p_0, p_1, \cdots, p_{n-1}$ denote the control signal logic states from the first to the last 2/3 counter. Using n 2/3 counter cells in cascade, the possible division ratios range from 2^n (if all $p_k = 0$) to $2^{n+1} - 1$ (if all $p_k = 1$).

The block diagram of a second-order $\Delta\Sigma$ modulator is depicted in Figure 6.43. Generally, the data word length in the modulator should be long enough to prevent the occurrence of an overflow in the first accumulator. Based on simulations, it was shown in [27] that the parameters K and M can be adequately chosen as $-0.5M < K < 0.5M$ and $M < 2^b/2.5$, where b is the accumulator word length. The computation resolution, which is initially set to b bits, is increased to $b+1$ bits in certain steps to accommodate possible numerical overflows in the adders. The quantization operation is reduced to the overflow of the second adder, the sign bit of which is used as the modulator output.

Using a linear model of the modulator, the z-transform of the output can be expressed as

$$P_k(z) = \frac{H(z)}{1 + M \cdot H(z)} K(z) + \frac{1}{1 + M \cdot H(z)} E_Q(z) \qquad (6.84)$$

where E_Q is the quantization error and

$$H(z) = \frac{2 - z^{-1}}{(1 - z^{-1})^2} \qquad (6.85)$$

At dc, or $z = 1$, and $M \cdot H(z) \gg 1$, we can obtain

$$P_k \simeq \frac{K}{M} + \frac{E_Q}{M \cdot H} \tag{6.86}$$

For a slow-varying input signal, the time average of the binary output sequence produced by the modulator is a high-resolution representation of the value K/M.

A problem generally associated with $\Delta\Sigma$ fractional-N frequency synthesizers is the increased level of in-band spurs. Ideally, the PFD and charge-pump circuits should deliver an output signal that is proportional to the phase difference between the reference and feedback signals. However, their practical characteristics are not fully linear. In the lock condition, the phase difference can still take different values due to the changing division ratio in the feedback path. This in turn stimulates nonlinearities of the PFD and charge-pump circuits, generating a noise floor that boosts the in-band phase noise. Note that the use of a higher-order $\Delta\Sigma$ modulator helps reduce the in-band quantization noise, but at the price of an increased level of the out-of-band noise.

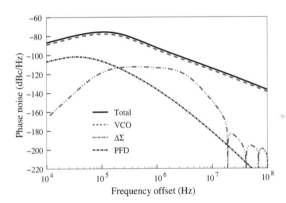

FIGURE 6.44
Noise figure of narrow-band PLL.

The typical ratio of the PLL bandwidth to the phase detector frequency is less than $1/100$ in wireless and wireline applications, but only less than $1/15$ in digital system applications such as the clock signal generation for a microprocessor and high-speed I/O link systems.

Figures 6.44 and 6.45 show the noise contribution of each block in a narrow-band PLL and a wide-band PLL, where the PLL bandwidth is about 100 kHz and 1 MHz, respectively, and the clock frequency is assumed to be 25 MHz. The VCO noise determines the overall phase noise performance in the narrow-band PLL, while quantization noise of the $\Delta\Sigma$ modulator is dominant at high frequencies in the wide-band PLL. The choice of the loop bandwidth in the design of a wide-band $\Delta\Sigma$ fractional-N PLL is eventually limited by

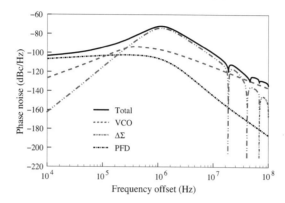

FIGURE 6.45

Noise figure of wide-band PLL.

the quantization noise and can involve a trade-off between the minimization of the quantization noise and VCO noise.

Several techniques can be used for quantization noise reduction either in the analog or in the digital domain. DAC-based noise canceling and phase rotator-based fractional division are two approaches to reduce the quantization noise. However, the level of noise reduction with both techniques can be limited by the linearity and matching requirements of analog components. Alternatively, in order to suppress the quantization noise at certain frequencies, a digital finite-impulse-response filter can be inserted along the path of the control bits generated by the $\Delta\Sigma$ modulator.

6.4.2 Clock and data recovery

In serial data transmissions, the signal is generally distorted by the transmission channel (coaxial cable, optic fiber). Because it is generally impractical to transmit the necessary sampling clock signal separately from data, the timing information is usually derived from the transmitted data, which are asynchronous and noisy. This is realized using a clock recovery circuit, which can be based on a PLL structure. With reference to Figure 6.46, the incoming data should be retimed by the D flip-flop, which is synchronized by the recovered clock signal in such a way that the sampling clock edge is aligned with the middle of the data bit period in order to minimize the bit-error rate.

A clock and data recovery (CDR) circuit, which relies on a PLL consisting of a phase detector, a charge-pump circuit, a filter, and a VCO, is generally unable to capture the incoming data if the free-running frequency of the VCO is more than a few hundred parts-per-million (ppm) from the frequency of the input data.

With the VCO free-running frequency exhibiting a wide variation over the IC process, temperature, and supply voltage, it is impossible for the PLL

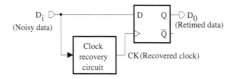

FIGURE 6.46
Principle of a clock and data recovery circuit.

to lock under all circumstances without relying on a frequency acquisition aid. Hence, the PLL used for clock and data recovery applications can be configured to include a frequency loop and a phase loop. The first one achieves the clock signal frequency acquisition by reducing the difference between the free-running frequency of the VCO and a reference frequency, while the second one is enabled for the phase locking of the clock signal with the incoming data signal. A PLL with a dual-loop configuration may be implemented with or without an external reference clock signal.

TABLE 6.1
CDR Structure Comparison

| | CDR circuit based on | | |
	Dual loop	PI	GVCO
Wide bandwidth	Good	Not good	Very good
Power efficiency	Not good	Good	Good
Fast locking	Not good	Not good	Very good
Capture range	Wide	Narrow	Narrow

In addition to dual-loop PLLs, other structures can be used for the implementation of CDR circuits. They can be based on a phase interpolator (PI) and a gated voltage-controlled oscillator (GVCO). Table 6.1 provides a comparison of their essential characteristics, especially for multi-lane applications.

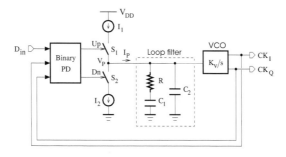

FIGURE 6.47
Block diagram of a single-loop binary CDR.

A single loop that can be used for phase tracking in a CDR circuit is shown in Figure 6.47. It consists of a binary phase detector, a charge pump, a loop filter, and a quadrature VCO. In the lock state, the edge of the clock signal, CK_I, should be aligned with the center of the data eye.

Stability in the z-domain of the single-loop CDR based on a binary PD

To take into account the sampling effects, it is necessary to perform the z-domain analysis of the loop [23,24]. Using the impulse invariant transformation [25], the response in continuous-time can be related to one in the discrete-time as follows,

$$\hat{g}(n) = [p_{\tau_d}(t) * g(t)]|_{t=nT} \tag{6.87}$$

$$= \left. \int_{-\infty}^{\infty} p_{\tau_d}(t)g(t-\tau)d\tau \right|_{t=nT} \tag{6.88}$$

or equivalently

$$\mathcal{Z}^{-1}\{\hat{G}(z)\} = \mathcal{L}^{-1}\{P_{\tau_d}(s)G(s)\}|_{t=nT} \tag{6.89}$$

where $p_{\tau_d}(t)$ and $g(t)$ are the impulse responses associated to the transfer functions $P_{\tau_d}(s)$ and $G(s)$, respectively.

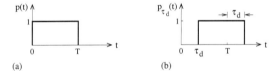

(a)   (b)

FIGURE 6.48
(a) Ideal PD pulse; (b) PD pulse with delay.

In the ideal case, the PD output sequence is supposed to be generated without a delay. However, in a practical implementation, it exhibits a delay, τ_d, due to the switching speed of gates and transistors, as shown in Figure 6.48. To take into account the effect of the delay, τ_d, on the stability in the z-domain, the loop delay can be put into the form,

$$T_d = q \cdot T + \tau_d \tag{6.90}$$

where T is the clock signal period, $0 \le \tau_d < T$, $q = \lfloor T_d/T \rfloor$, and $\lfloor x \rfloor$ denotes the largest integer not greater than x.

The PLL open-loop transfer function, $G(s)$, can then be converted from the s-domain to the z-domain using the impulse-invariant transformation. By performing the partial expansion

of $G(s)$, we obtain

$$G(s) = \frac{K_p K_v}{N \cdot C_T} \left[\frac{1}{s^2} + \frac{\omega_p - \omega_z}{\omega_z \omega_p} \left(\frac{1}{s} - \frac{1}{s + \omega_p} \right) \right] \qquad (6.91)$$

Using the s-to-z domain equivalences given in Table 6.2, the impulse-invariant transform of the PLL transfer function can be obtained as,

$$\hat{G}(z) = \frac{K_p K_v T^2}{2 \cdot N \cdot C_T} \frac{\alpha z^3 + \beta z^2 + \gamma z + \delta}{z^{q+1}(z - 1)(z - z_0)} \qquad (6.92)$$

where

$$\alpha = \left(1 - \frac{\tau_d}{T} \right)^2 + \frac{2}{T} \left(\frac{\omega_p - \omega_z}{\omega_z \omega_p} \right) \left(1 - \frac{\tau_d}{T} + \frac{z_0 z_1 - 1}{\omega_p T} \right) \qquad (6.93)$$

$$\begin{aligned} \beta &= 1 - z_0 - \frac{(2 + z_0)\tau_d^2}{T^2} + \frac{2(1 + z_0)\tau_d}{T} \\ &\quad - \frac{2}{T} \left(\frac{\omega_p - \omega_z}{\omega_z \omega_p} \right) \left(1 + z_0 - \frac{(2 + z_0)\tau_d}{T} - \frac{2 + z_0 - 3z_0 z_1}{\omega_p T} \right) \end{aligned} \qquad (6.94)$$

$$\begin{aligned} \gamma &= -z_0 + \frac{(1 + 2z_0)\tau_d^2}{T^2} - \frac{2z_0 \tau_d}{T} \\ &\quad + \frac{2}{T} \left(\frac{\omega_p - \omega_z}{\omega_z \omega_p} \right) \left(z_0 - \frac{(1 + 2z_0)\tau_d}{T} - \frac{1 + 2z_0 - 3z_0 z_1}{\omega_p T} \right) \end{aligned} \qquad (6.95)$$

$$\delta = z_0 \left[-\frac{\tau_d^2}{T^2} + \frac{2}{T} \left(\frac{\omega_p - \omega_z}{\omega_z \omega_p} \right) \left(\frac{\tau_d}{T} + \frac{1 - z_1}{\omega_p T} \right) \right] \qquad (6.96)$$

$$z_1 = \exp(\omega_p \tau_d) \qquad (6.97)$$

and

$$z_0 = \exp(-\omega_p T) \qquad (6.98)$$

FIGURE 6.49
PLL linearized model in the discrete-time domain.

Based on the PLL linearized model of Figure 6.49, the closed-loop transfer function can be written as,

$$\frac{\theta_0(z)}{\theta_i(z)} = \frac{\hat{G}(z)}{1 + \hat{G}(z)} \qquad (6.99)$$

TABLE 6.2

Impulse Invariant Transformation: s-to-z Domain Equivalences

s domain	z domain
$\dfrac{1}{s}$	$\dfrac{T}{z^q} \cdot \dfrac{z(1 - \tau_d/T) + \tau_d/T}{z(z - 1)}$
$\dfrac{1}{s + \omega_p}$	$\dfrac{1}{\omega_p z^q} \cdot \dfrac{z(1 - z_0 z_1) + z_0 z_1 - z_0}{z(z - z_0)}$ where $z_1 = \exp(\omega_p \tau_d)$ and $z_0 = \exp(-\omega_p T)$
$\dfrac{1}{s^2}$	$\dfrac{T^2}{z^q} \cdot \dfrac{z^2(1 - \tau_d/T)^2 + z(1 + 2\tau_d/T - 2\tau_d^2/T^2) + \tau_d^2/T^2}{2z(z - 1)^2}$

In general, the stability condition[2] in the z-domain is more constraining than that in the s-domain. The parameters of the charge-pump PLL should be chosen such that the roots of the characteristic equation, $1 + \hat{G}(z)$, remain inside the unit circle defined by $|z| = 1$ in the z plane.

A useful stability condition can be derived by relating the analysis of the locus behavior to the radius of curvature (see Appendix B.1) [26]. The open-loop transfer function, $\hat{G}(z)$, has a pole of order $q + 1$ at $z = 0$, a pole at $z = r_0$ (along the real axis and inside the unit circle), and two poles at $z = 1$. For the charge-pump PLL to be stable, it is required that the root locus departing at $z = 1$ should first migrate inside the unit circle before it can cross the unit circle. That is,

$$RC_1 > (1 + C_2/C_1)(T_d + T/2) \qquad (6.100)$$

This last condition is more conservative than the requirement to have all poles inside the unit circle, but an obvious advantage is its simplicity due to the linear behavior.

---

[2]Based on Jury test, a charge-pump PLL characterized in the z-domain by a third-order ($q = 0$) closed-loop transfer function with a denominator of the form,

$$D(z) = a_3 z^3 + a_2 z^2 + a_1 z + a_0$$

is stable provided

$$D(1) > 0, \qquad D(-1) < 0, \qquad |a_0| < a_3$$

and

$$|a_0^2 - a_3^2| > |a_0 a_2 - a_1 a_3|$$

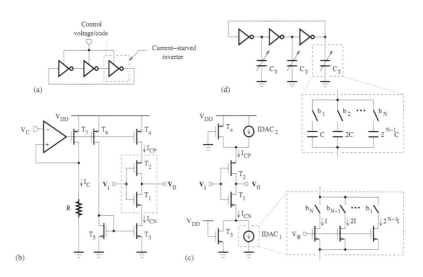

FIGURE 6.50
(a) Ring VCO based on current-starved inverters; (b) current-starved inverter with variable bias voltage; (c) current-starved inverter using a digital tuning code; (d) VCO based on inverters with variable load capacitors.

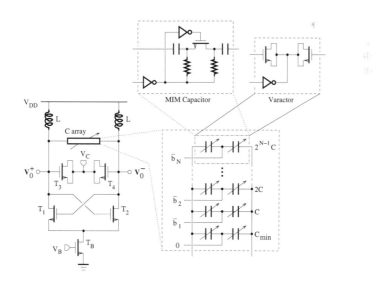

FIGURE 6.51
LC VCO with coarse and fine tunings.

PLL and CDR circuits can be implemented using either ring or LC oscillators. Ring oscillators, as shown in Figure 6.50(a), consist of delay stages that can easily be designed. They are scalable, exhibit a wide tuning range, and can inherently provide multi-phase output signals.

A current-starved inverter can be tuned by varying the bias voltage, as illustrated in Figure 6.50(b). Assuming that transistors T_6 and T_7 operate in the saturation region and $V_{GS_6} = V_{GS_7}$, we obtain $I_{D_6}/(W_6/L_6) = I_{D_7}/(W_7/L_7)$, provided that the threshold voltages of both transistors are identical. Due to the high gain of the amplifier, $V^+ \simeq V^- = V_C$. Hence, $I_{D_7} = I_C = V_C/R$.

Another tuning approach consists of using switched current-sources controlled by a digital code, as shown Figure 6.50(c).

Inverters can also be tuned by means of variable load capacitors, as shown in Figure 6.50(d).

However, VCOs based on ring oscillators feature a poor phase noise performance compared to LC VCOs. Figure 6.51 shows the circuit diagram of the LC VCO. Coarse tuning can be performed using either switched metal-insulator-metal (MIM) capacitor arrays, or MOS varactors that are switched between accumulation and depletion regions.

The accuracy of MIM capacitor arrays can be limited by the parasitic components of switches. On the other hand, MOS varactors are known to exhibit a large capacitance spread (or ratio between the capacitances when a varactor is turned on and off, respectively), and can then also be employed for fine tuning. The bias-dependent capacitance is proportional to the number of gate fingers used in the layout, the gate length and width. The capacitance spread is maximized when transistors are configured with minimum finger length.

6.4.2.1 Dual-loop CDR

A dual-loop CDR is first locked to the frequency of an external oscillator that is close to a fraction of the data rate, and then the phase is adjusted to appropriately align the clock signal edges and data outputs.

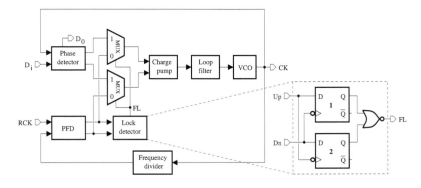

FIGURE 6.52
Block diagram of a dual-loop PLL requiring an external reference clock signal.

Referring to Figure 6.52, the block diagram shows a dual-loop PLL [29] requiring a reference clock signal. Before the closing of the data-acquisition loop by the lock detector, the frequency-acquisition loop should bring the VCO free-running frequency near the desired operating range, if necessary. The data-acquisition loop minimizes the remaining frequency error and aligns the phase of the VCO for an optimal sampling of the incoming data. The lock detector continuously monitors that the difference between the reference signal frequency and the divider output frequency remains within a predetermined range.

An approach to perform the lock detection can consist of monitoring the Up and Dn pulses that are generated by the three-state PFD [30]. This can simply be realized, as shown in the inset of Figure 6.52, by a lock detector structure whose components are two D flip-flops, inverters, and a NOR gate. The output signal FL is set to the high logic state indicating the frequency-locked mode only in the case where the Up and Dn pulses are at the low logic state. When the Up and Dn signals are in different logic states due to the fact that one of the PFD inputs is leading or lagging the other, the output signal FL is maintained to the low logic state. Depending on the signal propagation delay, the resulting error in the detection of phase alignment between the Up and Dn pulses is within 5 to 15% of the full-scale phase deviation of the PFD.

Another scheme for implementing the lock detector relies on the use of a counter-based frequency comparator to monitor the reference clock frequency and the divider output frequency [31]. The choice of Gray counters may help reduce the occurrence of latch metastability due to the fact that the reference clock and the divider output signal can be asynchronous with each other. After a given time period, the lock detector determines whether or not the frequency difference between the two signals is within a predetermined range. The counters should then be reset before the beginning of the next counting interval.

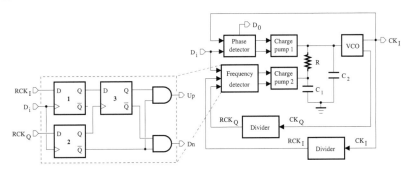

FIGURE 6.53
Block diagram of a dual-loop PLL without a reference clock signal.

A dual-loop PLL can also be designed to operate without a reference clock signal, as shown in Figure 6.53. It requires two separate charge pumps driven by a phase detector and a frequency detector, respectively. In one implementation of the frequency detector, three D flip-flops and two AND gates are necessary [32]. During the frequency acquisition, the frequency detector delivers Up and Dn signals with a frequency that is equal to the difference between the input data frequency and quadrature clock signal frequency. When the free-running frequency of the VCO is set within the loop capture range, the Up and Dn signals are maintained at a low logic level. It is then not necessary to disable the frequency detector, which can continuously track the frequency changes without affecting the operation of the data acquisition loop. Because the VCO is equivalent to a gain element in the frequency domain, a zero is not needed in the transfer function of the filter included in the frequency loop. Hence, the charge-pump circuit controlled by the frequency detector is connected to the node between the resistor R and capacitor C_1.

With V_c being the control voltage delivered by the loop filter in response to the charge-pump current I_p, the filter impedance function can be expressed as

$$Z(s) = \frac{V_c(s)}{I_p(s)} = \frac{N(s)}{sC_2(s + \omega_p)} \tag{6.101}$$

where $\omega_p = (C_1 + C_2)/RC_1C_2$. Assuming that $\omega_z = 1/RC_1$, the numerator is given by $N(s) = s + \omega_z$ for the data acquisition loop or $N(s) = 1$ for the frequency acquisition loop.

The signal CK_Q is 90° out of phase with the signal CK_I. By implementing the VCO as a four-stage ring oscillator, the quadrature signals, CK_I and CK_Q, are available at the outputs of two of the stages. The input range of the FD can be numerically predicted to be on the order of 25% of the desired VCO free-running frequency.

Dual-loop CDRs are not power efficient because an individual PLL is required for each lane. The PI-based CDR improves the power efficiency by using a single PLL that is shared by all lanes, while the GVCO-based CDR uses a single PLL that sets the control signal for all lanes. In a PI-based CDR, the PI resolution is generally limited due to the conflicting design trade-offs between linearity, area, and power consumption. As a result, the phase step that is produced when the PI switches from one phase to another can contribute to the jitter generation. In contrast to PI-based structures, GVCO-based CDRs take advantage of the injection to simultaneously detect and adjust the phase in relation to the incoming data, resulting in a wide bandwidth and a small locking time. However, the capture range (or the jitter tolerance) of GVCO-based CDRs can be limited by mismatches between GVCOs.

6.4.2.2 Phase interpolator-based CDR circuit

A CDR circuit can also be based on phase interpolator (PI), as shown in Figure 6.54. This architecture is generally adopted in high-speed serial links to avoid the need of using a PLL for each of the input/output pins [33]. A single PLL then generates in-phase and quadrature clock signals that are distributed to all channels. It is connected via the voltage-controlled oscillator to the data recovery circuit section. The PI converts the generated clock signals into a signal with the optimal phase for data recovery.

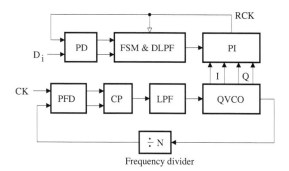

FIGURE 6.54
Phase interpolator-based CDR.

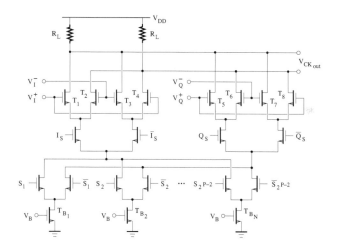

FIGURE 6.55
Phase interpolator (PI).

The PI-based CDR consists of a PLL that generates a quadrature clock signal at the data rate of the input data and a data recovery circuit that aligns the clock signal phases to the transitions on the incoming data stream. The data recovery loop includes a phase detector (PD), a finite-state machine (FSM) and a digital lowpass filter (DLPF), and a PI. A binary PD (or bang-bang PD) provides signals that are related to the sign of the sampling phase error between an interpolated version of the clock signals and transitions of the input data. The following FSM includes demultiplexers (or deserializer),

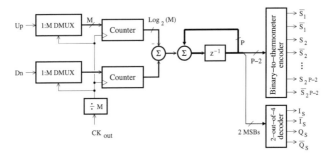

FIGURE 6.56

Finite-state machine and digital lowpass filter.

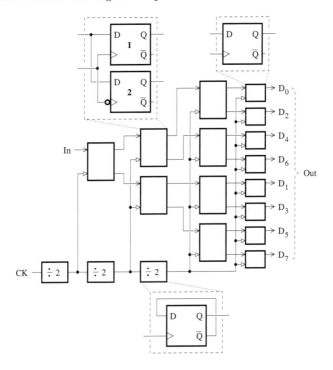

FIGURE 6.57

Circuit diagram of a 1-to-8 demultiplexer.

roll-over counters, and a summer. Its output is applied to a digital lowpass filter, that is based on an integrator, to generate a P-bit binary code. This latter is used to configure the phase interpolator so that one out of 2^P phases is selected each time to align the sampling clock to the center of the input data. Circuit diagrams of the PI, demultiplexer, FSM, and DLPF are shown in Figures 6.55, 6.56, and 6.57, respectively.

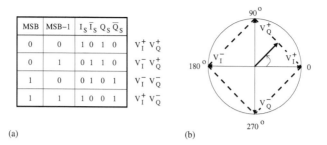

(a)   (b)

FIGURE 6.58
(a) MSB decoding table; (b) quadrant illustration on the clock phase circle.

The PI produces an output clock signal that is a weighted sum of the quadrature clock signals, V_I and V_Q. Hence,

$$V_{CK_{out}} = \alpha V_I + (1-\alpha)V_Q \tag{6.102}$$

where $V_I = V_I^+ - V_I^-$, $V_Q = V_Q^+ - V_Q^-$, and $0 \leq \alpha \leq 1$ and α represents a linear weighting coefficient. By using a P-bit digital code for the PI control, the phase quadrant (or the I/Q polarity) can be selected by two most significant bits (MSBs), while each quadrant is subdivided into 2^{P-2} steps that are associated to the combinations of other $P-2$ bits. Figure 6.58 shows the MSB decoding table and quadrant illustration on the clock phase circle. A 2-bit selection of the I/Q polarity allows for a complete 360° phase rotation with the phase step determined by a weighting digital-to-analog converter with a resolution of $P-2$ bits. By choosing the weighting coefficients as α and $1-\alpha$, the PI can generate arbitrary phase shifts from 0° to 360°. To simplify the implementation, the phase interpolation is based on weighting coefficients that are linear functions of the clock phase, instead of sinusoidal functions generally associated to a phasor rotating in a circle. The resulting signal amplitude variation has no detrimental effect on the data sampling that essentially depends on zero-crossings of the clock signal.

6.4.2.3 CDR circuit based on a gated VCO

CDR circuits based on PLL are characterized by a wide capture range and a good jitter rejection performance, and generally find applications in systems (serial links) operating in continuous mode. A fast acquisition time is often not required because a steady and uninterrupted stream of bits is transmitted.

For burst-mode systems, such as asynchronous transfer mode (ATM) networks and local area networks (LANs), where data are transmitted only for a limited period of time, the acquisition time should be minimized in order to meet the low latency requirement. In this case, suitable CDR architectures can be based on an injection locked oscillator or gated oscillator [34].

The circuit diagram of a CDR based on a gated VCO is shown in Fig-

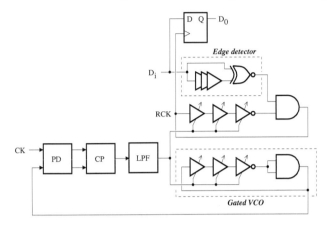

FIGURE 6.59
CDR based on a gated VCO.

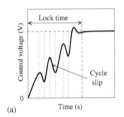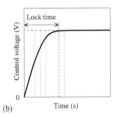

FIGURE 6.60
Representation of the oscillator control voltage in a CDR (a) without phase reset and (b) with phase reset.

ure 6.59. It is composed of a PLL section which tracks the clock signal frequency and a data recovery section that achieves the phase acquisition (or aligns the clock transition to the center of the data eye). The PLL section includes a binary phase detector (PD), a charge-pump (CP) circuit, a lowpass filter (LPF), and a gated VCO (GVCO). The data recovery section consists of an edge detector and a GVCO. The recovered clock signal (RCK) is used to trigger the D flip-flop that provides the recovered data, D_{out}. By using a GVCO, the phase of the recovered clock can be periodically reset to prevent a cycle slip. The delay of the edge detector is typically sized so that a rising edge of the clock signal is located at the center of the data eye whenever the oscillation resumes.

The PD is designed to track phase variations and its performance remains limited in the presence of frequency variations. In this latter case and during the *pull-in* process, the VCO control voltage is adjusted to minimize the difference between the VCO output frequency and the data rate at the input. To minimize the frequency error in a CDR without phase reset, the VCO control voltage can alternatively be changed in the right and wrong directions, leading to a cycle slip, as shown in Figure 6.60(a). The cycle slip process is due to the PD periodic output. On the other hand, in a CDR with phase reset, the VCO control voltage tends to keep moving in the same direction, preventing a cycle slip and resulting in a reduced lock time, as shown in Figure 6.60(b).

The GVCO oscillation frequencies should match the data rate. Once the GVCO control voltage reaches the steady state, the GVCO is no longer reset and the CDR circuit can now operate normally by aligning the clock signal with data sampling instants.

However, in practice, the tolerance of the CDR to continuous identical digits can be affected by mismatches in oscillation frequencies among gated oscillators that can lead to a drift in clock phases.

A CDR based only on a GVCO with a symmetric circuit topology can used to relax the timing requirements.

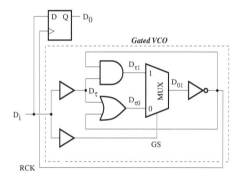

FIGURE 6.61
Reference-free CDR circuit based only on a gated VCO.

In applications (high-speed access networks such as LANs), where the fine delay for timing adjustment has to be set in accordance with a known target clock frequency, the data recovery can be implemented as shown in Figure 6.61.

When implemented using differential building blocks, the data recovery circuit features symmetric loops that operate complementarily and help avoid the timing mismatch between the signals from both loops [35]. As a result, an edge of the recovered clock signal can be aligned with the middle of the data eye. The differential circuit for the AND gate implementation is shown in Figure 6.62(a). The realized logic function is of the form,

$$Z = X \cdot Y \tag{6.103}$$

where $X = X^+ - X^-$, $Y = Y^+ - Y^-$, and $Z = Z^+ - Z^-$. To implement the OR gate, the following Boolean logic identities should be exploited:

$$\overline{Z} = \overline{X} \cdot \overline{Y} \tag{6.104}$$

$$= \overline{X + Y} \tag{6.105}$$

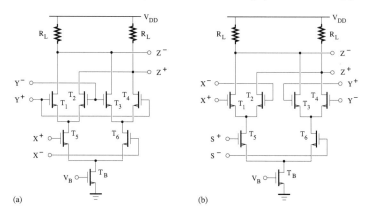

FIGURE 6.62
(a) Differential circuit for the AND and OR gate implementations; (b) differential implementation of the multiplexer.

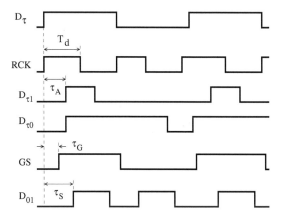

FIGURE 6.63
Waveforms of the CDR based on a gated VCO.

or equivalently

$$Z = X + Y \tag{6.106}$$

where $\overline{X} = X^- - X^+$, $\overline{Y} = Y^- - Y^+$, and $\overline{Z} = Z^- - Z^+$. The AND and OR gates are realized using the same circuit configuration so that their propagation delays can be as close as possible.

Figure 6.62(b) presents the differential implementation of the 2:1 multiplexer.

The waveforms of the data recovery circuit based on a gated VCO are represented in Figure 6.63.

The gated VCO oscillates at a frequency f_0 given by

$$f_0 = 1/(2T_d) \qquad (6.107)$$

where T_d is half of the oscillation period and is set by one of both oscillator loops. The oscillation frequency is then determined by the propagation delays of buffer circuits, the logic gate (AND or OR), and the 2:1 multiplexer.

To design the data recovery circuit so that the signal edge can be detected by the 2:1 multiplexer, the delay of the gating signal, τ_G, must be smaller than the propagation delay of the AND and OR logic gates, τ_A. Hence, $\tau_G < \tau_A$. On the other hand, the difference, $\tau_A - \tau_G$, should be small enough to avoid any unnecessary increase of the jitter.

At a data rate of 12.5 Gb/s, a value of 6.9 ps can be attributed to τ_G, while τ_A is on the order of 7 ps.

6.4.2.4 Reference-less dual-loop CDR circuit

A dual-loop CDR can also be designed without an external reference clock signal, as shown in Figure 6.64(a). It consists of a binary phase detector (PD), a rotational frequency detector (FD), two charge-pump circuits, CP_1 and CP_2, a loop filter, and a VCO. The circuit diagram of the half-rate binary PD is depicted in Figure 6.64(b), where D_0 and D_{180} represent the recovered data.

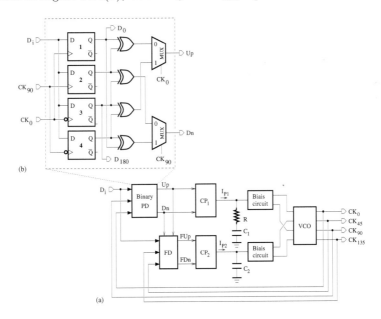

FIGURE 6.64
Dual-loop CDR with a half-rate binary PD.

A reference-less dual-loop CDR [36–38] extracts the clock frequency di-

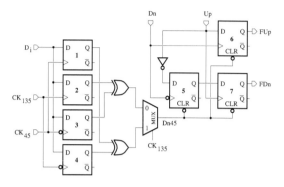

FIGURE 6.65
Rotational frequency detector (FD).

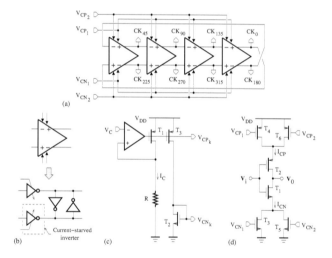

FIGURE 6.66
(a) VCO; (b) pseudo-differential inverter stage; (c) voltage-controlled bias circuit; (d) current-starved inverter.

rectly from the input data stream, but its frequency acquisition range is limited to about $\pm 50\%$ of the data rate due to the effects of the sampling process on the frequency detector. The VCO is driven by a first-order frequency loop towards the frequency lock state by directly extracting the frequency error from the incoming data stream, D_i. A second-order phase loop including the PD then takes over to appropriately align the data bits and the sampling edges of the clock signal, thus minimizing the phase error. The phase loop has a capture range that is smaller than the one of the frequency loop.

  The half-rate binary phase detector of Figure 6.64(b) samples the input data using the clock signals CK_0, CK_{90}, CK_{180}, and CK_{270}. The output signal of each of the XOR gates is valid for the phase range of 270°. Specifically for the topmost multiplexer, the inputs selected by the low and high logic states are valid from 90° to 360° and from 270° to 360° + 180°, respectively, while for bottommost multiplexer, the inputs selected by the low and high logic states are valid from 180° to 360° + 90° and from 360° to 360° + 2700°, respectively. By using the clock signals CK_0 and CK_{90} as selection signals for the multiplexers, only the valid 180° portions of each of the XOR outputs can be transferred to the PD outputs. When the phase loop is in the locked state, the edges of quadrature clock signals are aligned with the data transitions.

The rotational FD is represented in Figure 6.65. It generates the signals FUp and FDn using the clock signals CK_{45} and CK_{135}, and the PD output signals, Up and Dn. The rotational FD generates an FUp pulse if a data transition occurs between the clock phases, 0° and 45°, and the next between 135° and 180°. It also produces an FUp pulse if a data transition occurring between the clock phases, 180° and 225°, is followed by another one between 315° and 360°. For a data transition that takes place between the clock phases, 315° and 0°, and the next between 180° and 225°, or for a data transition that takes place between the clock phases, 135° and 180°, and the next between 0° and 45°, a FDn pulse is generated. In the locked state, data transitions can only occur around the edges of the clock signals CK_{90} and CK_{270}. Consequently, there is no data transition taking place between 315° and 45°, and between 135° and 225°. That is, the signal $Dn45$ is set to the low logic state. Because the rotational FD is reset using the signal $Dn45$, it remains disabled as long as the frequency loop is the locked state.

In the phase domain, the phasor representing the phase difference between data transitions and clock signal edges rotates in a clockwise direction if the clock frequency is lower than the data rate, and an anticlockwise direction if the clock frequency is higher than the data rate. It remains stationary if the phase difference is equal to zero.

A four-stage ring oscillator that produces quadrature outputs is represented in Figure 6.66(a). It consists of pseudo-differential inverter stages, as shown in Figure 6.66(b). The oscillation frequency can be adjusted using the voltage-controlled bias circuit of Figure 6.66(c), where $k = 1, 2$. The circuit diagram of a current-starved inverter is represented in Figure 6.66(d).

Each delay stage produces an output current that not only flows into its own parasitic load capacitors, but also the parasitic interstage coupling capacitor. As a result, the delay stage matching can be improved due to phase error averaging. Assuming that $\triangle\phi$ is the phase difference between adjacent delay stages and N is the number of stages, the phase shift around the loop that, by taking into account the feedback phase inversion, is of the form, $N\triangle\phi + \pi$, should be equal to 2π.

6.4.2.5 Reference-less single-loop CDR circuit using a linear phase detector

A reference-less CDR circuit can also be designed, as shown in Figure 6.67, by exploiting the single-sided frequency detection capability of the half-rate linear PD [39]. It is based on a single-loop structure that is composed of a PD, a CP circuit, a filter, and a VCO. The circuit diagram of a half-rate linear PD is represented in Figure 6.68, where flip-flops are triggered either by the recovered clock signal RCK or by its complementary version $\overline{RCK}$, and D_0 is the recovered data.

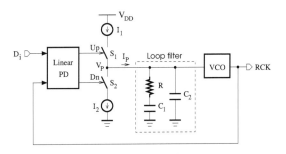

FIGURE 6.67
Single-loop CDR circuit using a linear PD $(I_1 = 2I_2)$.

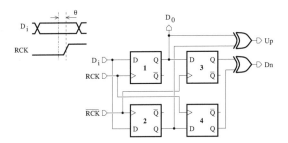

FIGURE 6.68
Half-rate linear PD.

The timing diagram of Figure 6.69, where $\oplus$ is the exclusive OR operation, illustrates the CDR operation when the clock frequency is equal to half the data rate. It was assumed that the duration of one data bit is 2π and the clock signal frequency, f_{RCK}, is half the data frequency, f_{D_i}. In the case where the sampling edges of the clock signal are exactly aligned with the data eye (or center), the Up and Dn pulses take the widths of π and 2π, respectively. Assuming that $I_1 = 2I_2 = I$ and θ is the phase difference between the center of the data and the rising edge of the clock signal, the average current delivered

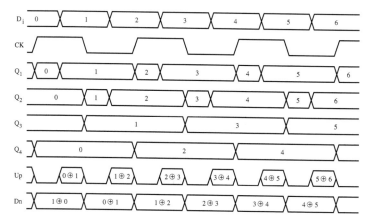

FIGURE 6.69
Timing diagram of a half-rate linear PD.

by the CP circuit can be expressed as

$$\bar{I}_P = \alpha I_1 \times \frac{\theta + \pi}{2\pi} - \alpha I_2 \times \frac{2\pi}{2\pi} \tag{6.108}$$

$$= \alpha I \frac{\theta}{2\pi} \quad \text{for} \quad -\pi < \theta < \pi \tag{6.109}$$

where α is the bit transition density.

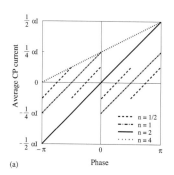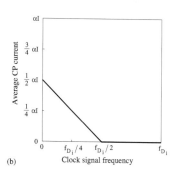

FIGURE 6.70
Phase (a) and frequency (b) characteristics of a half-rate linear PD.

General expressions of the average CP current can be derived by analyzing the PD timing diagram under various operating conditions.

When $f_{RCK} = f_{D_i}/n$, where n is an integer greater than 2, and the sampling edges of the clock signal are exactly aligned with the data eye, the widths of the Up and Dn pulses are $(n-1)\pi$ and $n\pi$ per every $n/2$ bits, respectively.

The average CP current is given by

$$\bar{I}_P = \alpha I \times \frac{\theta + (n-1)\pi}{n\pi} - \alpha\frac{I}{2} \times \frac{n\pi}{n\pi} \tag{6.110}$$

$$= \alpha I \frac{\theta + (n-2)\pi/2}{n\pi} \quad \text{for} \quad -\pi < \theta < \pi \tag{6.111}$$

When $f_{RCK} = f_{D_i}/n$, where n is an odd integer greater than or equal to 3, and $\theta = \pm\pi/2$, the widths of the Up and Dn pulses are $2(n-1)\pi$ and $2n\pi$ per every n bits, respectively. Depending on the value of θ, the average CP current can be written as

$$\bar{I}_P = \alpha I \times \frac{\theta + \pi/2 + (n-1)\pi}{n\pi} - \alpha\frac{I}{2} \times \frac{n\pi}{n\pi} \tag{6.112}$$

$$= \alpha I \frac{\theta + (n-1)\pi/2}{n\pi} \quad \text{for} \quad -\pi < \theta < 0 \tag{6.113}$$

and

$$\bar{I}_P = \alpha I \times \frac{\theta - \pi/2 + (n-1)\pi}{n\pi} - \alpha\frac{I}{2} \times \frac{n\pi}{n\pi} \tag{6.114}$$

$$= \alpha I \frac{\theta + (n-3)\pi/2}{n\pi} \quad \text{for} \quad 0 < \theta < \pi \tag{6.115}$$

The phase characteristic is represented in Figure 6.70(a) for various values of n.

The average CP current with respect to the phase is defined as

$$\bar{I}_P = \frac{1}{2\pi} \int_{-\pi}^{\pi} I_P \, d\theta \tag{6.116}$$

Because it also shows a dependence on the clock signal frequency, the frequency characteristic of the half-rate linear PD can be plotted as shown in Figure 6.70(b). Data points from the frequency characteristic can be interpolated to derive an equation of the following form:

$$\bar{I}_P = \begin{cases} \dfrac{\alpha I}{2}\left(1 - \dfrac{2f_{RCK}}{f_{D_i}}\right) & \text{if} \quad f_{RCK} \leq \dfrac{f_{D_i}}{2} \\ 0 & \text{if} \quad f_{RCK} > \dfrac{f_{D_i}}{2} \end{cases} \tag{6.117}$$

The half-rate linear PD can then be used for frequency detection provided the clock signal frequency does not exceed half the frequency of the input data. After the frequency acquisition, the CDR enters the next operation mode that consists of minimizing the phase difference between the clock signal and the input data.

In addition to providing a phase detection capability, a full-rate linear PD can also be used to detect a frequency difference [40]. A reference-less CDR

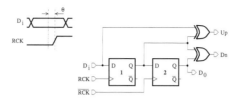

FIGURE 6.71
Full-rate linear PD.

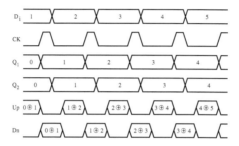

FIGURE 6.72
Timing diagram of a full-rate linear PD ($f_{RCK} = f_{D_i}$).

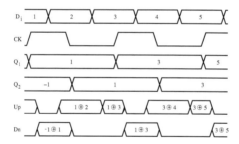

FIGURE 6.73
Timing diagram of a full-rate linear PD ($f_{RCK} = f_{D_i}/2$).

based on a single-loop architecture with $I_1 = I_2$ (see Figure 6.67) that includes a full-rate linear PD then has a frequency detection capability as long as the clock frequency is lower than the frequency of the incoming data.

The circuit diagram of a full-rate linear PD is shown in Figure 6.71. The retimed data is available at the D_0 terminal. Note that θ represents the phase difference between the center of the received data and the rising edge of the recovered clock. The timing diagram of Figures 6.72 and 6.73 illustrate the operation of a full-rate linear PD, when $f_{RCK} = f_{D_i}$ and $f_{RCK} = f_{D_i}/2$, respectively. Note that f_{RCK} is the frequency of the recovered clock signal and f_{D_i} is the frequency of the received data.

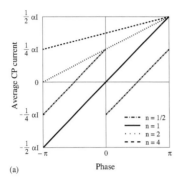

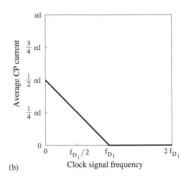

(a) Phase   (b) Clock signal frequency

FIGURE 6.74
Phase (a) and frequency (b) characteristics of a full-rate linear PD.

Assuming that $f_{RCK} = f_{D_i}/n$, where n is a number, and the duration of one bit of the received data is 2π, the pulse width at the Dn output is $n\pi$ per every n bits, while the pulse width at the Up output is $(2n-1)\pi$ per every n bits. The Dn pulse is generated by computing the XOR logic function of a given bit and only the adjacent bit at the n-th position, while for the generation of the Up pulse, it requires an estimate of the XOR logic function of a given bit and each of the consecutive neighboring bits from the first to the n-th position. Hence, assuming that $I_1 = I_2 = I$, the average CP current can be expressed as

$$\bar{I}_P = \alpha I \times \frac{\theta + (2n-1)\pi}{2n\pi} - \alpha I \times \frac{n\pi}{2n\pi} \tag{6.118}$$

$$= \alpha I \times \frac{\theta + (n-1)\pi}{2n\pi} \quad \text{for} \quad -\pi < \theta < \pi \tag{6.119}$$

Simulation results show that the average CP current is always positive regardless of θ if $f_{RCK} \leq f_{D_i}$, whereas it is always equal to zero if $f_{RCK} > f_{D_i}$. The expression of the average CP current as a function of the clock frequency is given by

$$\bar{I}_P = \begin{cases} \dfrac{\alpha I}{2}\left(1 - \dfrac{f_{RCK}}{f_{D_i}}\right) & \text{if} \quad f_{RCK} \leq f_{D_i} \\ 0 & \text{if} \quad f_{RCK} > f_{D_i} \end{cases} \tag{6.120}$$

Therefore, the full-rate linear PD has a single-sided frequency detection capability. The phase and frequency characteristics of a full-rate linear PD are shown in Figures 6.74(a) and (b), respectively.

On the other hand, it should be noted that a binary PD does not exhibit a frequency detection capability.

6.5 Delay-locked loop

In applications (multi-phase clock generation and synchronization, memory interface) where the desired clock frequency is already available, a voltage-controlled delay line (VCDL) can be used instead of a VCO. Hence, while the PLLs generate a new clock signal, the delay-locked loops (DLLs) actually retard the incoming clock signals in such a way that the delay in the output clock signals is greatly reduced.

The block diagram of a DLL is shown in Figure 6.75. The output signal is generated by delaying the input clock. The phase difference between the VCDL output and reference clock signals is provided by a phase detector whose output is applied to the loop filter, which generates the VCDL control voltage. The delay cell is actually equivalent to a gain stage and the stabilization of the loop can be achieved using a first-order lowpass filter.

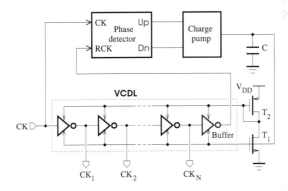

FIGURE 6.75
Block diagram of a DLL.

The closed-loop phase transfer function is given by

$$T(s) = \frac{\Theta_0(s)}{\Theta_i(s)} = \frac{1}{1 + \dfrac{1}{K_d K_p H(s)}} \tag{6.121}$$

where K_d is the gain of the VCDL and H is the transfer characteristic of the loop filter, which is realized here by a charge-pump circuit loaded by a capacitor and is then similar to an integrator.

A VCDL can be implemented as shown in Figure 6.76(a) [18]. The biasing circuit uses a current mirror to source and sink current corresponding to the dc level of the control voltage, V_c. Transistors $T_1 - T_2$ operate as a current source and the inversion of the input signal is achieved by $T_3 - T_4$. The time delay is related to the output load capacitance. The inverter structure depicted in Figure 6.76(b) can be adopted to improve the output driving capability.

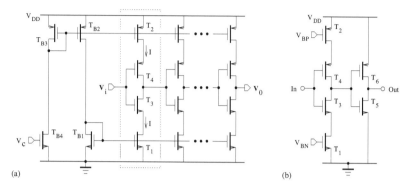

FIGURE 6.76
(a) Delay line-based oscillator using inverters; (b) inverter stage with an improved driving capability.

Generally, single-ended structures like the ones of Figure 6.76 are sensitive to the power supply noise, which affects the phase of the output signal.

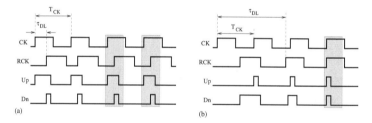

FIGURE 6.77
DLL signal waveforms during abnormal operation: (a) stuck false lock ($\tau_{DL} < T_{CK}$), (b) harmonic lock ($T_{CK} < \tau_{DL} < 2T_{CK}$).

DLLs with a wide operating range are required in various applications. In general, a DLL can lock the input clock period, T_{CK}, to the time delay, τ_{DL}, associated to the VCDL and replica buffer, only if the value of τ_{DL} is comprised between $T_{CK}/2$ and $3T_{CK}/2$. As a result, with $\tau_{DL} = N \cdot T_{CK}$, where N is an integer greater than or equal to one, the rising edge of the input clock is aligned with the one of the output generated by the VCDL. A wide-range DLL can feature a significantly large spread between the minimum and maximum values of τ_{DL}, and therefore, be subject to false lock problems.

A PFD is often used as a phase detector in a DLL because its operating range is not limited by the duty cycle ratio of the clock signal. But, its improper state setting, that can be expected to occur at the operation start, can cause a stuck false lock. For a correct operation under the condition $T_{CK} > \tau_{DL}$, the Dn signal should be high to allow that the delay τ_{DL} can increase until it reaches the value T_{CK}. Meanwhile, the Up signal may become

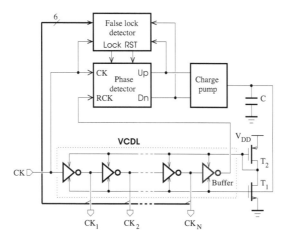

FIGURE 6.78
Block diagram of a wide-range DLL.

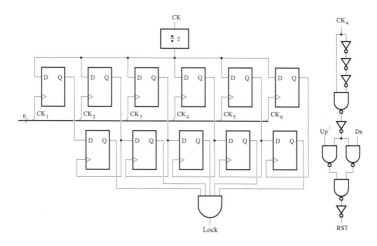

FIGURE 6.79
Circuit diagram of the false lock detector.

high as a result of RCK arriving later than CK. This leads to a stuck false lock, as shown in Figure 6.77(a). A measure for counteracting the stuck false lock can consist of resetting the PFD.

When the DLL is unable to distinguish the normal lock case ($N = 1$) from the other cases ($N > 1$), as shown in Figure 6.77(b), the false lock is said to be harmonic. An effective way to prevent this situation is to always keep the condition $\tau_{DL} < 2T_{CK}$ satisfied.

The frequency lock range of a conventional DLL appears to be very limited,

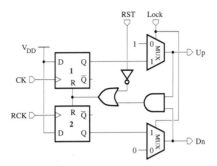

FIGURE 6.80
Circuit diagram of the phase detector.

and a solution to achieve a wide operation range then consists of using an extra control circuit to help meet the locking requirements set on the delay τ_{DL}.

The block diagram of a wide-range DLL is shown in Figure 6.78. It consists of a phase detector, a charge-pump circuit loaded by a capacitor, a multi-phase voltage-controlled delay line, and a false lock detector [41].

The circuit diagram of the false lock detector is depicted in Figure 6.79. It comprises a harmonic lock detector that helps maintain the delay τ_{DL} less than $2T_{CK}$ and a reset circuit that can correct the effect related to the improper state setting of the phase detector.

Figure 6.80 shows the circuit diagram of the phase detector, which is realized by adding two input signals, RST and Lock, to a conventional PFD.

In normal operation, the delay τ_{DL} can be reduced only when the rising edge of RCK occurs between the times T_{CK} and $(7/6)T_{CK}$. Hence, the Up signal can take the high level only during $(\alpha - 1)T_{CK}$, where $1 < \alpha < 7/6$. To correct a malfunction caused by the improper PD state setting, the state of the Dn signal is checked against CK_4, and if the Dn signal is high at the arrival of the rising edge of CK_4, the phase detector will be reset so that the normal operation can resume. On the other hand, the increase of the delay τ_{DL} can only take place provided the rising edge of RCK occurs before the duration T_{CK}. After an occurrence of the improper PD state setting, the phase detector should be reset. Here, this is achieved by checking if the state of the Up signal is high at the arrival of the rising edge of CK_4.

For normal operation of the phase detector, the signal Lock must be in the high state. When the signal Lock takes the low state, the Up output is set to the high level. Subsequently, the harmonic lock can be prevented by increasing the delay line control voltage until the delay τ_{DL} becomes less than $(7/6)T_{CK}$. In general, the maximum capture range is $(N + 1)T_{CK}$ for a false lock detector using a number N of multi-phase clocks.

A DLL is less prone to stability problems than a PLL whose loop bandwidth is often determined by a high-order narrowband filter requiring large components with a high sensibility to the IC process, voltage, and temperature variations. The requirement to use such a loop filter in a PLL also results in a longer acquisition time, usually in the 1/2 to 1 microsecond range, while a DLL can lock to the data rate in just a few clock cycles. By employing components with high values, a PLL occupies a large chip area compared to a DLL. Because a DLL adjusts the amount of delay or phase without affecting the frequency in order to achieve the desired synchronization, its operating frequency range is severely limited. Furthermore, a DLL may falsely lock into a harmonic frequency of the reference clock signal.

6.6 PLL with a built-in self-test structure

Typically, functional tests of PLLs consist of verifying parameters such as loop gain, loop bandwidth, lock time, capture range, lock range, and jitter. These tests are realized by applying a stimulus to the PLL and then observing the corresponding response. But, their design and execution may be time-consuming. The lock range, for instance, is determined only after the phase lock has been achieved and by gradually increasing or decreasing the stimulus frequency until the phase locking condition is no longer met. Furthermore, a tester allowing a high precision control of signal transition timing is required.

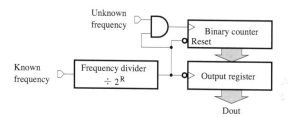

FIGURE 6.81
BIST circuit for the frequency measurement.

By designing a PLL with a built-in self-test (BIST) structure [42–45], the number of test devices required after the manufacturing phase is reduced. The BIST circuit is preferably designed to operate without affecting the internal signal path of the PLL. It can then be connected to usual PLL outputs, some of which are: VCO output, frequency divider output, and phase lock indicator output.

A simple BIST circuit is shown in Figure 6.81. It is based on pulse counting using a gated binary counter and a reference frequency, and can be used to verify the VCO center frequency or any PLL frequency range. The output count is the digital representation of the quantity $2^{R-1} f_{unknown}/f_{known}$, where f_{known} and $f_{unknown}$ are the known and unknown frequencies, respectively, and 2^R is the frequency divider factor.

6.6.1   Gain, capture and lock range, and lock time

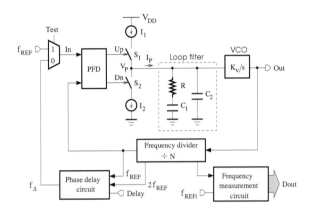

FIGURE 6.82
Circuit diagram of a PLL with BIST for the measurement of gain, capture, and lock characteristics.

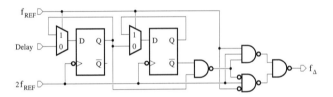

FIGURE 6.83
Phase delay circuit.

In general, the PLL open-loop gain can be written as,

$$G_{OL} = \frac{K_p K_v H(s)}{N \cdot s} \tag{6.122}$$

where $H(s)$ is the filter transfer function, K_p denotes the gain of the phase detector and charge-pump section, K_v is the VCO gain, and N is the frequency divider factor. The open-loop gain is a suitable characteristic to be tested because it is dependent on parameters that determine the PLL behavior.

The verification of the open-loop gain G_{OL}, the capture and lock range, and the lock time can be performed by using the BIST structure of Figure 6.82, that is based on phase shifting. To determine the open-loop gain, the PLL loop is open by preventing the feedback signal applied at the PFD input to be dependent on the VCO oscillation frequency. The frequency divider output is used to generate an input to the PLL at the same frequency as the feedback signal. As a result, no phase difference will be detected by the PFD and the charge-pump circuit will not source nor sink current. Ideally, the output of the

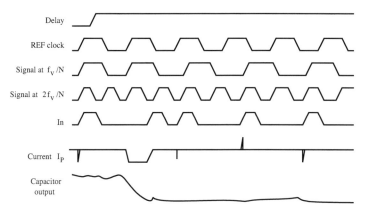

FIGURE 6.84
PLL waveforms during a test.

loop filter will then remain unchanged. From this last state, the application at the PLL inputs of a phase shift lasting only one cycle of the feedback signal, whose frequency equals f_{REF}, affects the VCO output frequency that first changes quickly and then remains constant as soon as the phase shift becomes zero again. Next, the VCO output frequency change can be quantified to ascertain the open loop gain.

Using the phase delay circuit of Figure 6.83 in the implementation of the PLL BIST, a 25% duty cycle output pulse can be delayed by half a clock cycle relative to the frequency f_v/N in response to the detection of the high logic state just taken by the control signal Delay.

The PLL waveforms during a test are shown in Figure 6.84. Initially, the PLL is locked to align the rising edges of the reference and feedback clock signals, such that the frequency f_v/N is equal to f_{REF}. Following the introduction of a phase shift, the PLL begins to adjust the VCO frequency in order to re-align the reference and feedback clocks.

Especially, the application of a phase shift, $\triangle \theta_i$, that lasts ρ clock cycles, to the PLL inputs leads to a VCO frequency variation, $\triangle \omega_v$. The VCO can be described by the next equation,

$$\triangle \omega_v = K_v V_{tune} \tag{6.123}$$

where V_{tune} is the tuning voltage generated at the filter output. Assuming that the loop filter is reduced to a single capacitor, the instantaneous current and voltage can be written as,

$$i_c = C \frac{dv_{tune}}{dt} \tag{6.124}$$

Considering only variations of average current and voltage, we arrive at

$$\overline{I_c} = C \frac{V_{tune}}{\triangle t} \tag{6.125}$$

where $\triangle t = \rho/f_{REF}$, $\overline{I_c} = K_p \triangle \theta_i$, and $K_p = I_p/(2\pi)$. Hence,

$$\triangle \omega_v = \frac{\rho K_v I_p \triangle \theta_i}{2\pi f_{REF} C} \tag{6.126}$$

The feedback phase change due to the VCO frequency step can be derived as,

$$\triangle \theta_{fb} = \frac{\triangle \theta_v}{N} = \frac{\triangle \omega_v|_{\rho=1} \cdot \triangle t}{N} = \frac{K_v I_p \triangle \theta_i}{2\pi f_{REF}^2 NC} \tag{6.127}$$

The open-loop gain is then given by

$$G_{OL} = \left| \frac{\triangle \theta_{fb}}{\triangle \theta_i} \right| = \frac{K_v I_p}{2\pi f_{REF}^2 NC} \tag{6.128}$$

Using the fact that $\triangle \theta_i = 2\pi f_i$ and $\triangle \omega_v = 2\pi \triangle f_v$, where f_i and $\triangle f_v$ denote the input frequency and the VCO frequency variation, respectively, and combining (6.126) and (6.128) yield

$$G_{OL} = \left| \frac{\triangle \theta_{fb}}{\triangle \theta_i} \right| = \frac{\triangle f_v}{\rho f_{REF} f_i N} \tag{6.129}$$

The value of the open-loop gain can be determined by performing frequency measurements. To reduce the effect of possible duty-cycle distortions or mismatch between the sink and source charge-pump currents, the measurements can be realized for an input phase lead and phase lag, $\pm f_i$, and then be averaged.

The capture range is the range of input frequencies over which the PLL can still lock onto the incoming signal, while the lock range is the input frequency range within which the PLL is able to remain in a previously acquired lock state. In the specific case of PLLs using edge-sensitive PFDs, the capture range and lock range seem equal.

One measurement approach is to continuously apply a frequency shift to induce a slow and controlled change of the VCO output frequency towards its maximum (or minimum) value so that several measurements can be performed during the frequency transition. This is realized by connecting the PLL's input to a version of the frequency divider output with a frequency equal to twice (or half) the internal feedback frequency. The output frequency is monitored continuously until no further significant change can take place within a specified time interval, that is, the minimum (or maximum) value is reached, and the last frequency count is retained.

The lock time is the amount of time required by the PLL to acquire a locked state with an input signal within its capture range after the power-up or a frequency change.

When a lock detector is part of the PLL circuit, its indication can be exploited for the lock-time measurement. Alternatively, the lock time can also be measured by first driving the VCO to its maximum frequency and then counting the number of reference clock cycles until the lock state is reached within some defined margin.

6.6.2   Jitter

The PLL clock jitter is another characteristic that can be measured using the BIST approach. It is a fundamental limitation in high-speed applications.

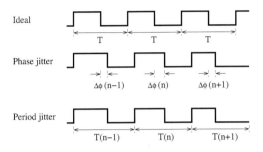

FIGURE 6.85
Phase and period jitter.

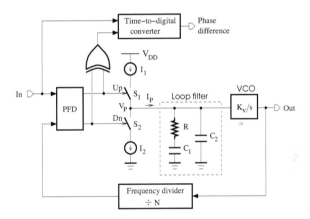

FIGURE 6.86
Circuit diagram of a PLL with time-to-digital converter-based BIST for jitter measurement.

Jitter can be classified as phase jitter, period jitter (see Figure 6.85), or cycle-to-cycle jitter.

Phase jitter refers to the deviation of rising or falling edges of a clock signal as compared to the ideal clock signal. It can be defined as,

$$J_\phi(n) = \Delta\phi(n) - \overline{\Delta\phi} \tag{6.130}$$

where $J_\phi(n)$ is the phase jitter during the cycle n, and $\overline{\Delta\phi}$ is the average value of the phase offset.

Period jitter is used to describe the random period fluctuation with respect

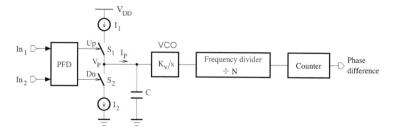

FIGURE 6.87
Circuit diagram of a time-to-digital converter.

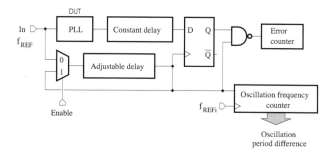

FIGURE 6.88
Circuit diagram of a PLL with delay-based BIST for cycle-to-cycle jitter measurement.

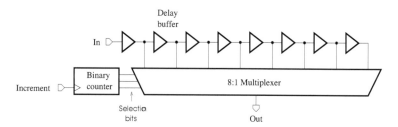

FIGURE 6.89
Circuit diagram of an adjustable delay.

to the ideal or average period. Its definition for the clock cycle n can be stated as,

$$J_T(n) = T(n) - T \qquad (6.131)$$

where $T(n)$ is the period jitter during the cycle n, and T is the ideal period of the clock signal.

Cycle-to-cycle jitter is defined as the deviation in period length of adjacent cycles of a signal. It is calculated by using the following formula,

$$J_{cc}(n) = T(n) - T(n-1) \qquad (6.132)$$

where $J_{cc}(n)$ denotes the cycle-to-cycle jitter for the cycle n.

Jitter is better described by taking many measurement samples that can be processed to determine peak-to-peak and RMS values.

Let N be the number of samples obtained by measuring the phase offset $\triangle\phi(n)$ or the period values $T(n)$. The peak-to-peak jitter can be defined as the largest swing in phase offset or period values and is given by

$$J_{\phi,PP} = \max\{\triangle\phi(1), \triangle\phi(2), \cdots, \triangle\phi(N)\}$$
$$- \min\{\triangle\phi(1), \triangle\phi(2), \cdots, \triangle\phi(N)\} \qquad (6.133)$$

or

$$J_{T,PP} = \max\{T(1), T(2), \cdots, T(N)\}$$
$$- \min\{T(1), T(2), \cdots, T(N)\} \qquad (6.134)$$

The RMS jitter can be expressed as

$$J_{\phi,RMS} = \sqrt{\frac{1}{N} \sum_{n=1}^{N} (\triangle\phi(n) - \overline{\triangle\phi})^2} \qquad (6.135)$$

or

$$J_{T,RMS} = \sqrt{\frac{1}{N} \sum_{n=1}^{N} (T(n) - \overline{T})^2} \qquad (6.136)$$

where

$$\overline{\triangle\phi} = \frac{1}{N} \sum_{n=1}^{N} \triangle\phi(n) \qquad (6.137)$$

and

$$\overline{T} = \frac{1}{N} \sum_{n=1}^{N} T(n) \qquad (6.138)$$

represent the average phase offset and period, respectively. Note that the period jitter can be linked to the phase jitter by an equation.

Several methods can be adopted to implement BIST circuits for PLL jitter measurements. They generally exploit the concept of time interval evaluation.

The PLL BIST circuit of Figure 6.86 is based on a time-to-digital converter that can detect the phase difference between the reference and feedback clock signals. The time-to-digital converter, as shown in Figure 6.87, is composed of a phase and frequency detector (PFD), a charge-pump circuit loaded by a capacitor C, a voltage-controlled oscillator (VCO), a frequency divider, and a counter.

The PFD is enabled only at each sampling instant to detect the phase difference between the reference and feedback clock signals that is assumed to be proportional to the time delay $\triangle T$. The corresponding voltage variation, $\triangle V$, generated at the output of the charge-pump circuit can be written as,

$$\triangle V = \frac{I_p}{C} \cdot \triangle T \tag{6.139}$$

where I_p is the charge-pump current. The change of the VCO oscillation frequency, $\triangle f_v$, is translated into the frequency variation $\triangle f$ at the frequency divider output. Hence,

$$\triangle f = \frac{\triangle f_v}{N} = \frac{K_v \triangle V}{N} \tag{6.140}$$

where N is the frequency divider factor, and K_v is the gain of VCO. The signal frequency is attenuated by the frequency divider to allow a proper operation of the counter. When its evaluation by the counter lasts T_c, the count difference can be obtained as

$$\triangle \text{Count} = \triangle f \cdot T_C \tag{6.141}$$

Combining (6.139), (6.140), and (6.141) yields

$$\triangle T = \frac{N \cdot C}{K_v I_P T_C} \cdot \triangle \text{Count} \tag{6.142}$$

The count difference is proportional to the phase difference between the reference and feedback clock signals. An initial test with a known $\triangle T$ can be exploited for the determination of the coefficient $N \cdot C/(K_v I_P T_C)$. Note that the measurement accuracy can be limited by mismatch non-linearities and power-supply noise. The charge-pump current should then be sized as large as possible to reduce the measurement error to about a few picoseconds.

A jitter measurement method that uses constant and adjustable delay lines, a phase detector, and counters, as shown in Figure 6.88, makes it possible to estimate the auto-correlation function of the reference and feedback clock signals around the signal period. Cycle-to-cycle jitter is measured by counting all events where the signal and its delayed version differ. The output of the D flip-flop is compared with the expected logic state, and the error counter is incremented each time an error is found out. When the maximum count, E, is reached, the counter output is latched and then reset. Figure 6.89 shows the circuit diagram of an adjustable delay line that consists of a counter, a multiplexer, and delay buffers.

To measure the RMS jitter, the adjustable delay line is first set to exhibit the minimum time delay guaranteeing that the error count can remain equal to zero for each group of E cycles of the PLL input signal. Next, the delay line counter is incremented after each E cycles to adjust the time delay, and its counts for which the error count is, for instance, 15.9% and 84.1% of E are recorded. This can be interpreted as the standard deviations ($\pm\sigma$) of a random variable (jitter) with normal distribution.

For the peak-to-peak jitter measurement, it is required to store the delay line counter contents for which the error count is first equal to one and first equal to E (or any other magnitude range boundaries). The adjustable delay line is operated in a self-oscillating mode to determine the delay in both cases, so that the delay difference can then be computed.

Vernier delay lines [46] can also be used to measure the period or phase error in a PLL with a high accuracy. However, in addition to the large area overhead, the long latency of delay-line-based measurements limits the application to relatively low frequencies.

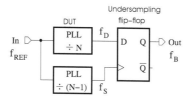

FIGURE 6.90
Circuit diagram of a PLL with undersampling-based BIST for jitter measurement.

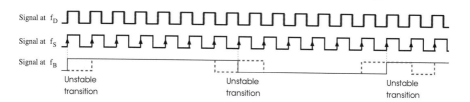

FIGURE 6.91
Waveforms illustrating the undersampling principle.

An undersampling method [47] can be adopted to ease the speed and area requirements of the PLL BIST circuit. Figure 6.90 shows the circuit diagram of a PLL with undersampling-based BIST for jitter measurement. The D flip-flop receives the output signal that is generated by the PLL under-test at the frequency f_D. Its sampling frequency f_S (with $f_S = f_D(N-1)/N$) is set by a second PLL (or any reference source). Jitter characteristics can then be determined by monitoring the output samples of the D flip-flop to detect signal transitions occurring after a relatively long stopover at the same logic level and capturing amplitude variations that are caused by jitter.

By sampling at a rate, f_S, slightly lower than the transmitted data rate, f_D, a signal is produced at a much slower frequency, f_B, called the beat frequency. This is related to the fact that both frequencies, f_S and f_D, are very close to each other so that a sequence of data samples at the same logic level will first be acquired before switching to the other logic level. Due to fluctuations of data edges induced by jitter, amplitude variations are produced around the rising and falling edges of the signal f_B when the sampling takes place around rising or falling edges of the data sequence f_D. Figure 6.91 shows waveforms illustrating the undersampling principle. The timing information of signal sequences with amplitude variations can be collected and analyzed as data samples with normal distribution. The accuracy of the jitter determination is improved by increasing the size of collected data samples.

The undersampling method is capable of evaluating jitter with sub-picosecond resolution in applications requiring gigabit data transfers.

6.7   PLL specifications

A PLL will operate at the free-running frequency, ω_0, when it is not locked to an input signal. This frequency is determined by the VCO components. A nonlinear model of the PLL is necessary for the characterization of the acquisition process.

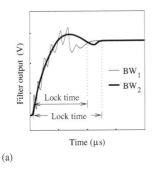
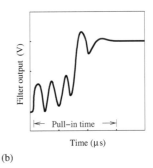

(a)   (b)

FIGURE 6.92
Transient response of the VCO control voltage due to a step signal in the PLL input: Illustration of the (a) lock time and (b) pull-in time.

A PLL is generally designed to meet a specified lock range, lock time, pull-in range, pull-in time, pull-out range, and loop noise bandwidth.

Let the frequency range of the PLL input signal be in the frequency range from $\omega_0 - \triangle\omega$ to $\omega_0 + \triangle\omega$.

Lock range, $\pm\triangle\omega_L$, is the angular frequency range over which the PLL acquires a lock within the first cycle of the input signal or within a given frequency tolerance. The lock range is affected by the difference between the input and free-running frequencies. Note that $\triangle\omega_L \geq \triangle\omega$. The parameter $\triangle\omega_L$ is centered at the free-running angular frequency and determines the PLL operating frequency range. It decreases with increasing loop bandwidth. However, a significant increase of the loop bandwidth may cause loop instability.

Lock time, t_L, is the time after which the PLL can lock in one single-beat note between the input reference frequency and the output frequency. The lock time, as illustrated in Figure 6.92(a)) for two loop bandwidths, is primarily a function of the loop characteristics.

Pull-out range, $\pm\triangle\omega_{po}$, is the angular frequency range over which the PLL operation is stable. The pull-out range is smaller than or equal to the pull-in range.

Pull-in range, $\pm\triangle\omega_{pi}$, is the angular frequency range over which the PLL would naturally become locked after the cycle slipping occurrence, as shown in Figure 6.92(b), where the frequency of the input step is assumed to lie outside the lock range. Cycle slipping occurs when the PFD is unable to accurately follow the phase error variations caused by large frequency offsets, for instance. In the event of a cycle slip, the PLL momentarily loses the frequency lock before settling at the actual output frequency.

Hold range, $\pm\triangle\omega_H$, is the input frequency range over which the PLL can maintain static lock. It then represents the maximum static tracking range. When the frequency offset of the input signal is less than the holding range but larger than the pull-in range, the PLL is said to be conditionally stable. Theoretically, the hold range can be high enough. However, in practice, it can be limited by the operating or tuning range of PLL components, especially the voltage-controlled oscillator.

Note that $\triangle\omega_H > \triangle\omega_{pi} > \triangle\omega_{po} > \triangle\omega_L$.

Pull-in time, t_p, is the transient time required by the PLL to always become locked. The pull-in process has a nonlinear nature and can be slow and unreliable. Hence, the pull-in time is dependent on the PLL initial conditions (frequency and phase errors) and loop characteristics.

Loop noise bandwidth, B_n, is the one-sided bandwidth of a unity-gain and ideal lowpass filter, which transmits as much noise power as does a linear PLL model. The signal components located outside B_n are greatly attenuated.

After the lock is achieved, the accuracy of the acquisition loop can still be limited by the effects of the jitter, phase offset, and step-size errors. The jitter is a random variation of the clock signal transitions due to noise. The phase offset, which is constant in nature, is essentially caused by mismatches between circuit components and timing misalignment. The step-size errors are associated with the minimum resolution of the delay or VCO control signal.

6.8   VCO-based analog-to-digital converter

A VCO-based analog-to-digital converter (ADC) is an alternative architecture for analog-to-digital conversion. It relies on time resolution that is improved with the continued down-scaling of CMOS process technology and is in the order of tens of picoseconds for a 130-nm CMOS process.

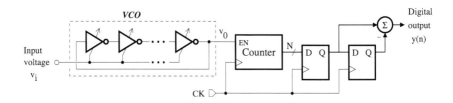

FIGURE 6.93
Block diagram of an open-loop VCO-based ADC.

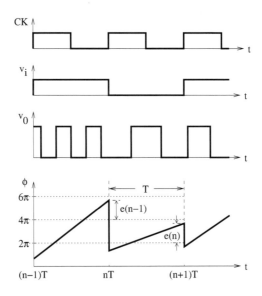

FIGURE 6.94
Waveforms of the open-loop VCO-based ADC.

The block diagram of an open-loop VCO-based ADC is shown in Figure 6.93. It includes a ring VCO, a counter, and a differentiator. The VCO consists of delay (or inverter) stages and its oscillation frequency is set by the input voltage, v_i. A 2^N-modulo counter determines the total number of rising transitions in the VCO output during every period of the clock signal, $T = 1/f_s$, with f_s being the clock signal frequency. The resulting output digital code is proportional to the value of the input voltage that determines the propagation delay of each VCO inverter.

Note that the phase increment of the VCO output must remain lower than 2π during a sampling interval, otherwise the sampled value will not contain the complete information about the phase variation. The clock signal frequency should then be kept greater than the highest VCO oscillation frequency to avoid phase roll-over. This corresponds to an operation in the oversampling mode.

Waveforms illustrating the operation principle of the open-loop VCO-based ADC are represented in Figure 6.94.

By modeling the VCO as an ideal integrator with an input voltage signal, v_i, and an output phase signal, ϕ, the relationship between the VCO output phase and input voltage can be written as:

$$\phi(n) = K_v \int_{nT}^{(n+1)T} v_i(n) \, dt + \phi_i(n) \tag{6.143}$$

$$= K v_i(n) + \phi_i(n) \tag{6.144}$$

where $K = K_v T$, the VCO gain is represented by K_v, and ϕ_i is the initial phase. It is assumed that the sampling frequency is much higher than the signal bandwidth so that the input signal can be considered constant between two consecutive sampling instants.

The counter is supposed to play the role of a quantizer characterized by the following function:

$$q\left(\frac{\phi}{2\pi}\right) = \frac{\phi}{2\pi} - \text{mod}_{2^N}\left(\frac{\phi}{2\pi}\right) \tag{6.145}$$

where $\text{mod}_{2^N}(\phi/2\pi)$ denotes the remainder of the division of $\phi/2\pi$ by 2^N. In comparison to $\phi/2\pi$, the quantizer output $q(\phi/2\pi)$ is shifted down by half the quantization step size. As a result, the quantization error, given by $e = q(\phi/2\pi) - \phi/2\pi$, is comprised between -1 and 0. The output code generated by the differentiator can be put into the following form:

$$y(n) = \frac{1}{2\pi}(\phi(n) + e(n-1)) \tag{6.146}$$

$$= \frac{1}{2\pi}(K v_i(n) - e(n) + e(n-1)) \tag{6.147}$$

Using z-transforms, it can be shown that

$$Y(z) = \frac{1}{2\pi}(KV_i(z) - E(z) + z^{-1}E(z)) \tag{6.148}$$

$$= \frac{1}{2\pi}(KV_i(z) - (1 - z^{-1})E(z)) \tag{6.149}$$

Hence, the VCO-based ADC achieves a first-order noise shaping. An output decimation stage is required to filter out unwanted out-of-band quantization noise.

6.9   PLL based on time-to-digital converter

PLL can also be designed using time-to-digital converters (TDCs) that generally find applications in time difference measurements. It has the advantage of requiring a simple digital lowpass filter instead of the conventional PLL loop filter that demands large and leaky integrating capacitors.

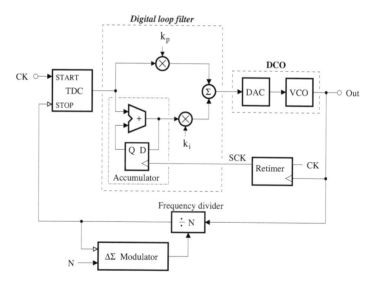

FIGURE 6.95
Fractional-N PLL based on TDC.

A fractional-N PLL based on a TDC is shown in Figure 6.95, where the programmable multi-modulus frequency divider is controlled by a delta-sigma modulator to achieve a fractional frequency resolution, and SCK is the system clock signal available at the output of the retimer flip-flops that synchronize the reference (or feedback) clock signal, RCK, with the incoming clock signal,

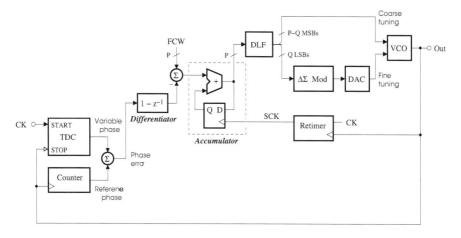

FIGURE 6.96
PLL based on counter-assisted TDC.

CK. Note that SCK serves also as a clock signal for building blocks with synchronization and purely digital logic functions.

A fractional-N PLL can also be designed by using a counter-assisted TDC [48], as shown in Figure 6.96. In the phase-locked state, the desired PLL output frequency is equal to the reference frequency multiplied by the value of the frequency command word (FCW). By defining the FCW as the fractional frequency division ratio, the fine frequency resolution is set by the FCW word length, P.

The phase error quantizer, which consists of a counter, a TDC, and an adder, compares the variable phase of the output signal with the phase of the reference clock signal (CK). The counter can be considered to have a resolution of 2π because it is incremented by one at every rising edge of the output signal. To achieve a fine resolution, the TDC measures the fractional phase, or time interval between the reference edge and the preceding rising edge of the output signal.

The TDC should include output synchronization flip-flops triggered by the PLL output signal because the TDC and counter operations are synchronized to the edges of the reference clock signal and PLL output signal, respectively, instead of the same signal.

The digital phase error is differentiated to obtain a frequency word, which is then compared with the specified frequency command word (FCW). The resulting frequency errors are accumulated to generate a phase error sequence that is used to drive the digital loop filter (DLF).

The most-significant bits (MSBs) of the DLF output are directly used to control the coarse capacitor array of the VCO, while the least-significant bits (LSBs) are first applied to a delta-sigma modulator ($\Delta\Sigma$ mod) followed by a

DAC and the resulting signal is then used to tune the fine capacitor array of the VCO.

The DLF can consist of a proportional path and an integration path and be designed using a transfer function of the form

$$H(z) = H_{PI}(z) \tag{6.150}$$

where

$$H_{PI}(z) = k_p + k_I \frac{1}{1 - z^{-1}} \tag{6.151}$$

with k_p and k_I being constant numbers that can be related to the loop characteristics (bandwidth and phase margin). To suppress out-of-band quantization noise, the DLF transfer function can alternatively be chosen as

$$H(z) = H_{PI}(z)H_{LPF}(z) \tag{6.152}$$

where

$$H_{LPF}(z) = \prod_{i=1}^{4} \frac{\lambda_i}{1 - (1 - \lambda_i)z^{-1}} \tag{6.153}$$

and λ_i are constants.

However, TDC-based PLLs can exhibit phase noise and spurious tone performances that are limited due to the finite resolution of TDC and DCO.

6.9.1 Flash TDC

The classical approach of realizing a TDC consists of counting the number of sequential inverter delays that occur between two rising edges of START and STOP signals. Figure 6.97(a) shows a Flash TDC that comprises a chain of delay elements (or buffers), D flip-flops, and a thermometer-to-binary encoder.

The rising edge of the START signal is successively delayed by a series of delay elements and applied to the inputs of flip-flops that are triggered at the rising edges of the STOP signal. A thermometer code, that corresponds to the number of signal transitions that have occurred within the measurement interval, $\triangle T$, is then generated at the flip-flop outputs.

Figure 6.97(b) shows the waveforms illustrating the operation of a TDC using 4 flip-flops.

In general, the generation of an N-bit output code requires the use of 2^N flip-flops. Hence,

$$\tau = T_S/2^N \tag{6.154}$$

and the measured time difference, $\triangle T$, is given by

$$\triangle T = \tau \sum_{k=1}^{N} 2^{k-1} b_k \tag{6.155}$$

where τ is the propagation delay of each buffer, T_S is the period of the STOP

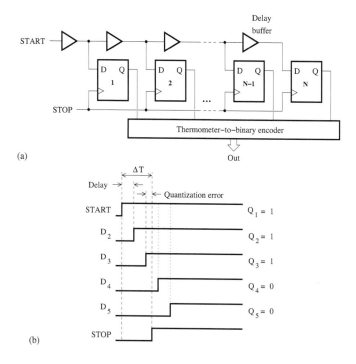

FIGURE 6.97
(a) Flash TDC; (b) waveform representation.

signal, b_k is the k-th bit of the output code, and N is the number of bits of the output code.

Based on CMOS implementations, the resulting resolution is generally limited to a few hundred picoseconds.

6.9.2 Vernier TDC

The improvement of the resolution can be achieved by adopting a Vernier TDC structure, as shown in Figure 6.98, where two chains of delay elements are used.

Assuming that the propagation delay of a buffer in the upper chain, tau_1, is slightly greater than the one of a buffer in the lower chain, τ_2, the time difference between the START and the STOP signals is successively decreased at each TDC stage by $\tau = \tau_1 - \tau_2$, until it becomes less than τ. The time difference, $\triangle T$, is measured with the resolution τ and then encoded as a binary number. Hence,

$$\triangle T = \tau \sum_{k=1}^{N} 2^{k-1} b_k \qquad (6.156)$$

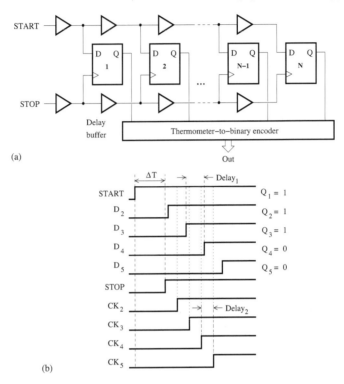

FIGURE 6.98
Vernier TDC.

where b_k is the k-th bit of the output code. In practice, the resolution can be limited by component noise and mismatches. Vernier TDCs can exhibit a resolution as low as 25 picoseconds.

TDC architectures based on chains of delay elements are simple. However, their dynamic range, $DR = 2^N \tau$, can be extended only by augmenting the number of delay elements, which, in turn, leads to an increase of the power consumption and a decrease of the maximum sampling rate. An improvement can be achieved by adopting TDCs based on switched ring oscillators.

6.9.3   Switched ring oscillator TDC

A switched ring oscillator (SRO) TDC [49] is an alternative architecture to achieve sub-gate-delay resolution. It relies on ring VCOs that are switched between two frequencies to achieve a first-order noise shaping of the quantization error. The input signal is integrated by VCOs, whose output phases are quantized and then differentiated. The quantization error is first-order shaped because its value at the end of a measurement period is transferred to the next measurement period. The TDC resolution can be improved by choosing high oversampling ratios.

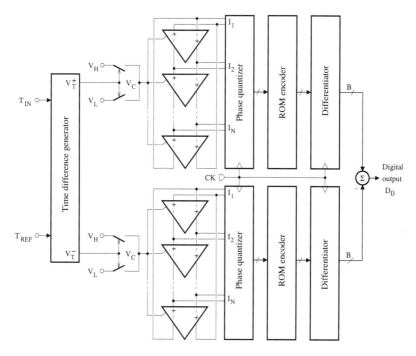

FIGURE 6.99
Block diagram of an SRO TDC.

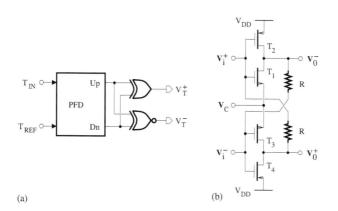

FIGURE 6.100
(a) Block diagram of the time difference generator; (b) circuit diagram of the VCO delay stage.

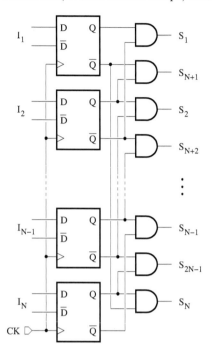

FIGURE 6.101
Circuit diagram of the phase quantizer.

TABLE 6.3
ROM Encoder for $N = 16$

Address	Data Out
S_1	00001
S_2	00010
S_3	00011
S_4	00100
$\vdots$	$\vdots$
S_{28}	11100
S_{29}	11101
S_{30}	11110
S_{31}	11111

The block diagram of the SRO-TDC is shown in Figure 6.99. It consists of a time difference generator, VCOs, phase quantizers, read-only memory (ROM) encoders, digital differentiators, and a B-bit subtractor. The time difference generator provides a continuous-time pulse-width modulated signal that corresponds to the time difference between the input signal, T_{IN}, and reference signal, T_{REF}, and is used to switch the VCO control voltages between two reference levels, V_H and V_L. The oscillation frequency of VCOs can be set either at F_H or at F_L, depending on the (high or low) value of the control voltages. The VCO phase accumulation is achieved at a faster rate for F_H than for F_L, because F_H is higher than F_L. The VCO output phases roll over to zero when they reach 2π. They are quantized before being applied to ROM encoders. The ROM encoder outputs are differentiated and then applied to the subtractor that generates the TDC output code. The phase wrapping around 2π is taken into account by using modulo (two's complement) arithmetic operations in the differentiators.

Note that the phase accumulation will not exceed π radians in one clock period, if F_H is selected to be less than $f_s/2$, with f_s being the sampling frequency.

The block diagram of the time difference generator is represented in Figure 6.100(a). It is realized using a phase-frequency detector (PFD) and XOR gates (or a single differential XOR gate).

The VCO is based on the delay stage, whose circuit diagram is depicted in Figure 6.100(b). To obtain a pseudo-differential structure, resistors are inserted in feedforward paths, and the sources of all NMOS transistors are connected to the control voltage.

Figure 6.101 shows the circuit diagram of the phase quantizer. Sense-amplifier D flip-flops are used to sample the VCO output phase, while a transition detector consisting of AND gates compares adjacent sampled phases to detect phase transitions.

The content of a ROM encoder for $N = 16$ is represented in Table 6.3. The transition decoder outputs are used to address the ROM encoder that converts the thermometer-coded values of the quantized VCO phases into a 5-bit binary code with 31 levels.

The following figure-of-merit (FOM) can be used to characterize noise-shaping TDCs:

$$\text{FOM} = \frac{P}{2^{\text{ENOB}} \times \min(2f_i, f_s)} \qquad (6.157)$$

where f_i represents the input signal frequency, f_s is the sampling frequency, P denotes the power consumption, ENOB is the effective number of bits that can be obtained as $\text{ENOB} = (\text{SNDR} - 1.76)/6.02$, and SNDR is the signal-to-noise-plus distortion ratio.

6.10 High-speed input/output link transceiver

Clock and data recovery (CDR) circuits are generally required in the design of a high-speed input/output (I/O) bus based on standards such as Peripheral

Component Interconnect Express (PCIe), QuickPath Interconnect (QPI), and Double Data Rate (DDR).

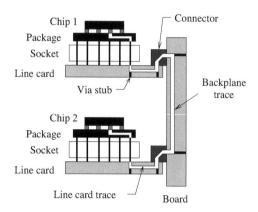

FIGURE 6.102
Representation of a line card/backplane link.

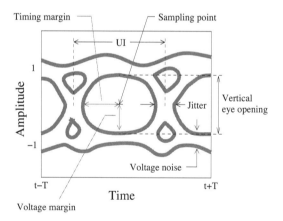

FIGURE 6.103
Outline of a binary eye diagram.

A typical backplane/line card application is shown in Figure 6.102. The I/O link transceiver goal is to achieve a multi-gigabit-per-second signaling rate for the communication from the chip on one line card to the chip on another line card and with the lowest power consumption. However, the undesirable electrical characteristics of the transmission channel (or the long trace on the line card and backplane) and crosstalk between neighboring channels can degrade the signal waveform. Specifically, the transmitted signal is affected by skin effects and dielectric losses of the channel and reflections due to via stubs on circuit boards and other impedance discontinuities of the chip packages and connectors. This can result in signal attenuation of about 20-30 dB for 10-Gb/s data and inter-symbol interference (ISI) that can be seen from an eye diagram.

The dispersive nature of the transmission channel can lead to a significant spreading of the data pulse. That is, adjacent symbols can overlap into each other, resulting in inter-symbol interference and possible bit errors.

The bit error rate (BER) is the likelihood of a bit misinterpretation due to component imperfection. In the presence of a noise source characterized by a Gaussian distribution with zero mean, the BER expression is given by

$$\text{BER} = Q\left(\frac{V_{z,pp}}{2\sqrt{\overline{V_{z,n}^2}}}\right) \tag{6.158}$$

where $V_{z,pp}$ and $\overline{V_{z,n}^2}$ are the peak-to-peak voltage swing (or vertical eye opening) and noise voltage variance at the input of the decision device, respectively, and the Q-function is defined as

$$Q(x) = \frac{1}{\sqrt{2\pi}} \int_x^\infty \exp\left(-\frac{u^2}{2}\right) du \tag{6.159}$$

When the comparator used in the implementation of the decision device is assumed to have a decision threshold voltage, V_{thr}, and the probability of the transmitted signal taking one of both logic states is the same, or say equal to $1/2$, the BER expression can be rewritten as

$$\text{BER} = \frac{1}{2}Q\left(\frac{V_{z,pp}/2 - V_{thr}}{\sqrt{\overline{V_{z,n}^2}}}\right) + \frac{1}{2}Q\left(\frac{V_{z,pp}/2 + V_{thr}}{\sqrt{\overline{V_{z,n}^2}}}\right) \tag{6.160}$$

where V_{thr} may be affected by the comparator offset voltage, V_{off}. With the voltage V_{thr} set to 0, for instance, the value of $V_{z,pp}$ required for a BER of 10^{-12} is such that

$$V_{z,pp} \geq 14\sqrt{\overline{V_{z,n}^2}} \tag{6.161}$$

because $Q(7) \simeq 10^{-12}$. Note that the BER can also be obtained by transmitting a pseudo-random bit stream to the channel and recording the number of transmission errors.

To display the eye diagram on an oscilloscope, the signal of interest is repetitively sampled and applied to the vertical input, while the horizontal time base is synchronized to the symbol rate. An eye diagram, as represented in Figure 6.103, can indicate the best point for sampling, the amount of jitter, and voltage noise. With increasing data rates, the effect of the ISI that makes the width of a data bit exceed one data period (also known as a unit interval, UI) becomes dominant. It is caused by the superposition of delayed and distorted versions of the signal and results in the reduction of both the voltage and timing margins in which the signal can be sampled, so that the bit error rate (BER) is improved. In practice, the ISI effect is mitigated using equalizers with an appropriate number of coefficients (generally less or equal to 16).

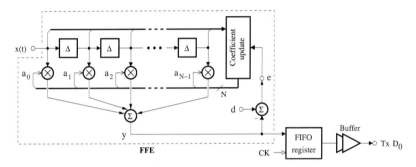

FIGURE 6.104
Block diagram of the transmitter section with a feedforward equalizer.

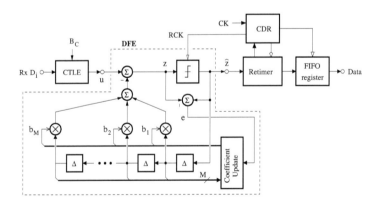

FIGURE 6.105
Block diagram of the receiver section with a decision feedback equalizer.

I/O circuits are composed of transmitter (Tx) and receiver (Rx) sections.
The Tx block diagram is shown in Figure 6.104. It consists of a feedforward equalizer (FFE), first-in first-out (FIFO) registers, and buffers.
A FFE is used to realize the transmit pre-emphasis, or the pre-distortion of the transmitted signal to anticipate the channel distortion and loss. Espe-

cially, it adjusts the magnitude of high-frequency signal components that are supposed to be attenuated by the channel.

The FFE error signal is of the form, $e = d - y$, where d represents the desired signal and y is the equalizer output computed as follows:

$$y(k) = \sum_{n=0}^{N-1} a_n x(k-n) \tag{6.162}$$

To minimize the error function $E(e^2)$ approximated by the instantaneous estimate e^2, the equalizer coefficients are updated in a direction opposite to the gradient of the error signal power given by

$$\frac{\partial e^2(k)}{\partial a_n} = -2e(k)x(k-n) \tag{6.163}$$

In the discrete-time domain, the equalizer coefficients are estimated by means of the least-mean square (LMS) algorithm according to

$$a_n(k) = a_n(k-1) - \mu \frac{\partial e^2(k)}{\partial a_n} \tag{6.164}$$

$$= a_n(k-1) + 2\mu e(k)x(k-n) \qquad n = 0, 1, \cdots, N-1 \tag{6.165}$$

where μ is the adaptation step size. In the continuous-time domain, the coefficient update equations are equivalently expressed as

$$\frac{da_n(t)}{dt} = 2\mu e(t)x(t - n\triangle) \qquad n = 0, 1, \cdots, N-1 \tag{6.166}$$

To reduce the hardware complexity, a sign-sign LMS algorithm, where the error e and samples of the input signal x are replaced by their respective signs, can be adopted.

The Rx block diagram is represented in Figure 6.105. It is composed of a continuous-time linear equalizer (CTLE), a decision feedback equalizer (DFE), a clock data recovery (CDR) circuit, retimer flip-flops, and FIFO registers.

A CTLE can consist of several cascaded gain stages, as shown in Figure 6.106. It is designed to provide a digitally selected amplification to high-frequency signal components.

A peaking gain stage can be based on the source-degenerated differential pair structure shown in Figure 6.107(a). Assuming that the gain stage is loaded by capacitors C_L and modeling the input transistors as a simple transconductance g_m, we can obtain the following transfer function:

$$G(s) = G_0 \frac{1 + s/\omega_z}{(1 + s/\omega_{p_1})(1 + s/\omega_{p_2})} \tag{6.167}$$

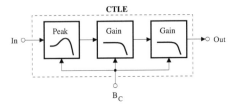

FIGURE 6.106
Block diagram of a CTLE.

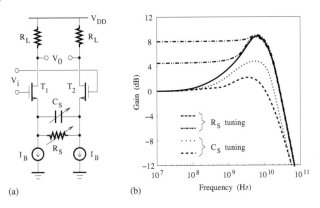

FIGURE 6.107
Peaking gain stage (a) and its magnitude frequency response (b).

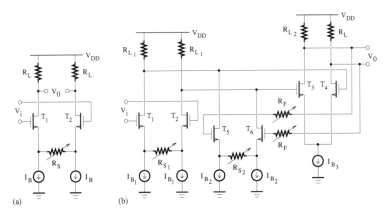

FIGURE 6.108
(a) Conventional gain stage; (b) gain stage with active feedback.

where

$$G_0 = \frac{g_m R_L}{1 + g_m R_S} \tag{6.168}$$

$$\omega_z = 1/R_S C_S \tag{6.169}$$

$$\omega_{p1} = \frac{1 + g_m R_S/2}{R_S C_S} \tag{6.170}$$

and

$$\omega_{p_2} = 1/R_L C_L \qquad (6.171)$$

By implementing R_S and C_S as digitally programmable resistor and capacitor arrays, respectively, the peaking gain stage should provide a high-frequency boost, whose peak can be controlled by a digital code. Figure 6.107 shows the effect of tuning either the resistor R_S, or the capacitor C_S, on the magnitude frequency response. The level of the high-frequency boost is modified essentially by varying the capacitor C_S, while variations of the resistor R_S affect both the high-frequency boost and *dc* gain.

The amplification provided by the peaking gain stage may be insufficient. Other amplifier stages are then required to maximize both the gain and bandwidth. Figure 6.108(a) shows a conventional gain stage, while Figure 6.108(b) presents a gain stage with active feedback.

The multistage CTLE can provide a peaking in the order of 25 dB at Nyquist frequency corresponding to the data bit rate.

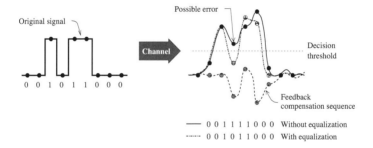

FIGURE 6.109
Signal waveforms illustrating the operation of a DFE.

A DFE [50] is required to mitigate the channel ISI. It makes use of previous decisions to cancel the interference from symbols that are subsequently received. The feedback filter helps cancel the post-cursor ISI. The difference between the input, z, and output, $\hat{z}$, of the decision device constitutes an error signal, $e = \hat{z} - z$, which can then be minimized to update the equalizer coefficients. In the discrete-time domain, this translates into the following equations:

$$b_m(k) = b_m(k-1) - \mu \frac{\partial e^2(k)}{\partial b_m} \qquad (6.172)$$

$$= b_m(k-1) - 2\mu e(k)\hat{z}(k-m) \qquad m = 1, 2, \cdots, M \qquad (6.173)$$

where $\partial e^2(k)/\partial b_m = 2e(k)\hat{z}(k-m)$ and μ is the adaptation step size. The implementation complexity can be reduced by using the sign-sign LMS algorithm. That is:

$$b_m(k) = b_m(k-1) - 2\mu\,\text{sign}(e(k))\text{sign}(\hat{z}(k-m)) \qquad m = 1, 2, \cdots, M \quad (6.174)$$

In the continuous-time domain, the coefficient update equations can equivalently be expressed as:

$$\frac{db_m(t)}{dt} = -2\mu e(t)\hat{z}(t - m\triangle) \qquad m \doteq 1, 2, \cdots, M \qquad (6.175)$$

The optimum equalizer coefficients are such that the ISI contributions of the previous symbols are canceled from the current decision.

Figure 6.109 shows signal waveforms illustrating the operation of a DFE. Due to ISI, when the data sequence 001011000 is transmitted without equalization, the received data sequence 001111000 contains a possible error due to the effect of the channel ISI. With the use of an appropriate DFE, the margin between data bits is now restored so that the received data sequence is likely to be sampled correctly. Note that the DFE can be considered as a non-linear equalizer because of the non-linear characteristic of the decision device.

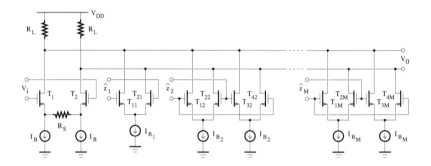

FIGURE 6.110
Circuit diagram of the DFE input summer stage.

The filter section of the DFE consists of delay elements that can be implemented using D flip-flops or sample-and-hold circuits and an input summer stage. Let u be the input signal of the DFE provided by the CTLE. The signal at the input of the decision device is given by

$$z(k) = u(k) - \sum_{m=1}^{M} b_m(k)\hat{z}(k - m) \qquad (6.176)$$

where the equalizer coefficients and samples of the decision device output signal are represented by b_m and $\hat{z}(k - m)$, respectively. Figure 6.110 shows the circuit diagram of the DFE input summer stage. The summation in current mode generally offers the advantage of a simple implementation and high

operation speed over its voltage mode counterpart. Adjustable equalizer coefficients (or post-cursor taps) are implemented by using transconductance
stages with variable tail currents, I_{B_j}, where $j = 1, \cdots M$, that can be controlled by digital-to-analog converters (DACs) with resolution of at least 6
bits. Hence, the magnitude of the feedback tap coefficient currents are set
by DAC-controlled current sources that are driven by digital DFE update
circuits. The first post-cursor tap is set to a negative polarity, while other
post-cursor taps are designed with a bipolar polarity to take into account the
effect of reflections that can occur in some channels. The summer settling
performance can be limited by the RC time constant set by the output load
resistance and capacitance. To accurately cancel the post-cursor ISI from the
most recent bit, all operations performed by the DFE must be completed in
one UI, or say before the arrival of the next symbol.

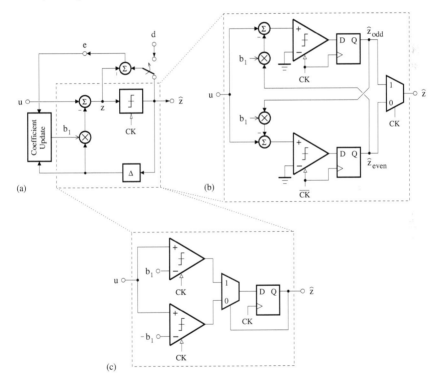

FIGURE 6.111
(a) Block diagram of a one-tap equalizer; (b) half-rate implementation; (c)
implementation using loop unrolling.

For the specific case of a 10-Gb/s binary link, the UI of 100 ps may not be
sufficient to allow the DFE operating at the full rate to settle as required. The
critical timing constraints can be alleviated by adopting time-interleaved (half-
rate, quarter-rate or eighth-rate) architectures and using DFE loop unrolling

(or coefficient speculation) [51, 52], but at the expense of increased power consumption and chip area. Due to the fact that the number of parallel slicing paths grows as a power of two of the number of speculative coefficients, the loop unrolling is often limited to the first coefficient) [53] and helps eliminate the potential feedback delay associated with the first coefficient.

The block diagram of a one-tap equalizer is shown in Figure 6.111(a), while the corresponding half-rate and unrolled implementations are depicted in Figures 6.111(b) and (c), respectively. A half-rate DFE still has the same timing constraint for comparators and latches as the full-rate DFE, but specifications of the clock buffer are relaxed.

In the one-tap DFE implementation using loop unrolling, the timing constraint of the feedback loop is alleviated by making two speculative decisions each cycle. Hence, it is necessary to use two comparators with thresholds set by the first coefficient and its negative value, respectively. The decision of one comparator is made as if the previous output was in the high logic state, and the decision of the other comparator as if the previous output bit was in the low logic state. The multiplexer then selects one of both decisions once the logic state of the previous bit is known.

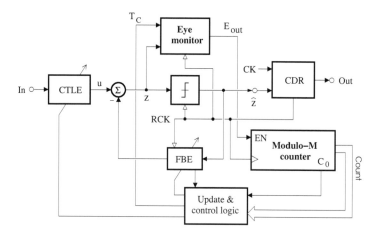

FIGURE 6.112

Block diagram of an eye monitor-based equalizer.

At high BERs, an adaptive equalizer designed to minimize the comparator decision error can exhibit a performance that is degraded due to the channel noise and loading effects of the internal equalizer nodes. In this case, the adaptation approach relying on the error estimated by an eye monitor can contribute to a performance improvement.

The block diagram of an eye monitor-based equalizer is shown in Figure 6.112. The eye monitor provides an error signal that can be used as a figure of merit for monitoring the equalizer performance, and for adapting the feedback equalizer (FBE) coefficients and tuning the CTLE. The counter has

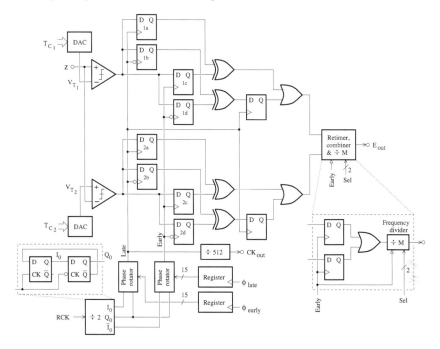

FIGURE 6.113
Block diagram of an eye monitor.

a modulo M. It is reset whenever the count reaches $M - 1$, and the carry-out signal C_0 is then set to 1. The CDR generates the sampling clock by aligning the phase of the reference clock signal CK to the transitions of the incoming data stream. The recovered data is then obtained by sampling the incoming data with a recovered clock.

An eye monitor can be implemented as shown in Figure 6.113. It uses a rectangular mask to characterize the opening of an eye diagram. An error is detected whenever a data transition passes inside the mask. The eye monitor output can be related to the mask error rate (MER), or say, the number of data transitions that fall inside the defined mask normalized by the whole number of transitions during the same time period. Hence,

$$\text{MER} = \frac{N \times f_{out}}{BR} \qquad (6.177)$$

where f_{out} is the frequency of the output error signal, N is the total divide ratio inserted in the signal path, and BR is the input bit rate.

The operation of the eye monitor is illustrated in Figure 6.114. The vertical opening of the mask is specified by two reference voltages, V_{T_1} and V_{T_2}, while the horizontal opening is set by two phases of the sampling clock signal, CK_{Early} and CK_{Late}. The eye trace labeled with a circled 1 is error-free

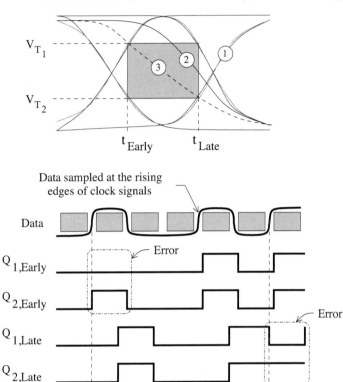

FIGURE 6.114
Illustration of the eye monitor operation.

while eye traces labeled with a circled 2 and a circled 3 are related to a mask violation. Signals obtained by comparing the incoming data with V_{T_1} and V_{T_2} are sampled at instants determined by both CK_{Early} and CK_{Late}. For each sampling clock signal, the difference between the sampled values is detected using an XOR logic gate that can generate a pulse signal whenever a violation occurs for each side of the mask. An OR logic gate is then required to combine the resulting error signals. One violation for each side of the mask can be observed in the timing diagram of Figure 6.114. Note that the errors associated to the left and right sides of the mask can also be counted separately.

The CDR can also be designed to be decision-directed by the eye monitor [54].

Output drivers can be realized as a limiting amplifier or a linear amplifier that can drive various output loads and appropriately set the differential voltage output swing, say from 800 mV to 1200 mV, for instance.

6.11 Relaxation oscillator

Relaxation oscillators can be used to produce triangular and clock signals. They find applications in the design of class D power amplifiers, switching power supplies, dual-slope analog-to-digital converters, and function generators.

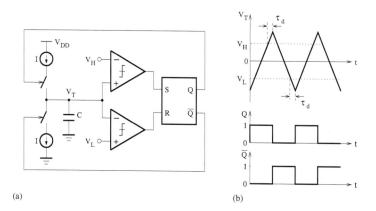

(a)   (b)

FIGURE 6.115

(a) Block diagram of a relaxation oscillator; (b) waveforms.

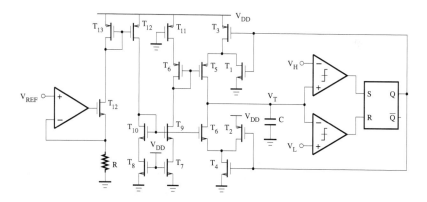

FIGURE 6.116

Circuit diagram of a relaxation oscillator.

The block diagram of a relaxation oscillator is represented in Figure 6.115(a). The constant currents I_1 and I_2 that are used to charge and discharge the capacitor C are provided by a reference generator. Two comparators determine whether the voltage, V_T, across the capacitor C exceeds the threshold voltages V_H and V_L, and then set or reset an SR latch. When the voltage V_T rises above V_H, the output of the upper comparator changes to the high state, and the outputs Q and $\overline{Q}$ then become 1 and 0, respectively. Consequently, the switch S_1 is open while S_2 is closed, leading to the discharge of the capacitor C through the current sink I_2. On the other hand, when the voltage V_T falls below V_L, the output of the lower comparator goes to the high state, and the outputs Q and $\overline{Q}$ become 0 and 1, respectively. The capacitor C can then be charged by the current source I_1, because the switch S_1 is closed while S_2 is open. Figure 6.115(b) shows waveforms that illustrate the operation of the relaxation oscillator.

The period of the resulting oscillation can be expressed as

$$T = \frac{1}{f} = T_c + T_d = C(V_H - V_L)\left(\frac{1}{I_1} + \frac{1}{I_2}\right) + 4\tau_d \qquad (6.178)$$

where T_c and T_d are the charging and discharging times, respectively, τ_d is the propagation delay from the comparator inputs to the SR latch outputs, and f is the oscillation frequency. The oscillation period is sensitive to process fluctuations because it is a function of process-dependent characteristics (I_1, I_2, and τ_d). The adoption of process variation-aware design techniques can help improve the accuracy of the oscillation period.

The circuit diagram of the relaxation oscillator is represented in Figure 6.116. Transistors T_1–T_4 operate as switches. Note that transistors T_3 and T_4 help reduce the effect of charge sharing that can be associated to parasitic capacitances. For oscillations in the kilohertz range, the capacitor C should be in the order of few picofarads.

6.12 Class D amplifier

A class D power amplifier is basically a non-linear switching amplifier. It relies on the principle of pulse-width modulation (PWM) to generate a rectangular wave signal that is used to drive switching output stages in either single-ended or bridge-tied-load configurations. The efficiency of a class D amplifier can reach 100% in the ideal case, but it is in the order of 90% in practical designs.

In comparison with open-loop structures, closed-loop class D power amplifiers offer the advantages of reducing the non-ideal effects that can be caused by switching dead times, finite on-resistances of switches, and supply voltage noises [55].

The circuit diagram of a class D amplifier with a second-order loop filter is represented in Figure 6.117. It includes a loop filter, two comparators, two drivers, four power MOSFET switches, LC filters, and a speaker. For audio applications, the input frequency is typically comprised between 20 Hz and 20

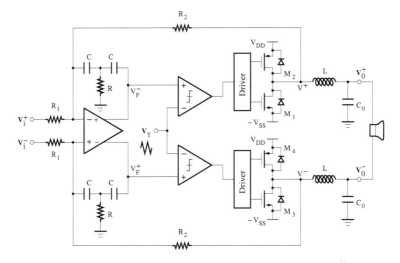

FIGURE 6.117
Circuit diagram of a class D amplifier using a triangular reference signal.

kHz. The difference of the loop filter output signals is compared with a high-frequency triangular signal to generate a pulse-width modulated signal that is used to drive the switching output stage. The duty cycle of the pulse-width modulated signal is such that its average content corresponds to the input analog signal. Figure 6.118 shows the waveforms illustrating the operation of a class D amplifier. The outputs V^+ et V^- are not two complementary signals of equal amplitude but opposite phase. In addition to the differential voltage, $V_d = V^+ - V^-$, there is thus a significant common-mode component, V_c, available at the outputs.

In the time domain, the filter output can be expressed as

$$V_F(t) = h(t) * [V_i(t) + V(t)] \tag{6.179}$$

where $*$ is the convolution product, $h(t)$ represents the loop filter impulse response, $V_F(t) = V_F^+(t) - V_F^-(t)$ and $V(t) = V^+(t) - V^-(t)$. More specifically, we have

$$V^+(t) = \begin{cases} 1 & \text{if} \quad V_F^+(t) > V_T(t) \\ 0 & \text{if} \quad V_F^+(t) < V_T(t) \end{cases} \tag{6.180}$$

and

$$V^-(t) = \begin{cases} 1 & \text{if} \quad V_F^-(t) > V_T(t) \\ 0 & \text{if} \quad V_F^-(t) < V_T(t) \end{cases} \tag{6.181}$$

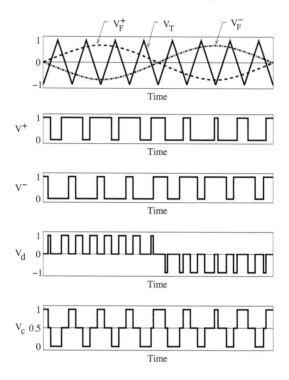

FIGURE 6.118
Class D amplifier waveforms.

The differential rectangular-wave PWM signal $V(t)$ can take each of the values 1, 0, and -1. It is fed back to the loop filter input, providing the negative feedback.

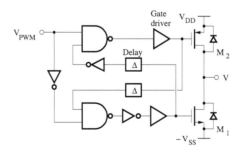

FIGURE 6.119
Half bridge with the switch driver.

Switch drivers are required in order to quickly turn on and off the power MOSFETs, because comparators are unable to support large currents. They can be realized based on the operation principle of a two-phase nonoverlapping clock signal generator and a level shifter. Figure 6.119 shows half bridge with the switch driver that can help control the delay between turn-on times, as required to reduce the total harmonic distortion. The gate driver and delay can be implemented by cascading an even number of inverters.

A pulse-width modulated waveform can be generated by comparing a signal with a reference (carrier) signal. The carrier is chosen as a sawtooth waveform and a triangle waveform to perform a single-sided modulation (leading or trailing edge) and a double-sided modulation, respectively [56]. The spectral characteristics of the signal obtained by modulating a sine wave can be determined using double Fourier series (see Appendix B.2).

Consider an input sine wave of the form, $M \cdot A \cdot \pi \cos(\omega t)$, with A representing the amplitude. In the case of a double-sided modulation, it can be shown that

$$V_{PWM}(t) = M \cdot A \cos(\omega t)$$

$$- 4A \sum_{\substack{m=1}}^{+\infty} \sum_{\substack{n=-\infty \\ n \neq 0}}^{+\infty} \frac{J_n\left(\frac{m \cdot M}{2}\pi\right)}{m\pi} \sin\left(\frac{m+n}{2}\pi\right) \sin\left(\frac{n}{2}\pi\right) \sin\left(m\omega_c t + n\omega t - \frac{n}{2}\pi\right)$$

$$(6.182)$$

where M is the modulation index, ω_c is the angle frequency of the carrier signal, J_n represents the Bessel function of n-th order, n is the index to the harmonics of the input signal, and m denotes the index to the harmonics of the carrier signal.

It can be noted that only intermodulation components at $m\omega_c t \pm n\omega t$ are present in the spectrum. Because the modulated signal does not contain any harmonics of the carrier frequency, the intermodulation components with the lowest angle frequencies are located around $2\omega_c$.

On the other hand, the double-sided modulation produces a signal that exhibits a common-mode component, and is then also known as a three-level (class BD) pulse-width modulated signal.

Generally, a post filter is required at the amplifier output for the intermodulation component suppression.

An LC filter is required for the reduction of high-frequency undesired components related to the modulation process. Its frequency response varies with the speaker load impedance. The equivalent circuit of the LC filter is represented in Figure 6.120(a). For differential signals, the load resistance, R_L, is equal to half the speaker resistance. The transfer function of the LC filter can then be put into the form,

$$H(s) = \frac{V_{0d}(s)}{V_i(s)} = \frac{\omega_0^2}{s^2 + s(\omega_0/Q) + \omega_0^2} \tag{6.183}$$

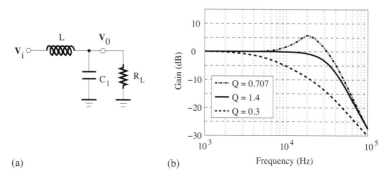

FIGURE 6.120
LC filter (a) equivalent circuit and (b) gain response.

where $\omega_0 = 1/\sqrt{LC}$ and $Q = R_L\sqrt{C/L}$. The LC filter frequency response is overdamped for $Q < 1/\sqrt{2}$, critically damped when $Q = 1/\sqrt{2}$, and underdamped for $Q > 1/\sqrt{2}$. The effect of Q on the frequency response is illustrated in Figure 6.120(b). The critically damped response is preferred because a high peaking can cause the amplifier to malfunction due to the operation in over-current condition, while the overdamped response can be related to the loss of some high-frequency components of the audio signal. A typical LC filter has a Butterworth response with a cut-off frequency of about 40 kHz and can exhibit a maximally flat passband. Hence, $Q = 1/\sqrt{2}$ and the values of the filter components can be derived as follows:

$$C = 1/(\sqrt{2}R_L\omega_0) \quad \text{and} \quad L = \sqrt{2}R_L/\omega_0 \qquad (6.184)$$

The load resistance, R_L, is equal to half the speaker resistance which is generally in the range from 4 Ω to 8 Ω.

For common-mode signals, the load resistance R_L can be removed. The transfer function of the resulting lossless LC filter is given by

$$H(s) = \frac{V_{0c}(s)}{V_i(s)} = \frac{\omega_0^2}{s^2 + \omega_0^2} \qquad (6.185)$$

where $\omega_0 = 1/\sqrt{LC}$.

The choice of Butterworth filter response for 8 Ω speakers results in inductors L of 33 μH that can be very large and bulky for portable devices. A solution in this case can consist of using a class D amplifier based on filterless modulation schemes.

For input frequencies up to 20 kHz, a class D amplifier designed with a carrier frequency of 315 kHz can achieve a maximum power efficiency of about 90%. The total harmonic distortion (THD) computed by taking into account all harmonics within the audio range (that is, from 20 Hz to 20 kHz) is in the order of −76 dB when the input frequency is equal to 2 kHz. It is increased to −66 dB for an input frequency of 6 kHz.

To improve the power efficiency, a class D amplifier should be designed with comparators exhibiting a low propagation delay in the order of a few tens of nanoseconds and output power MOSFETs featuring low on-resistances.

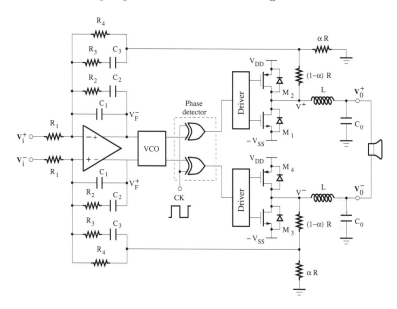

FIGURE 6.121
Circuit diagram of a class D amplifier using a rectangular reference signal.

For applications requiring a bandwidth up to 500 kHz and a high switching frequency, such as power-line communications, the class D amplifier of Figure 6.121, that requires a rectangular reference signal, may be more suitable [57]. It consists of a loop filter, a voltage-controlled oscillator (VCO), a phase detector (PD), an output driver, and a filter. The difference between the input and output signals is filtered to generate the control voltage of the VCO. The PWM signal is delivered at the output of the PD that operates by comparing the phase of the VCO output signal with that of a rectangular reference signal. It is then used to drive the amplifier output stage.

The VCO can be realized using a pseudo-differential architecture consisting of current-starved inverters and cross-coupled inverters. In the locked state, the average of the VCO control signal should ideally be equal to zero. Because the difference between the input and output signals also has an average value of zero, the output signal must be identical to the inverted-phase version of the input signal in the passband frequency range of the loop filter. This is required so that the loop can track the input signal variations.

A class D amplifier designed with a reference frequency of 20 MHz can exhibit a power efficiency in the order of 80% and a total harmonic distortion (THD) of −67 dB for input signal frequencies up to 60 kHz. The power efficiency is reduced to about 70% and the THD is −60 dB when the reference and input frequencies are set to 60 MHz and 500 kHz, respectively. The maximum power efficiency is achieved when the duty cycle of the reference signal is equal to 50%.

6.13 Summary

Circuits for clock signal generation and synchronization were described at the behavioral and architectural level. Specifically, they are used to reduce the voltage and timing errors in the clocking or data transmission networks.

In addition to a low jitter performance, it is important to achieve a low power consumption and die area in portable equipment. Either passive or active methods can be used to reduce the clock jitter induced by the power supply noise. The first method employs filters on the power supply lines, while the second one regulates the power supply by relying on a feedback control loop that keeps the voltage constant.

Circuit architectures and techniques that can be used in the design of input/output transceivers for high-speed serial links, relaxation oscillators, and class D power amplifiers, were also reviewed.

6.14 Circuit design assessment

1. **Phase and frequency detector**

   The phase and frequency detector of Figure 6.122 [2] possess the advantage of not having a dead zone. Use SPICE simulations to obtain the transient response of the circuit and show that the error detection range is limited in the range from $-\pi$ to π.

2. **Single-ended VCO based on inverters**

   Consider the VCO structure shown in Figure 6.123(a). It consists of N inverters (see Figure 6.123(b)), capacitors, and transistors operating as switches controlled by V_c.

   Why must the number of stages, N, be odd?

   With the assumption that each stage can be modeled with a first-order transfer function of the form

   $$\frac{V_0}{V_i}(s) = \frac{A_0}{1 + \dfrac{s}{\omega_0}} \tag{6.186}$$

   where A_0 is the dc gain and ω_0 is the -3 dB bandwidth, find the oscillation frequency of the VCO.

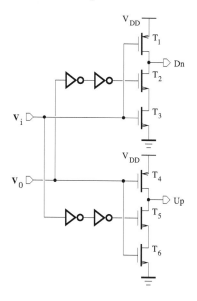

FIGURE 6.122
Phase and frequency detector.

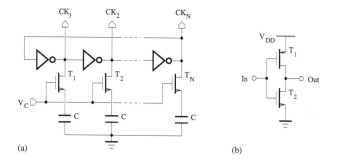

FIGURE 6.123
(a) VCO structure; (b) circuit diagram of the inverter.

3. Ring oscillator

A ring oscillator [58], as shown in Figure 6.124, can consist of N delay stages in a unity feedback loop. For a single-ended configuration, the number N of stages should be odd so that the total phase shift can satisfy the required oscillation condition.

Assuming that all transistors operate in the saturation region, verify that the application of Kirchhoff's current law at the drain node of

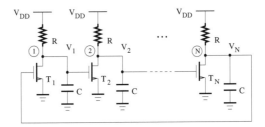

FIGURE 6.124

Circuit diagram of a ring oscillator.

each delay stage leads to:

$$\text{Node 1} \qquad \frac{V_1 - V_{DD}}{R} + C\frac{dV_1}{dt} + K(V_N - V_T)^2 = 0 \qquad (6.187)$$

$$\text{Node 2} \qquad \frac{V_2 - V_{DD}}{R} + C\frac{dV_2}{dt} + K(V_1 - V_T)^2 = 0 \qquad (6.188)$$

$$\vdots \qquad\qquad \vdots$$

$$\text{Node N} \qquad \frac{V_N - V_{DD}}{R} + C\frac{dV_N}{dt} + K(V_{N-1} - V_T)^2 = 0 \quad (6.189)$$

where $K = (1/2)\mu_n C_{ox}(W/L)$, $C = C_L + C_p$, C_L and C_p are the load capacitance and the total parasitic capacitance at the drain node of each delay stage, respectively, and V_T is the transistor threshold voltage.

The output of the ring oscillator can be considered to be a sinusoidal waveform of the form,

$$V_0(t) = B + A\cos\omega t \qquad (6.190)$$

To find the amplitude, A, the *dc* level, B, and the signal frequency, $f = \omega/2\pi$, the phase relationship can be exploited to write:

$$V_1(t) = V_2\left(t + \left(\frac{T}{2} - \frac{T}{2N}\right)\right) \qquad (6.191)$$

$$= B + A\cos\left(\omega t + \frac{N-1}{N}\pi\right) \qquad (6.192)$$

where T is the signal period.

Show that the current equation for the node 2 can be rewritten as,

$$\frac{B + A\cos(\omega t) - V_{DD}}{R} - \omega C A \sin(\omega t)$$

$$+ K\left[B + A\cos\left(\omega t + \frac{N-1}{N}\pi\right) - V_T\right]^2 = 0$$

$$(6.193)$$

Combining each of the following sets of trigonometric relations,

(1) $\sin \omega t = -1$ and $\cos \omega t = 0$

(2) $\sin \omega t = 0$ and $\cos \omega t = 1$

(3) $\sin \omega t = 0$ and $\cos \omega t = -1$

with Equation (6.193), show that

$$
\begin{cases}
(1) & \dfrac{B - V_{DD}}{R} + \omega C A + K(B - A\sin(\pi/N) - V_T)^2 = 0 \\[2ex]
(2) & \dfrac{B + A - V_{DD}}{R} + K(B - A\cos(\pi/N) - V_T)^2 = 0 \\[2ex]
(3) & \dfrac{B - A - V_{DD}}{R} + K(B + A\cos(\pi/N) - V_T)^2 = 0
\end{cases}
\tag{6.194}
$$

Deduce that

$$
B = V_T + \frac{1}{2RK\cos(\pi/N)}
\tag{6.195}
$$

$$
A = \sqrt{\dfrac{\dfrac{V_{DD} - B}{R} - K(B - V_T)^2}{K\cos^2(\pi/N)}}
\tag{6.196}
$$

$$
f = \frac{\dfrac{B - V_{DD}}{R} + K(B - A\sin(\pi/N) - V_T)^2}{2\pi C A}
\tag{6.197}
$$

Verify that all transistors operate in the saturation region, provided $V_1(t) \geq V_N(t) - V_T$, $V_k(t) \geq V_{k-1}(t) - V_T$, where $k = 2, \cdots, N$, and $B - A \geq V_T$.

4. **Analysis of a second-order PLL**

    Consider the linear equivalent model of the second-order PLL shown in Figure 6.125.

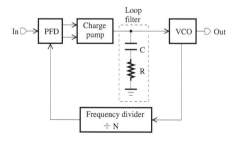

FIGURE 6.125
Block diagram of a second-order PLL.

Estimate the closed-loop phase transfer function, T, of the PLL. Show that T can be put into the following form:

$$T(s) = \frac{\Theta_0(s)}{\Theta_i(s)} = \frac{N\left(1 + 2\dfrac{\zeta}{\omega_n}s\right)}{1 + 2\dfrac{\zeta}{\omega_n}s + \left(\dfrac{s}{\omega_n}\right)^2} \tag{6.198}$$

where

$$\omega_n = \sqrt{\frac{K_v I_P}{NC}} \tag{6.199}$$

$$\zeta = \frac{1}{2}\sqrt{\frac{K_v I_P C R^2}{N}} \tag{6.200}$$

By applying an angular frequency step to the PLL, the input phase angle, Θ_i, can be written as

$$\Theta_i(t) = \begin{cases} 0 & \text{for } t < 0 \\ \Delta\omega t & \text{for } t > 0 \end{cases} \tag{6.201}$$

Prove that the phase error is given in the s-domain and time domain by

$$\Theta_e(s) = (1 - T(s))\frac{\Delta\omega}{s^2} \tag{6.202}$$

and

$$\Theta_e(t) = \begin{cases} \dfrac{N\Delta\omega}{\omega_n\sqrt{1-\zeta^2}}\sin(\sqrt{1-\zeta^2}\,\omega_n t)\exp(-\zeta\omega_n t) & \text{for } \zeta < 1 \\ N\Delta\omega t\exp(-\omega_n t) & \text{for } \zeta = 1 \\ \dfrac{N\Delta\omega}{\omega_n\sqrt{\zeta^2-1}}\sinh(\sqrt{\zeta^2-1}\,\omega_n t)\exp(-\zeta\omega_n t) & \text{for } \zeta > 1 \end{cases} \tag{6.203}$$

respectively.

The noise bandwidth of the PLL is defined as

$$B_n = \int_0^{+\infty} |T(j\omega)|^2 d\omega \tag{6.204}$$

Verify that

$$B_n = N^2\omega_n\frac{1 + 4\zeta^2}{8\zeta} \tag{6.205}$$

The parameter ζ is generally chosen around the value ζ_m, which minimizes B_n. Find ζ_m.

5. **Frequency detection capability of PDs**

A second-order PLL can be modeled using the block diagram shown in Figure 6.126(a). Practical circuit implementations of full-rate binary and linear phase detectors (PDs) are represented in Figures 6.126(b) and (c), respectively. The phase difference between the center of the data and the rising edge of the clock signal is represented by θ.

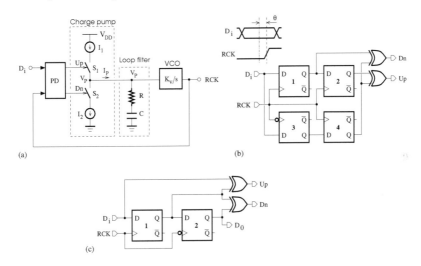

FIGURE 6.126

(a) PLL block diagram; (b) full-rate binary PD; (c) full-rate linear PD.

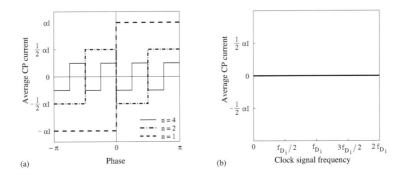

FIGURE 6.127

Phase (a) and frequency (b) characteristics of a full-rate binary PD.

To show that the full-rate binary PD does not have a frequency detection capability, verify that:

– the phase and frequency characteristics of the full-rate binary PD can be obtained as shown in Figures 6.127(a) and (b), respectively.

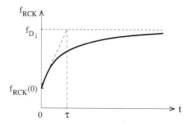

FIGURE 6.128
Transient response of the frequency acquisition loop.

– the average CP output current, $\bar{I}_P$, is always equal to zero regardless of θ if $f_{RCK} = f_{D_i}/2^k$, where k is an integer.

For restricted frequencies of the clock signal, the full-rate linear PD can detect a frequency difference, and can then be used in the implementation of clock and data recovery circuits based on a single-loop architecture, as shown in Figure 6.126(a). Its average CP current as a function of the clock frequency is given by

$$\bar{I}_P = \begin{cases} \dfrac{\alpha I}{2}\left(1 - \dfrac{f_{RCK}}{f_{D_i}}\right) & \text{if} \quad f_{RCK} \le f_{D_i} \\ 0 & \text{if} \quad f_{RCK} > f_{D_i} \end{cases} \tag{6.206}$$

where α is the bit transition density.

Use the following equations

$$\frac{dV_P(t)}{dt} = \frac{\bar{I}_P(t)}{C} + R\frac{d\bar{I}_P(t)}{dt} \tag{6.207}$$

$$f_{RCK}(t) = K_v V_P(t) \tag{6.208}$$

to show that

$$\frac{df_{RCK}(t)}{dt} + Af_{RCK}(t) = B \tag{6.209}$$

where

$$A = \frac{K_v \alpha I/2C f_{D_i}}{1 + K_v R\alpha I/2 f_{D_i}} \tag{6.210}$$

$$B = \frac{K_v \alpha I/2C}{1 + K_v R\alpha I/2 f_{D_i}} \tag{6.211}$$

Deduce that the frequency acquisition loop has a first-order transient behavior from the initial state to the final state, as shown in Figure 6.128, that can be described by:

$$f_{RCK}(t) = f_{D_i} - (f_{D_i} - f_{RCK}(0))\,e^{-t/\tau} \tag{6.212}$$

where the time constant can be written as

$$\tau = RC\left(1 + \frac{2f_{D_i}}{K_v R \alpha I}\right) \tag{6.213}$$

The following loop parameters can be used for simulations: $R = 1$ kΩ, $C = 160$ pF, $\alpha = 0.5$, $I = 100$ μA, $K_v = 2$ GHZ/V, $f_{D_i} = 2$ GHz, $f_{RCK}(0) = 1$ GHz.

6. **Jitter-peaking-free PLL**

Consider the PLL of Figure 6.129(a) that consists of a phase detector (PD), a charge pump circuit, a loop filter, and a voltage-controlled oscillator (VCO). The loop filter transfer function is given by

$$H(s) = \frac{V_f(s)}{I_p(s)} = R + \frac{1}{sC} = \frac{1 + s/\omega_z}{sC} \tag{6.214}$$

where $\omega_z = 1/RC$.

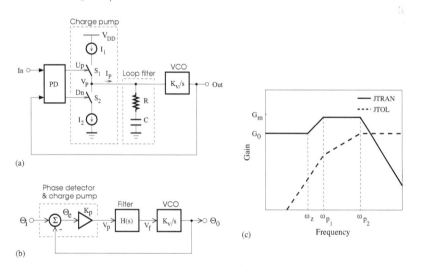

(a)

(b)

(c)

FIGURE 6.129
PLL block diagram (a) and its linear equivalent model (b); PLL with a cascaded VCDL (c).

Use the equivalent linear model of Figure 6.129(b) to show that:

– the input-output phase transfer function, also known as the jitter transfer function (JTRAN), can be obtained as follows

$$G(s) = \frac{\Theta_0(s)}{\Theta_i(s)} = \frac{\omega_n^2(1 + s/\omega_z)}{s^2 + 2\zeta\omega_n s + \omega_n^2} \tag{6.215}$$

where $\omega_n = \sqrt{K_P K_v/C}$, $\zeta = RC\omega_n/2$, and $\omega_z = 1/RC$;

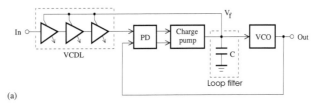

(a)

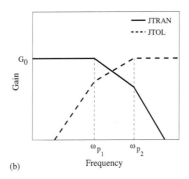

(b)

FIGURE 6.130
Jitter transfer function and tolerance for a second-order PLL (a) and a PLL with a cascaded VCDL (b).

— the error transfer function, $E(s)$, is inversely proportional to the jitter tolerance (JTOL) characteristic and can be written as

$$E(s) = \frac{\Theta_e(s)}{\Theta_i(s)} = 1 + \frac{\Theta_0(s)}{\Theta_i(s)} = 1 - G(s) = \frac{s^2}{s^2 + 2\zeta\omega_n s + \omega_n^2} \quad (6.216)$$

Typically, the damping factor ζ is selected between 0.5 and 2. When ζ is greater than one, the transfer functions G and E exhibit two real poles, ω_{p_1} and ω_{p_2}, at $(-\zeta \pm \sqrt{\zeta^2 - 1})\omega_n$, where ω_n is the undamped natural frequency.

Verify that the magnitudes of the functions G and E can be represented as the asymptotic bode plots of Figure 6.129(c).

It can be shown that the -3 dB bandwidth (in rad) and the level of gain peaking (in dB) can be expressed as,

$$BW = \omega_n \sqrt{1 + 2\zeta^2 + 2\sqrt{\zeta^4 + \zeta^2 + 1/2}} \quad (6.217)$$

and

$$G_m = 20 \log_{10} |G(\omega_m)|$$

$$= 10 \log_{10} \left(\frac{8\zeta^4}{8\zeta^4 - 4\zeta^2 - 1 + \sqrt{1 + 8\zeta^2}} \right) \quad (6.218)$$

where

$$\omega_m = \frac{\omega_n}{2\zeta}\sqrt{\sqrt{1+8\zeta^2}-1} \tag{6.219}$$

Design a PLL assuming that the input frequency (or reference frequency) is 2 MHz, the VCO gain is $K_v = 0.375$ MHz/V, and using the following steps:

− set the level of the gain peaking to 1.4 dB to determine the damping factor, ζ;

− set the loop bandwidth, BW, to one tenth of the reference frequency due to the stability requirement associated to the PD discrete-time nature to determine the undamped natural frequency, ω_n;

− compute R and C assuming that $K_p = 10/2\pi$ mA/rad.

Due to the presence of a closed-loop zero at a frequency lower than that of the poles, the magnitude of $G(s)$ exhibits a gain in excess of unity, known as jitter peaking. In the PLL circuit of Figure 6.130(a), the elimination of the closed-loop zero contributes to the improvement of the jitter filtering by lowering the bandwidth of the jitter transfer function.

• Assuming a linear phase detector and using

$$V_f = (\Theta_i - K_d V_f - \Theta_0)\frac{1}{Cs} \tag{6.220}$$

$$= \Theta_e \frac{1}{Cs} \tag{6.221}$$

$$\Theta_0 = \frac{K_v}{s}V_f \tag{6.222}$$

show that

$$G(s) = \frac{\Theta_0(s)}{\Theta_i(s)} = \frac{\omega_n^2}{s^2 + 2\zeta\omega_n s + \omega_n^2} \tag{6.223}$$

and

$$E(s) = \frac{\Theta_e(s)}{\Theta_i(s)} = \frac{s^2}{s^2 + 2\zeta\omega_n s + \omega_n^2} \tag{6.224}$$

where $\omega_n = \sqrt{K_v/C}$ and $\zeta = (K_d/2)\sqrt{1/K_vC}$.

Verify that the PLL will not exhibit jitter peaking, as illustrated in the asymptotic bode plots of Figure 6.130(b), if the PLL components are selected so that the damping factor ζ is equal to $\sqrt{2}/2$ (second-order Butterworth response).

The −3 dB bandwidth (in rad) of the PLL is given by

$$BW = \omega_n\sqrt{1 - 2\zeta^2 + 2\sqrt{\zeta^4 - \zeta^2 + 1/2}} \tag{6.225}$$

Design a PLL with a bandwidth, BW, of 40 kHz, a VCO gain, K_v, of 0.375 MHz/V, and using the following steps:

– set the damping factor to $\sqrt{2}/2$ and determine the undamped natural frequency, ω_n;

– compute C using $K_p = 10/2\pi$ μA/rad and deduce K_d (in rad/V).

• When a binary phase detector is used, the phase error becomes almost equal to zero, $\Theta_e \simeq 0$, and the output phase, Θ_0, is related to the control voltage, V_f, by the VCO transfer function. Use

$$\Theta_i = \Theta_0 + K_d V_f \tag{6.226}$$

$$\Theta_0 = \frac{K_v}{s} V_f \tag{6.227}$$

to show that

$$G(s) = \frac{\Theta_0(s)}{\Theta_i(s)} = \frac{1}{1 + s/\omega_p} \tag{6.228}$$

where $\omega_p = K_v/K_d$.

7. **Design of a PLL for clock recovery applications**

Consider the phase-locked loop (PLL) shown in Figure 6.131 [21]. It includes a phase and frequency detector (PFD) based on two XOR gates, a charge-pump (CP) circuit, a loop filter, a voltage-controlled oscillator (VCO) with four differential delay stages, and a D flip-flip used for retiming the recovered clock signal.

With the assumption that I_1 and I_2 are nominally equal to I_P, the CP provides the charge $\pm I_P$ to the capacitor C_1.

Show that the transfer function of the section including the PFD, CP, and loop filter is given by

$$\frac{V_c}{\triangle\phi}(s) = \frac{I_P}{2\pi} \frac{1 + RC_1 s}{(C_1 + C_2)s + RC_1C_2 s^2} \tag{6.229}$$

where V_c is the VCO control voltage and $\triangle\phi$ is the phase difference between the PD inputs.

Determine the closed-loop transfer function of the PLL in the s-domain and z-domain.

Use the SPICE program to perform the transient analysis of the PLL.

8. **Fourth-order charge-pump PLL**

Consider the fourth-order charge-pump PLL shown in Figure 6.132. The frequency division by N is performed by cascading six divide-by-two stages, each of which is based on a D flip-flop whose inverted output node is connected back to the data input node. The

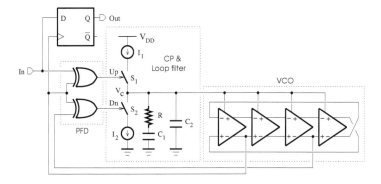

FIGURE 6.131
Circuit diagram of a third-order charge-pump PLL.

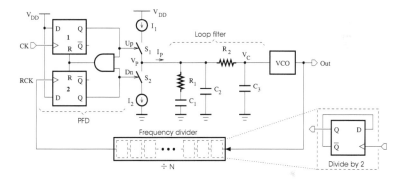

FIGURE 6.132
Circuit diagram of a fourth-order charge-pump PLL.

circuit diagrams of a charge-pump circuit and a VCO are depicted in Figures 6.133(a) and (b), respectively. Assuming that the pole frequency due to R_2 and C_3 is at least ten times higher than the loop bandwidth, and $C_3 \leq C_2/10$, the third-order loop filter can be considered a second-order filter section followed by a first-order filter section providing an attenuation α of the spurious sidebands at the multiples of the reference frequency. Hence,

$$C_1 \simeq \frac{K_p K_v}{N\omega_u^2} \cdot \frac{\eta}{\sqrt{1+\eta}} \tag{6.230}$$

$$R_1 \simeq \frac{\sqrt{1+\eta}}{\omega_u C_1} \tag{6.231}$$

$$C_2 \simeq \frac{C_1}{\eta} \tag{6.232}$$

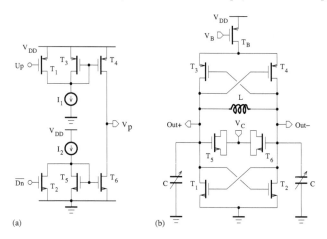

FIGURE 6.133
(a) Charge-pump circuit; (b) VCO circuit.

and

$$R_2 \simeq \frac{\sqrt{10^{\alpha/10} - 1}}{\omega_{ref} C_3} \qquad (6.233)$$

where

$$\eta = 2\tan^2 \phi_M + 2\tan \phi_M \sqrt{1 + \tan^2 \phi_M} \qquad (6.234)$$

Here, $K_p = I_P/(2\pi)$, $\omega_{ref} = 2\pi f_{ref}$, and f_{ref} is the reference frequency.

For a reference frequency of 37.5 MHz, a VCO free-running frequency of 2.4 GHz, $\omega_u/(2\pi) = 200$ kHz, $\phi_M = 65°$, $\alpha = 10$ dB, $K_v = 100$ MHz/V, $N = 64$, and $I_P = 200$ µA, first determine the initial values of the filter components and then use MATLAB® and SPICE simulations to size and fine-tune the values of the loop components in a given CMOS technology such that the charge-pump PLL can exhibit a loop bandwidth of 200 kHz.

9. **Relaxation oscillator**
   A relaxation oscillator can be implemented as shown in Figure 6.134(a). The capacitors C_1 and C_2 are alternatively charged by the current source I up to the reference voltage V_{REF} and then discharged to the ground depending on the logic states of the RS latch outputs. Transistors $T_1 - T_2$ and $T_3 - T_4$ operate as CMOS inverters. Figure 6.134(b) shows the waveforms illustrating the oscillator operation. Complementary square-wave clock signals are generated at the RS latch outputs.

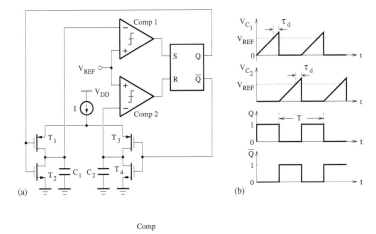

Comp

FIGURE 6.134
(a) Circuit diagram of a relaxation oscillator; (b) waveforms.

Verify that the oscillation period can be expressed as

$$T = (C_1 + C_2)V_{REF}/I + 2\tau_d \qquad (6.235)$$

where τ_d is the propagation delay of the latched comparator.

Taking into account the comparator offset voltages, the oscillation period becomes $T \pm \Delta T$. Show that the absolute accuracy of the period is of the form

$$\Delta T = (C_1 V_{off_1} + C_2 V_{off_2})/I \qquad (6.236)$$

where V_{off_1} and V_{off_2} represent the offset voltages of the first and second comparators, respectively.

Bibliography

[1] F. M. Gardner, "Charge-pump phase-lock loops," *IEEE Trans. on Communications*, vol. 28, pp. 1849–1858, Nov. 1980.

[2] H. O. Johansson, "A simple precharged CMOS phase frequency detector," *IEEE J. of Solid-State Circuits*, vol. 33, pp. 295–299, Feb. 1998.

[3] B. Razavi, "Challenges in the design of high-speed clock and data recovery circuits," *IEEE Communications Magazine*, vol. 40, pp. 94–101, Aug. 2002.

[4] D. W. Cline, "Noise, speed, and power trade-offs in pipelined analog to digital converters," PhD thesis, University of California, Berkeley, 1995.

[5] Seng-Pan U, R. P. Martins, and J. E. Franca, "A 2.5-V 57-MHz 15-tap SC bandpass interpolating filter with 320-Ms/s output for DDFS system in 0.35-μm CMOS," *IEEE J. of Solid-State Circuits*, vol. 39, pp. 87–99, Jan. 2004.

[6] I. W. Young, J. K. Greason, and K. L. Wong, "A PLL clock generator with 5 to 110 MHz of lock range for microprocessors," *IEEE J. Solid-State Circuits*, vol. 27, pp. 1599–1607, Nov. 1992.

[7] M. Mansuri, D. Liu, and C.-K. K. Yang, "Fast frequency acquisition phase-frequency detectors for Gsamples/s phase-locked loops," *IEEE J. of Solid-State Circuits*, vol. 37, pp. 1331–1334, Oct. 2002.

[8] C. R. Hogge Jr., "Signal detection apparatus," U.S. Patent 4,535,459, filed May 26, 1983; issued August 13, 1985.

[9] J. D. H. Alexander, "Clock recovery from random binary signals," *Electronics Letters*, vol. 11, pp. 541–542, Oct. 1975.

[10] J. Savoj and B. Razavi, "A 10-Gb/s CMOS clock and data recovery circuit with a half rate linear phase detector," *IEEE Journal of Solid-State Circuits*, vol. 36, pp. 761–768, May 2001.

[11] M. Fukaishi, K. Nakamura, H. Heiuchi, Y. Hirota, Y. Nakazawa, H. Ikeno, H. Hayama, and M. Yotsuyanagi, "A 20-Gb/s CMOS multichannel transmitter and receiver chip set for ultra-high-resolution digital displays," *IEEE J. of Solid-State Circuits*, vol. 35, pp. 1611–1618, Nov. 2000.

[12] C.-C. Huang, "Phase detector," U.S. Patent 6,259,278, filed June 14, 1999; issued July 10, 2001.

[13] A. Rezayee and K. Martin, "A 9-16 Gb/s clock and data recovery circuit with three-state phase detector and dual-path loop architecture," *Proc. of the 2003 European Solid-State Circuits Conf.*, pp. 683–686, Estoril, Portugal, Sept. 2003.

[14] A. Ong, S. Benyamin, J. Cancio, V. Condito, T. Labrie, Q. Lee, J. P. Mattia, D. K. Shaeffer, A. Shahani, X. Si, H. Tao, M. Tarsia, W. Wong, and M. Xu, "A 40-43-Gb/s clock and data recovery IC with integrated SFI-5 1:16 demultiplexer in SiGe technology," *IEEE J. of Solid-State Circuits*, vol. 38, pp. 2155–2168, Dec. 2003.

[15] P. Larsson, "A 2-1600-MHz CMOS clock recovery PLL with low-V_{dd} capability," *IEEE J. of Solid-State Circuits*, vol. 34, pp. 1951–1960, Dec. 1999.

[16] A. Waizman, "A delay line loop for frequency synthesis of de-skewed clock," *1994 IEEE ISSCC Digest of Technical Papers*, pp. 298–299, Feb. 1994.

[17] M. Rau, T. Oberst, R. Lares, A. Rothermel, R. Schweer, and N. Menoux, "Clock/data recovery PLL using half-frequency clock," *IEEE J. Solid-State Circuits*, vol. 32, pp. 1156–1159, July 1997.

[18] M. G. Johnson and E. L. Hudson, "A variable delay line PLL for CPU coprocessor synchronization," *IEEE J. Solid-State Circuits*, vol. 23, pp. 1218–1223, Oct. 1988.

[19] J. G. Maneatis, "Low-jitter process-independent DLL and PLL based on self-biased techniques," *IEEE J. of Solid-State Circuits*, vol. 31, pp. 1723–1732, Nov. 1996.

[20] J. G. Maneatis and M. A. Horowitz, "Precise delay generation using coupled oscillators," *IEEE J. of Solid-State Circuits*, vol. 28, pp. 1273–1282, Dec. 1993.

[21] D.-H. Kim and J.-K. Kang, "Clock and data recovery circuit with two exclusive-OR phase frequency detector," *Electronics Letters*, vol. 36, pp. 1347–1349, Aug. 2000.

[22] I. I. Novof, J. Austin, R. Kelkar, D. Strayer, and S. Wyatt, "Fully integrated CMOS phase-locked loop with 15 to 240 MHz locking range and ±50 ps jitter," *IEEE J. of Solid-State Circuits*, vol. 30, pp. 1259–1266, Nov. 1995.

[23] P. K. Hanumolu, M. Brownlee, K. Mayaram, and U. K. Moon, "Analysis of charge-pump phase-locked loops," *IEEE Trans. on Circuits and Systems–I*, vol. 51, pp. 1665–1674, Sept. 2004.

[24] Z. Wang, "An analysis of charge-pump phase-locked loops," *IEEE Trans. on Circuits and Systems–I*, vol. 52, pp. 2128–2138, Oct. 2005.

[25] J. Cherry and M. Snelgrove, "Clock jitter and quantizer metastability in continuous-time delta-sigma modulators," *IEEE Trans. on Circuits and Systems, Part II*, vol. 46, no. 4, pp. 376–389, Apr. 1999.

[26] J.-Y. Ihm, "Stability analysis of bang-bang phase-locked loops for clock and data recovery systems," *IEEE Trans. on Circuits and Systems, Part II*, vol. 60, no. 1, pp. 1–5, Jan. 2013.

[27] T. A. D. Riley, M. A. Copeland, and T. A. Kwasniewski, "Delta-sigma modulation in fractional-N frequency synthesis," *IEEE J. of Solid-State Circuits*, vol. 28, pp. 553–559, May 1993.

[28] C. S. Vaucher, I. Ferencic, M. Locher, S. Sedvallson, U. Voegeli, and Z. Wang, "A family of low-power truly modular programmable dividers in standard 0.35-μm CMOS technology," *IEEE J. of Solid-State Circuits*, vol. 35, pp. 1039–1045, July 2000.

[29] M. Meghelli, B. D. Parker, H. A. Ainspan, and M. Soyuer, "SiGe BiCMOS 3.3-V clock and data recovery circuits for 10-Gb/s serial transmission systems," *IEEE J. of Solid-State Circuits*, vol. 35, pp. 1992–1995, Dec. 2000.

[30] H. S. Hakkal and J. J. Hughes, "Circuit and method of a three state phase frequency lock detector," U.S. Patent 6,404,240, filed Oct. 30, 2000; issued June 11, 2002.

[31] M. B. Ghaderi and V. W. S. Tso, "Phase-frequency lock detector," U.S. Patent 5,870,002, filed June 23, 1997; issued Feb. 9, 1999.

[32] R. C. H. van de Beek, C. S. Vaucher, D. M. W. Leenaerts, E. A. M. Klumperink, and B. Nauta, "A 2.5-10-GHz clock multiplier unit with 0.22-ps rms jitter in standard 0.18-μm CMOS," *IEEE J. of Solid-State Circuits*, vol. 39, pp. 1862–1872, Nov. 2004.

[33] N. Kalantari and J. F. Buckwalter, "A multichannel serial link receiver with dual-loop clock-and-data recovery and channel equalization," *IEEE J. of Solid-State Circuits*, vol. 60, no. 11, pp. 2920–2931, Nov. 2013.

[34] R. Shivnaraine, M. S. Jalali, A. Sheikholeslami, M. Kibune, and H. Tamura, "An 8-11 Gb/s reference-less Bang-Bang CDR enabled by 'phase reset,'" *IEEE J. of Solid-State Circuits*, vol. 61, no. 7, pp. 2129–2138, July 2014.

[35] K. Kishine, H. Inaba, H. Inoue, M. Nakamura, A. Tsuchiya, H. Katsurai, and H. Onodera, "A multi-rate burst-mode CDR using a GVCO with symmetric loops for instantaneous phase locking in 65-nm CMOS," *IEEE J. of Solid-State Circuits*, vol. 62, no. 5, pp. 1288–1295, May. 2015.

[36] A. Rezayee and K. Martin, "A 9-16 Gb/s clock and data recovery circuit with three-state phase detector and dual-path loop architecture," in *Proc. of the IEEE ESSCIRC'03*, pp. 683–686, 2003.

[37] R. Inti, W. Yin, A. Elshazly, N. Sasidhar, and P. K. Hanumolu, "A 0.5-to-2.5 Gb/s reference-less half-rate digital CDR with unlimited frequency acquisition range and improved input duty-cycle error tolerance," *IEEE J. of Solid-State Circuits*, vol. 46, no. 12, pp. 3150–3162, Dec. 2011.

[38] J. Han, J. Yang, and H.-M. Bae, "Analysis of a frequency acquisition technique with a stochastic reference clock generator," *IEEE Trans. Circuits Syst. II*, vol. 59, no. 6, pp. 336–340, Jun. 2012.

[39] S. Byun, "A 400 MB/s-2.5 Gb/s referenceless CDR IC using intrinsic frequency detection capability of half-rate linear phase detector," *IEEE Trans. Circuits Syst. I*, vol. 63, no. 10, pp. 1592–1604, Oct. 2016.

[40] C. H. Son and S. Byun, "On frequency detection capability of full-rate linear and binary phase detectors," *IEEE Trans. Circuits Syst. II*, vol. 64, no. 7, pp. 757–761, Jul. 2017.

[41] Y.-H. Moon, I.-S. Kong, Y.-S. Ryu, and J.-K. Kang, "A 2.2-mW $20-135$ MHz false-lock-free DLL for display interface in 0.15 μm CMOS," *IEEE Trans. on Circuits and Systems, Part II*, vol. 61, no. 8, pp. 554–558, Aug. 2014.

[42] S. Kim and M. Soma, "An all-digital built-in self-test for high-speed phase-locked loops," *IEEE Trans. on Circuits and Systems*, vol. 48, pp. 141–150, Feb. 2001.

[43] C.-L. Hsu, Y. Lai, and S.-W. Wang, "Built-in self-test for phase-locked loops," *IEEE Trans. on Instrumentation and Measurement*, vol. 54, pp. 996–1002, June 2005.

[44] S. Sunter and A. Roy, "BIST for phase-locked loops in digital applications," in *Proc. of the International Test Conf.*, 1999, pp. 532–540.

[45] D. M. Fischette, A. L. S. Loke, R. J. DeSantis, and G. R. Talbot, "An embedded all-digital circuit to measure PLL response," *IEEE J. of Solid-State Circuits*, vol. 45, no. 8, pp. 1492–1503, Aug. 2010.

[46] A. Chan and G. Roberts, "A jitter characterization system using a component-invariant vernier delay line," *IEEE Trans. on Very Large Scale Integration Systems*, vol. 12, no. 1, pp. 79–95, Jan. 2004.

[47] J. L. Huang and K. T. Cheng, "An on-chip short-time interval measurement technique for testing high-speed communication links," in *Proc. of the VLSI Test Symposium*, Jan. 2001, pp. 380–385.

[48] R. B. Staszewski, J. L. Wallberg, S. Rezeq, C.-M. Hung, O. E. Eliezer, S. K. Vemulapalli, C. Fernando, K. Maggio, R. Staszewski, N. Barton, M.-C. Lee, P. Cruise, M. Entezari, K. Muhammad, and D. Leipold, "All-digital PLL and transmitter for mobile phones," *IEEE J. of Solid-State Circuits*, vol. 40, no. 12, pp. 2469–2482, Dec. 2005.

[49] A. Elshazly, S. Rao, B. Young, and P. K. Hanumolu, "A noise-shaping time-to-digital converter using switched-ring oscillators—Analysis, design, and measurement techniques," *IEEE J. Solid-State Circuits*, vol. 49, no. 6, pp. 1184-1197, May. 2014.

[50] D. D. Falconer, "Adaptive equalization of channel nonlinearities in QAM data transmission systems," *Bell System Technical Journal*, vol. 57, no. 7, pp. 2589–2611, Sep. 1978.

[51] K. K. Parhi, "High-speed architectures for algorithms with quantizer loops," in *Proc. of the IEEE International Symposium on Circuits and Systems*, vol. 3, May 1990, pp. 2357–2360.

[52] V. Stojanović et al, "Autonomous dual-mode (PAM2/4) serial link transceiver with adaptive equalization and data recovery," *IEEE J. Solid-State Circuits*, vol. 40, no. 4, pp. 1012–1026, Apr. 2005.

[53] T. O. Dickson, J. F. Bulzacchelli, and D. J. Friedman, "A 12-Gb/s 11-mW half-rate sampled 5-tap decision feedback equalizer with current-integrating summers in 45-nm SOI CMOS technology," *IEEE J. of Solid-State Circuits*, vol. 44, no. 4, pp. 1298–1305, Apr. 2009.

[54] H. Noguchi, N. Yoshida, H. Uchida, M. Ozaki, S. Kanemitsu, and S. Wada, "A 40-Gb/s CDR circuit with adaptive decision-point control based on eye-opening monitor feedback," *IEEE J. of Solid-State Circuits*, vol. 43, no. 12, pp. 2929–2938, Dec. 2008.

[55] S. M. Cox, J. Yu, W. L. Goh, and M. T. Tan, "Intrinsic distortion of a fully differential BD-modulated class-D amplifier with analog feedback," *IEEE Trans. on Circuits and Systems–I*, vol. 60, no. 1, pp. 63–73, Jan. 2013.

[56] K. Nielsens, "A review and comparison of pulsewidth modulation (PWM) methods for analog and digital input switching power amplifiers," in the *Proc. of the 102nd AES Convention*, Munich, Germany, Preprint 4446 (G4), March 1997.

[57] J. Lu, H. Song, and R. Gharpurey, "A CMOS class-D line driver employing a phase-locked loop based PWM generator," *IEEE J. of Solid-State Circuits*, vol. 49, no. 3, pp. 729–739, March 2014.

[58] P. M. Farahabadi, H. Miar-Naimi, and A. Ebrahimzadeh, "Closed-form analytical equations for amplitude and frequency of high-frequency

CMOS ring oscillators," *IEEE Trans. on Circuits and Systems, Part I,* vol. 56, no. 12, pp. 2669–2677, Dec. 2009.

A

Logic Building Blocks

CONTENTS

A.1 Boolean algebra

Boolean algebra is used to describe the relations between inputs and outputs of a digital circuit. The input and output signals are considered to be Boolean variables (X, Y, Z) whose values are either 0 (logic low level) or 1 (logic high level).

A.1.1 Basic operations

NOT	AND	OR
$\overline{0} = 1$	$0 \cdot X = 0$	$0 + X = X$
$\overline{1} = 0$	$1 \cdot X = X$	$1 + X = 1$
$\overline{\overline{X}} = X$	$X \cdot X = X$	$X + X = X$
	$X \cdot \overline{X} = 0$	$X + \overline{X} = 1$

- $X + X \cdot Y = X$
- $X \cdot Y + \overline{X} \cdot Z + Y \cdot Z = X \cdot Y + \overline{X} \cdot Z$
- $X + \overline{X} \cdot Y = X + Y$
- $(X + Y)(X + \overline{Y}) = X$
- $X(X + Y) = X$
- $(X + Y)(\overline{X} + Z) = X \cdot Z + \overline{X} \cdot Y$
- $X(\overline{X} + Y) = X \cdot Y$
- $(X+Y)(\overline{X}+Z)(Y+Z) = (X+Y)(\overline{X}+Z)$

A.1.2 Exclusive-OR and equivalence operations

- $X \oplus Y = X \cdot \overline{Y} + \overline{X} \cdot Y$
- $X(Y \oplus Z) = X \cdot Y \oplus X \cdot Z$ $\qquad$ $(X \oplus Y) \oplus Z = X \oplus (Y \oplus Z) = X \oplus Y \oplus Z$
- $X \odot Y = \overline{X \oplus Y} = X \cdot Y + \overline{X} \cdot \overline{Y} = \overline{X} \oplus Y = X \oplus \overline{Y}$
- $X \oplus 0 = X$ $\qquad$ $X \oplus X = 0$
- $X \oplus 1 = \overline{X}$ $\qquad$ $X \oplus \overline{X} = 1$
- $\overline{X \oplus Y} = X \oplus \overline{Y} = \overline{X} \oplus Y = X \oplus Y$
- $(X \cdot Y) \oplus (X + Y) = X \oplus Y$ $\qquad$ $(X \cdot Y) \oplus (\overline{X} \cdot \overline{Y}) = \overline{X \oplus Y} = X \odot Y$
- $X + Y = (X \oplus Y) \oplus (X \cdot Y)$
- $X \cdot Y = (X \oplus Y) \oplus (X + Y)$
- $X \cdot (Y \oplus Z) = (X \cdot Y) \oplus (X \cdot Z)$
- If $X \cdot Y = 0$, then $X + Y = X \oplus Y$

A.2 Combinational logic circuits

A.2.1 Basic gates

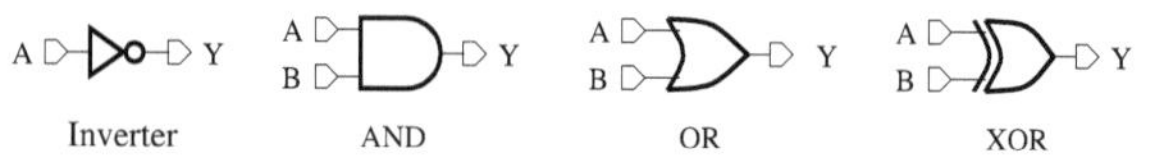

FIGURE A.1
NOT, AND, OR, and XOR gates.

Combinational circuits are digital circuits whose outputs depend only on the current states of inputs. Figure A.1 shows the NOT, AND, OR and XOR gates. Boolean expressions that relate the state of the output to the ones of inputs are given in Table A.1.

TABLE A.1
Logic Equations of NOT, AND, OR, and XOR Gates

Gate type	Inverter	AND	OR	XOR
Boolean function	$Y = \overline{A}$	$Y = A \cdot B$	$Y = A + B$	$Y = A \oplus B$

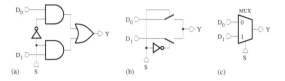

(a) (b) (c)

FIGURE A.2
2-to-1 Multiplexer: (a) circuit diagram, (b) switch-based implementation, (c) symbol.

A multiplexer permits the transmission of only one of many data inputs to the output. In general, a multiplexer with 2^k inputs and one output has k select bits that are used to choose the input to be connected to the output node. The circuit diagram, switch-based implementation, and the symbol of a 2-to-1 multiplexer are illustrated in Figure A.2. The Boolean expression for the output of the 2-to-1 multiplexer can be written as

$$Y = \begin{cases} D_0 & \text{if} \quad S = 0 \\ D_1 & \text{if} \quad S = 1 \end{cases} \tag{A.1}$$

or equivalently,

$$Y = D_0 \cdot \overline{S} + D_1 \cdot S \tag{A.2}$$

where D_0 and D_1 are the data inputs, and S denotes the selection signal.

A 4-to-1 multiplexer can be realized using logic gates as illustrated in Figure A.3(a), or using 2-to-1 multiplexers configured as shown in Figure A.3(b). Its symbol is depicted in Figure A.3(c). The logic equation of the 4-to-1 multiplexer output is given by

$$Y = \overline{S_1} \cdot \overline{S_0} \cdot D_0 + \overline{S_1} \cdot S_0 \cdot D_1 + S_1 \cdot \overline{S_0} \cdot D_2 + S_1 \cdot S_0 \cdot D_3 \tag{A.3}$$

where D_0, D_1, D_2, and D_3 represent the data inputs, and S_0 and S_1 are the selection inputs.

Decoding is necessary in applications such as interfacing, where it is often necessary to convert one digital format to another. In general, a binary decoder asserts one of 2^k output lines for each combination of the k input signals, provided it is enabled. It is implemented using k inverters and 2^k AND gates with $k + 1$ inputs.

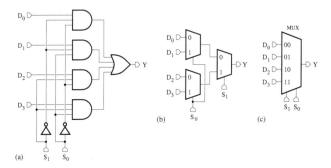

FIGURE A.3
4-to-1 Multiplexer based on logic gates (a) and 2-to-1 multiplexers (b); (c) symbol.

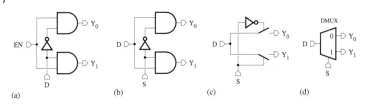

FIGURE A.4
(a) 1-out-of-2 Decoder with an enable signal; 1-to-2 demultiplexer: (b) circuit diagram, (c) switch-based implementation, (d) symbol.

The circuit diagram of the 1-out-of-2 decoder with an enable signal is shown in Figure A.4(a).

A demultiplexer is a logic circuit that switches a data input toward one of the outputs depending on the digital code applied at the selection inputs. The circuit diagram, switch-based implementation, and symbol of a 1-to-2 demultiplexer are represented in Figures A.4(b), (c), and (d), respectively.

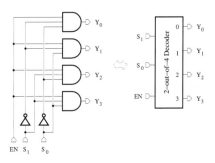

FIGURE A.5
2-out-of-4 Decoder: (a) Circuit diagram, (b) symbol.

Figure A.5 shows the circuit diagram and symbol of a 2-out-of-4 decoder, which consists of two inverters and four AND gates. The Boolean equations of the outputs can be derived as

$$Y_0 = EN \cdot \overline{S_1} \cdot \overline{S_0} \tag{A.4}$$

$$Y_1 = EN \cdot \overline{S_1} \cdot S_0 \tag{A.5}$$

$$Y_2 = EN \cdot S_1 \cdot \overline{S_0} \tag{A.6}$$

and

$$Y_3 = EN \cdot S_1 \cdot S_0 \tag{A.7}$$

where EN is the enable signal, and S_1 and S_0 represent the bits of the input code. For each input code, a different output line is asserted by the decoder. Note that each output line is numbered in accordance with the decimal equivalent of the input binary code.

Note that a 1-to-4 demultiplexer can be obtained from the 2-out-of-4 decoder by connecting the incoming data D to the enable input.

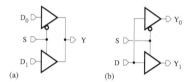

FIGURE A.6
2-to-1 Multiplexer (a) and 1-to-2 demultiplexer (b) based on tri-state buffers.

A multiplexer and demultiplexer can also be implemented using tri-state buffers, that operate as an open or closed switch depending on the logic level applied to selection input. Figures A.6(a) and (b) show the circuit diagrams of the 2-to-1 multiplexer and 1-to-2 demultiplexer, respectively.

A.2.2 CMOS implementation

TABLE A.2
Truth Table of CMOS Inverter

A	T_1	T_2	Y
0	OFF	ON	V_{DD}
V_{DD}	ON	OFF	0

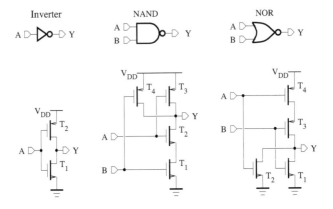

FIGURE A.7
CMOS implementations of (a) inverter, (b) NAND and (c) NOR gates.

A static CMOS inverter consisting of an nMOS transistor and pMOS transistor is shown in Figure A.7(a). It operates according to Table A.2. When the input is at V_{DD}, the nMOS transistor is on while the pMOS transistor is off. The output is then set to 0 V. On the other hand, when the input takes the value of 0 V, the nMOS transistor is off while the pMOS transistor is on. As a result, the output level equals V_{DD}. Assuming that the leakage current is negligible, an unloaded inverter should not dissipate a static power because the path between the supply voltage and ground is interrupted in the steady state.

TABLE A.3
Truth Table of CMOS NAND Gate

A	B	T_1	T_2	T_3	T_4	Y
0	0	OFF	OFF	ON	ON	V_{DD}
0	V_{DD}	ON	OFF	ON	OFF	V_{DD}
V_{DD}	0	OFF	ON	OFF	ON	V_{DD}
V_{DD}	V_{DD}	ON	ON	OFF	OFF	0

The circuit diagram of a two-input NAND gate is depicted in Figure A.7(b). By connecting two nMOS transistors in series between the output node and ground, the output will be set to 0 V only if the level of both inputs is V_{DD}. Due to the two parallel pMOS transistors connected between the supply voltage and the output node, the output level will equal V_{DD} if either the input A or B is 0 V. This is summarized in Table. A.3.

A CMOS NOR gate can be implemented as shown in Figure A.7(c). Its operation is illustrated by Table. A.4. By inserting two nMOS transistors in

TABLE A.4
Truth Table of CMOS NOR Gate

A	B	T_1	T_2	T_3	T_4	Y
0	0	OFF	OFF	ON	ON	V_{DD}
0	V_{DD}	ON	OFF	OFF	ON	0
V_{DD}	0	OFF	ON	ON	OFF	0
V_{DD}	V_{DD}	ON	ON	OFF	OFF	0

parallel between the output node and ground, the output will be set to 0 V if either the input A or B is at V_{DD}. The connection of two pMOS transistors in series between the supply voltage and ground forces the output to be at V_{DD} only if the level of both inputs is equal to 0 V.

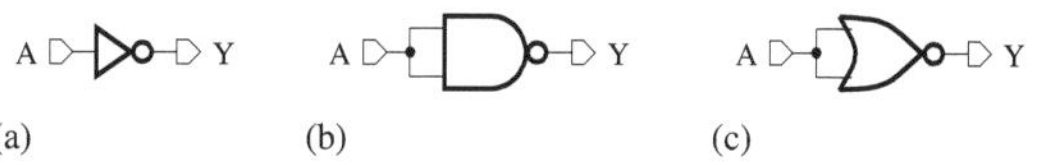

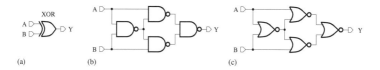

(a)   (b)   (c)

FIGURE A.8
(a) Inverter and its implementations based on (b) NAND and (c) NOR gates.

FIGURE A.9
(a) XOR gate and its implementations based on (b) NAND and (c) NOR gates.

The NAND and NOR gates are universal logic elements. Each of them can be used for the implementation of any logic function, as illustrated in Figures A.8 and A.9 in the case of the inverter and XOR gate, respectively.

Assuming that complementary signals are available, the XOR gate and multiplexer can also be realized as shown in Figure A.10. For the XOR gate, the pull-up transistors are configured based on the function $A \oplus B$, while the structure of the pull-down transistors is determined by the function $\overline{A \oplus B}$. The output is at the logic high level when only one of the inputs is set to the logic high level. When the logic level of both inputs is either high or low, that of the output is low. Considering the circuit diagram of the 2-1 multiplexer, which is also composed of pull-up and pull-down transistors, it can be shown that the implemented logic function is of the form $A \cdot S + B \cdot \overline{S}$. For any given

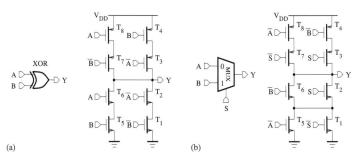

(a)   (b)

FIGURE A.10
CMOS implementation of (a) XOR gate and (b) 2-1 multiplexer.

logic state of the selection signal, S, the logic level of only one of the inputs
is transferred to the output.

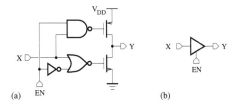

(a)   EN   (b)

FIGURE A.11
(a) Tri-state buffer and (b) its symbol.

Tri-state buffers allow multiple logic circuits to share a common data bus,
provided only one of the logic circuits is active at any given time. Figure A.11
shows the circuit diagram and symbol of a tri-state buffer. The output signal
of a tri-state buffer can be written as

$$Y = \begin{cases} X & \text{if} \quad EN = 1 \\ z & \text{if} \quad EN = 0 \end{cases} \tag{A.8}$$

where z denotes the high-impedance state. It can then assume one of three
states: logic low, logic high, or high impedance state. In the transfer mode,
that is, when $EN = 1$, the tri-state buffer is equivalent to a closed switch.
In the disconnect mode, that is, when $EN = 0$, the tri-state buffer output is
isolated from the input by a high impedance.

A.3 Sequential logic circuits

Sequential circuits are digital circuits whose outputs depend not only on the current state of inputs, but also on the previous state of outputs. Basic sequential circuits include latch and flip-flop. A latch is sensitive to either the high level or the low level of its inputs and is said to be level-sensitive or transparent, while a flip-flop can change its output state only at an edge of the clock signal and is said to be edge-triggered.

A.3.1 Asynchronous SR latch

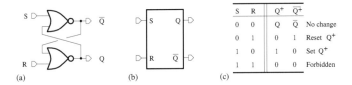

S	R	Q^+	$\overline{Q}^+$	
0	0	Q	$\overline{Q}$	No change
0	1	0	1	Reset Q^+
1	0	1	0	Set Q^+
1	1	0	0	Forbidden

(a)   (b)   (c)

FIGURE A.12
SR Latch: (a) Circuit diagram, (b) symbol, (c) truth table.

The asynchronous set-reset (SR) latch is one of the simplest sequential circuits. Figure A.12 shows the circuit diagram, symbol, and truth table of an SR latch implemented with two NOR gates. The characteristic equation of the SR latch is given by

$$Q^+ = \overline{R} \cdot S + \overline{R} \cdot Q = \overline{R} \cdot (S + Q) \tag{A.9}$$

where Q and Q^+ denote the current and next states of the output, respectively. In the case where the condition $(R = S = 1)$ is not supposed to occur, we have

$$Q^+ = S + \overline{R} \cdot Q, \quad \text{with the assumption that} \quad S \cdot R = 0 \tag{A.10}$$

A.3.2 Asynchronous $\overline{S}\,\overline{R}$ latch

Another simplest form of the latch circuit can be implemented using two NAND gates. Figure A.13 shows the circuit diagram, symbol, and truth table of the $\overline{S}\,\overline{R}$ latch, whose characteristic equation is of the form

$$Q^+ = S + \overline{R} \cdot Q \tag{A.11}$$

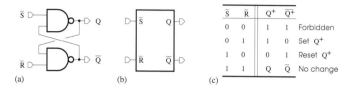

FIGURE A.13
$\overline{S}\,\overline{R}$ latch: (a) Circuit diagram, (b) symbol, (c) truth table.

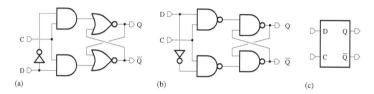

FIGURE A.14
D latch: (a) Implementation based on SR latch, (b) implementation based on $\overline{S}\,\overline{R}$ latch, (c) symbol.

A.3.3   D latch

A D (data) latch is used to capture the logic level of the signal at the data input when it is enabled by the control signal. Figure A.14 shows the circuit diagrams of D latches based on an SR latch and $\overline{S}\,\overline{R}$ latch, respectively, and the D-latch symbol. It can be found that the characteristic equation is

$$Q^+ = D \cdot C + \overline{C} \cdot Q \tag{A.12}$$

Hence, the latch output follows the D input when the control signal C is at the logic high level. When the C input is low, the latch output is maintained at the state previously acquired when the C input was high. Note that the outputs Q and $\overline{Q}$ of the D flip-flop are complementary.

A.3.4   D flip-flops

The operation of flip-flops is synchronized by either the rising or falling edge of a clock signal (CK). D flip-flops, whose outputs can only change at one edge of the clock signal, exhibit a characteristic equation of the form

$$Q^+ = D. \tag{A.13}$$

A D flip-flop can be implemented using a master-slave configuration, which consists of two D latches that are connected in series and controlled by inverted phases of the clock signal. Figure A.15 shows the circuit diagram, symbol, and truth table of the master-slave D flip-flop. The input data are captured when the clock signal is low, but its state is transferred to the output only at the

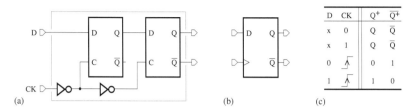

FIGURE A.15
D flip-flop with master-slave configuration: (a) Circuit diagram, (b) symbol, (c) truth table.

beginning of the next clock phase. Provided the setup and hold requirements are met, the aforementioned master-slave D flip-flop is referred to as a device triggered by the rising edge of the clock signal.

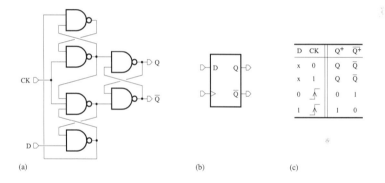

FIGURE A.16
D flip-flop: (a) Circuit diagram, (b) symbol, (c) truth table.

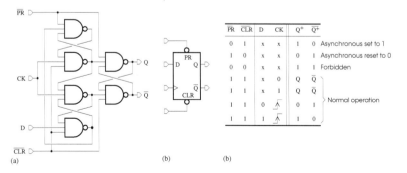

FIGURE A.17
D flip-flop with asynchronous (preset and clear) inputs: (a) Circuit diagram, (b) symbol, (c) truth table.

A D flip-flop can also be realized by relying on the use of an edge detector so that the input pulse can occur synchronously with a transition of the clock signal. Figure A.16 shows the circuit diagram, symbol, and truth table of a D flip-flop triggered by the rising edge of the clock signal. Six NAND gates are required in this implementation of the D flip-flop. The state of the data input is captured at the rising edge of the clock signal. It is then transferred to the output a short time after the edge occurrence due to the gate propagation delay.

In some applications, it may be necessary to asynchronously drive the flip-flop, thereby bypassing the clock signal control. Figure A.17 shows the circuit diagram, symbol, and truth table of a D flip-flop with asynchronous inputs. The D input is synchronous, while the preset and clear inputs, which are used to, respectively, set or reset the outputs regardless of the signal levels on the other input nodes (and especially the clock signal node), are asynchronous. The asynchronous set and reset functions are activated by the low level of signals, which are then denoted as $\overline{PR}$ and $\overline{CLR}$. Asynchronous inputs are required to determine the initial state of the flip-flop and are not used during the normal operation.

A.3.5   CMOS implementation

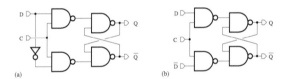

FIGURE A.18
Circuit diagrams of conventional D latches (a) with a single data input and (b) with complementary data inputs.

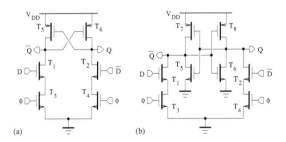

FIGURE A.19
CMOS implementations of (a) dynamic and (b) static D latches.

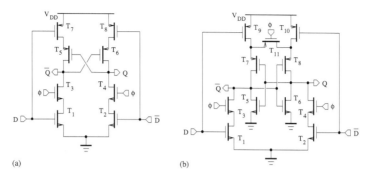

(a) (b)

FIGURE A.20

CMOS implementations of ratio-insensitive (a) dynamic and (b) static D latches.

D latches can be designed by substituting each logic gate of the circuit diagrams shown in Figure A.18 with its CMOS implementation. However, the resulting maximum operating frequency appears to be limited by the long delay of the critical path, especially at low supply voltages. An approach to reduce the critical path delay is to use latches with differential cascode structures [1], such as the ones depicted in Figure A.19. The first latch in Figure A.19(a) is a dynamic logic circuit based on cross-coupled transistors, which can track the changes of the input data only during one phase of the clock signal, while the second latch in Figure A.19(b), which uses cross-coupled inverters acting as a storage element, can be considered a static logic circuit. A static latch possesses an output storage node that remains connected to either the supply voltage or ground, while a dynamic latch exhibits a conducting path to the supply voltage or ground only during the evaluation of the input data. In general, dynamic CMOS logic circuits require less silicon area due to the number of transistors needed than their static counterparts. However, they can be affected by charge sharing at the output nodes. One solution to prevent variations in the logic level consists of using sufficiently large static inverters to isolate the output nodes of the dynamic latch stage.

The latches of Figure A.19 have the drawback of being sensitive to the aspect ratio between the nMOS and pMOS transistors, especially in the case where the input transistor is of the pMOS type. This is due to the fact that the current delivered by input section should overcome that from the cross-coupled transistors in order for the latch to switch, thereby increasing the length of the switching period.

Ratio-insensitive latches can be designed as shown in Figure A.20. Extra transistors are used so that the input node can be formed by connecting together the gate terminals of nMOS and pMOS transistors.

Flip-flops can be realized using two latches in a master-slave configuration, as shown in Figure A.21. The state of the input is acquired by the master latch when the clock signal becomes low. It is then transferred to the outputs of the slave latch when the clock signal goes high. The master latch is transparent,

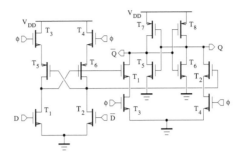

FIGURE A.21
CMOS implementation of a semi-static D flip-flop.

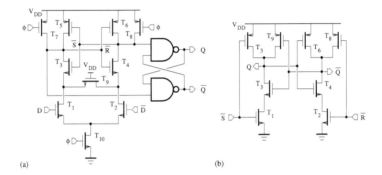

FIGURE A.22
CMOS implementations of (a) D flip-flop and (b) $\overline{S}\,\overline{R}$ latch.

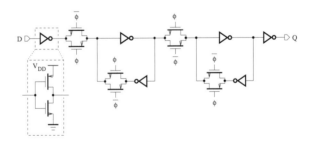

FIGURE A.23
Transmission gate flip-flop.

while the slave latch remains opaque — and vice versa. This flip-flop has the advantage of using only one clock signal, but it can be prone to a substantial voltage drop at the outputs due to the capacitive coupling effect between the master and slave latches.

Another approach to design flip-flop relies on the structure depicted in Figure A.22(a) [2, 3]. The first stage is based on a sense amplifier and the second stage is an $\overline{S}\,\overline{R}$ latch, which can be implemented as shown in Figure A.22(b). When the clock signal becomes low, the input differential transistor pair is in the cutoff region and the output nodes of the sense amplifier are precharged through transistors T_7 and T_8 to the high logic state. The $\overline{S}\,\overline{R}$ latch then holds its previous logic state. When the rising edge of the clock signal occurs, the sense amplifier is enabled and can track the input data to produce a transition from the high state to the low state on one of the outputs. The $\overline{S}\,\overline{R}$ latch updates its outputs according to the actual state of the sense amplifier. By applying the high state of the input data to the node D, T_3, T_1, and T_{10} provide a discharge path to the node $\overline{R}$, while T_4 and T_6 are biased in the cutoff and conduction region, respectively. On the other hand, when the high state of the input data occurs at the node $\overline{D}$, a discharge path is supplied to the node $\overline{S}$ by T_4, T_2, and T_{10}, while T_3 and T_5 operate in the cutoff and conduction region, respectively. After the completion of the state transition at one of the output nodes, the inputs become decoupled from the outputs and the sense amplifier then remains unaffected by any subsequent change of input data during the active clock phase. If the state of the input data changes, the discharge path will be interrupted, leaving the node at the low logic level floating. In order to prevent this floating node from being charged by the leakage or coupling currents, another discharge path should be provided by using the nMOS pass transistor T_9. This pass transistor also helps equalize the voltage values of the two differential branches during the pre-charge and reset the output nodes prior to the next sensing of the input data. The primary feature of the sense-amplifier flip-flop is the provision of high input impedance. However, the operation speed can be limited by the propagation delay introduced by the output latch.

A master-slave D flip-flop that is triggered by the rising edge of the clock signal can be realized, as illustrated in Figure A.23, by using two transmission gate-based latches enabled by complementary clock signals. It is particularly suitable for operation at high supply voltage. Its hold time may be limited by mismatches between the inverters on the clock and data paths.

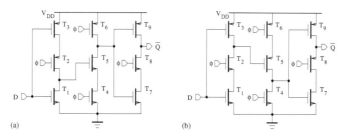

FIGURE A.24
CMOS implementations of (a) rising-edge and (b) falling-edge triggered D flip-flops.

More efficient circuits for D flip-flops can be derived using true single-

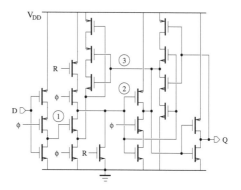

FIGURE A.25
Circuit diagram of a retentive TSPC D flip-flop.

phase clocking (TSPC) techniques [4]. TSPC D flip-flops operate with only one clock signal, and offer advantages such as a small circuit area for clock lines, a reduced clock skew, and high-speed operation. Figure A.24 shows the circuit diagrams of the rising-edge and falling-edge triggered D-flip-flops, which consist of three clocked inverting stages. The output signal is available in an inverted version. When the clock slope is not sufficiently steep, both nMOS and pMOS clocked transistors simultaneously operate in the conduction region during the clock transition. As a result, the logic level of internal signals may become undefined and a race condition may occur.

In comparison with conventional master-slave flip-flops, TSPC flip-flops have the advantages of reducing the load of the clock distribution network and the switching power dissipation. Moreover, by requiring only a single-phase clock signal, they are less affected by clock skews caused by process variations. However, the operation of TSPC flip-flops may be affected by data loss that is inherent to a dynamic operation. Figure A.25 shows the circuit diagram of a retentive TSPC D flip-flop [5]. The signal R can be used to reset the flip-flop. When the clock signal goes low, the actual logic state of the data is kept on the node 3. Meanwhile, the next logic state of the data is transferred to the node 1, while the node 2 is pre-charged to the supply voltage V_{DD} without effecting the retention process on the node 3 due to the pMOS transistor with the gate connected to the node 2. The data transfer begins at the rising edge of the clock signal. If the node 1 is charged to V_{DD}, the node 2 is discharged to ground by the nMOS transistor with the gate connected to the node 1 and a high logic state is held on the node 3. If, on the other hand, the node 1 is discharged to ground, the node 2 is not discharged and a low logic state is kept on the node 3. The forward-conditional feedback paths between the nodes 2 and 3 and between the nodes Q and 3 help avoid data loss on the node 3 during the data transfer phase by preventing the discharge of the node 2 due to leakage current, and during the retention phase by inducing the discharge of the node 3 to counteract possible contention transitions, respectively. For a high-speed and power-efficient design, transistors with a minimum width should preferably be used in the feedback paths.

A.4 Bibliography

[1] J. Yuan and C. Svensson, "New single-clock CMOS latches and flipflops with improved speed and power savings," *IEEE J. of Solid-State Circuits*, vol. 32, pp. 62–69, Jan. 1997.

[2] M. Matsui, H. Hara, Y. Uetani, L.-S. Kim, T. Nagamatsu, Y. Watanabe, A. Chiba, K. Matsuda, and T. Sakurai, "A 200 MHz 13 mm^2 2-D DCT macrocell using sense-amplifying pipeline flip-flop scheme," *IEEE J. of Solid-State Circuits*, vol. 29, pp. 1482–1490, Dec. 1994.

[3] J. Montanaro, R. T. Witek, K. Anne, A. J. Black, E. M. Cooper, D. W. Dobberpuhl, P. M. Donahue, J. Eno, G. W. Hoeppner, D. Kruckemyer, T. H. Lee, P. C. M. Lin, L. Madden, D. Murray, M. H. Pearce, S. Santhanam, K. J. Snyder, R. Stephany, and S. C. Thierauf, "A 160-MHz, 32-b, 0.5-W CMOS RISC microprocessor," *IEEE J. of Solid-State Circuits*, vol. 31, pp. 1703–1714, Nov. 1996.

[4] J. Yuan and C. Svensson, "High-speed CMOS circuit technique," *IEEE J. of Solid-State Circuits*, vol. 24, pp. 62–70, Jan. 1989.

[5] F. Stas and D. Bol, "A 0.4V 0.08fJ/cycle retentive true-single-phase-clock 18T flip-flop in 28nm FDSOI CMOS," in *Proc. of the IEEE ISCAS*, Baltimore, MD, USA, March 2017, pp. 2779–2782.

B

Notes on Circuit Analysis

CONTENTS

Proofs of some equations and results that were used to characterize PLL circuits are presented.

B.1   Radius of curvature

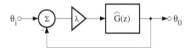

FIGURE B.1
Variable gain model of the PLL in the discrete-time domain.

Based on the variable gain model of Figure B.1, the closed-loop transfer function can be written as,

$$\frac{\theta_0(z)}{\theta_i(z)} = \frac{\lambda \hat{G}(z)}{1 + \lambda \hat{G}(z)} \tag{B.1}$$

where $1 + \lambda \hat{G}(z)$ is the characteristic equation, and λ represents the loop gain.

An approach that can be adopted to select a gain value for stability is the root locus, which is a plot of the characteristic equation roots as a function of the gain, λ.

Figure B.2 shows two examples of root loci for the charge-pump PLL [1] with a second-order lowpass filter.

In Figure B.2(a), $C_1 = 2$ pF and $C_2 = 0$. At the intersections between the unit circle and the root loci, $\lambda = \lambda_1$ or λ_2, with $\lambda_2 > \lambda_1$. When $0 < \lambda < \lambda_1$, all the roots are inside the unit circle, and the stability condition is met. When

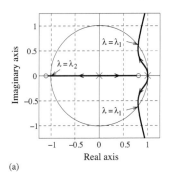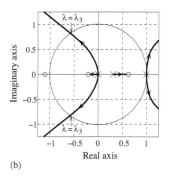

(a)　　　　　　　　　　　　　　　　　　(b)

FIGURE B.2
Root loci of the closed-loop transfer function poles: (a) $C_2 = 0$, (b) $C_2 \neq 0$.

$\lambda_1 < \lambda < \lambda_2$, the roots departing at $z = 1$ are located outside the unit circle while others remain inside the unit circle. Hence, θ_0 starts to grow, and then $\theta_i - \theta_0$ tends to increase, causing a decrease of λ. This process is repeated for several successive cycles until the roots outside the unit circle move back inside the unit circle, thus disclosing the existence of a stable limit cycle.

In Figure B.2(b), $C_1 = C_2 = 2$ pF. The root loci cross the unit circle when $\lambda = \lambda_3$. It can be verified that, in any case, there is no stable region of operation.

A useful stability criterion can be derived by relating the analysis of the locus behavior to the radius of curvature. Let the transfer function, $\hat{G}(z)$, be of the form,

$$\hat{G}(z) = \frac{N(z)}{D(z)} = \frac{\sum_{k=0}^{K} a_k z^k}{\sum_{l=0}^{L} b_l z^l} \tag{B.2}$$

where $z = x + jy$ and $j = \sqrt{-1}$. The Taylor series expansions of $N(z)$ and $D(z)$ about z are given by

$$N(x + jy) = \sum_{k=0}^{K} \frac{N^{(k)}(x)}{k!} (jy)^k \tag{B.3}$$

$$D(x + jy) = \sum_{l=0}^{L} \frac{D^{(l)}(x)}{l!} (jy)^l \tag{B.4}$$

where $N^{(k)}(x)$ and $D^{(l)}(x)$ represent the nth and lth derivatives of $N(x)$ and $D(x)$, respectively. Substituting (B.3) and (B.4) into (B.2) and rearranging the result in real and imaginary parts, and setting the imaginary part equal to zero, we arrive at

$$g(x, y) = Q_1(x) - y^2 Q_3(x) + y^4 Q_5(x) - \cdots = 0 \tag{B.5}$$

where

$$Q_r(x) = \sum_{r=0}^{R} (-1)^r \frac{N^{(r)}(x) D^{(R-r)}(x)}{r!(R-r)!} \tag{B.6}$$

In general, at a given point (x, y) on a curve implicitly defined by the equation

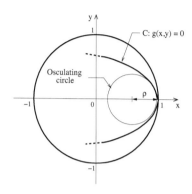

FIGURE B.3
Representation of the radius of curvature ρ at $(x, y) = (1, 0)$.

$C : g(x, y) = 0$, the curvature κ and the radius of curvature ρ can be estimated as,

$$\kappa(x, y) = \frac{g_{xx} g_y^2 - 2 g_{xy} g_x g_y + g_{yy} g_x^2}{(g_x^2 + g_y^2)^{3/2}} \tag{B.7}$$

$$\rho(x, y) = \frac{1}{|\kappa(x, y)|} \tag{B.8}$$

where for $u, v \in \{x, y\}$, $g_v = \partial g / \partial v$, $g_{uv} = \partial g_v / \partial u$, and $g_{uu} = \partial^2 g / \partial u^2$.

In Figure B.3, the curve $C : g(x, y) = 0$ is assumed to be a portion of the root locus with the double pole at $z = 1$. The radius ρ of the osculating circle is called the radius of curvature at the point $(x, y) = (1, 0)$ on the curve.

If the root locus has to first move toward the inside of the unit circle before it can cross the unit circle, then the next requirement should be satisfied,

$$\frac{1}{\kappa(1, 0)} < 1 \tag{B.9}$$

where $\kappa(1, 0)$ is assumed to be positive. From (B.5), it can be shown that $g_y(1, 0) = 0$. Furthermore, $g_x(1, 0) = Q_1'(1)$ and $g_{yy}(1, 0) = -2Q_3(1)$, where $Q_1'(x) = dQ_1(x)/dx$, and Equation (B.9) can be rewritten as,

$$-\frac{Q_1'(1)}{2 Q_3(1)} < 1 \tag{B.10}$$

where

$$Q'_1(1) = -\sum_{k=0}^{K}\sum_{l=0}^{L}(k-l)(k+l-1)a_k b_l \tag{B.11}$$

$$Q_3(1) = -\frac{1}{6}\sum_{k=0}^{K}\sum_{l=0}^{L}(k-l)[(k-l)^2 - 3(k+l)+2]a_k b_l \tag{B.12}$$

In the special case of the charge-pump PLL with a second-order lowpass filter, it can be shown that

$$Q'_1(1) = 4\hat{G}_0(1-r_0)^2 \tag{B.13}$$

$$Q_3(1) = -\hat{G}_0(1-r_0)^2\left[1 - 2\left(q + \frac{\tau_d}{T} - \frac{RC_1}{T} + \frac{RC_1 C_2}{T(C_1+C_2)}\right)\right] \tag{B.14}$$

where $\hat{G}_0 = K_p K_v T^2/(2 \cdot N \cdot C_T)$.

B.2   Spectral analysis of PWM signals

In general, the pulse-width modulated (PWM) waveform, that is generated by comparing an input signal with a carrier signal, is not periodic. Its spectral analysis requires the use of double Fourier series expansion [2] because the input signal frequency cannot be directly related to the carrier frequency.

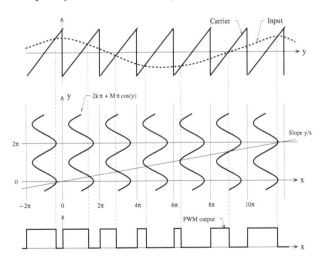

FIGURE B.4
Geometric surface for the spectral analysis of a PWM signal.

Let $x = \omega_c t$ and $y = \omega t$. The input signal can be written as,

$$u(y) = 2k\pi + M\pi \cos(y) \tag{B.15}$$

For a single-sided modulation, the carrier is a sawtooth signal. The spectral analysis of the PWM signal can be performed using a geometric surface, defined as $F(x, y)$ and illustrated in Figure B.4. Hence,

$$F(x, y) = \begin{cases} 1 & \text{if } (x - |x|_{2\pi}) \leq u(y) \\ 0 & \text{otherwise} \end{cases} \tag{B.16}$$

where $|x|_{2\pi}$ is the closed multiple of 2π, which is less than x. The function $F(x, y)$ remains identical in each 2π-square and is then periodic in two dimensions. The double Fourier series of $F(x, y)$ can be written as,

$$
\begin{aligned}
F(x, y) ={}& \frac{1}{2} A_{00} + \sum_{n=1}^{+\infty} [A_{0n} \cos(ny) + B_{0n} \sin(ny)] \\
&+ \sum_{m=1}^{+\infty} [A_{m0} \cos(mx) + B_{m0} \sin(my)] \\
&+ \sum_{m=1}^{+\infty} \sum_{\substack{n=-\infty \\ n \neq 0}}^{+\infty} [A_{mn} \cos(mx + ny) + B_{mn} \sin(mx + ny)]
\end{aligned} \tag{B.17}
$$

In the complex form, the Fourier coefficients of $F(x, y)$ can be expressed as

$$A_{mn} + jB_{mn} = \frac{1}{2\pi^2} \int_0^{2\pi} \int_0^{2\pi} F(x, y) e^{j(mx+ny)} \, dx \, dy \tag{B.18}$$

Taking into account the definition of $F(x, y)$ and assuming that $m, n \neq 0$, we have

$$A_{mn} + jB_{mn} = \frac{1}{2\pi^2} \int_0^{2\pi} \int_0^{2k\pi + M\pi \cos(y)} F(x, y) e^{j(mx+ny)} \, dx \, dy \tag{B.19}$$

$$= -\frac{j}{2\pi^2 m} \int_0^{2\pi} (e^{j[m(2k\pi + M\pi \cos(y)) + ny]} - e^{jny}) \, dx \, dy \tag{B.20}$$

$$= -\frac{j}{2\pi^2 m} e^{j2mk\pi} \int_0^{2\pi} e^{jmM\pi \cos(y)} e^{jny} \, dx \, dy \tag{B.21}$$

$$= -\frac{j}{\pi m} e^{j(2mk\pi + n\pi/2)} J_n(mM\pi) \tag{B.22}$$

where the Bessel function is given by

$$J_n(z) = \frac{j^{-n}}{2\pi} \int_0^{2\pi} e^{jz \cos(\phi)} e^{jn\phi} \, d\phi \tag{B.23}$$

Hence, for $m, n \neq 0$, we obtain

$$A_{mn} = \frac{j}{\pi m} \sin(2mk\pi + n\pi/2) J_n(mM\pi) \tag{B.24}$$

and

$$B_{mn} = -\frac{j}{\pi m} \cos(2mk\pi + n\pi/2) J_n(mM\pi) \tag{B.25}$$

For the special cases where $m, n = 0$, it can be shown that

$$A_{00} = 2k \tag{B.26}$$
$$A_{0n} = M/2 \tag{B.27}$$
$$B_{0n} = 0 \tag{B.28}$$
$$A_{m0} = \frac{J_0(mM\pi)}{m\pi} \sin(2mk\pi) \tag{B.29}$$
$$B_{m0} = \frac{1 - J_0(mM\pi)\cos(2mk\pi)}{m\pi} \tag{B.30}$$

By inserting the Fourier coefficients into Equation (B.17), we arrive at

$$
\begin{aligned}
F(x,y) = {} & k + M\cos(y) \\
& + 2\sum_{m=1}^{+\infty} \frac{\sin(mx)}{m\pi} - 2\sum_{m=1}^{+\infty} \frac{J_0(m\pi M)}{m\pi}\sin(mx - 2mk\pi) \\
& - 2\sum_{m=1}^{+\infty}\sum_{\substack{n=-\infty \\ n\neq 0}}^{+\infty} \frac{J_n(m\pi M)}{m\pi}\sin(mx + ny - 2mk\pi - n\pi/2)
\end{aligned}
\tag{B.31}
$$

By choosing to set k equal to $1/2$, then multiplying the resulting series by 2 (waveform with a pulse height of 2) and ignoring the dc component, the Fourier series can be written as,

$$
\begin{aligned}
F_{ADs}(x,y) = {} & M\cos(y) \\
& + 2\sum_{m=1}^{+\infty} \frac{\sin(mx)}{m\pi} - 2\sum_{m=1}^{+\infty} \frac{J_0(m\pi M)}{m\pi}\sin(mx - m\pi) \\
& - 2\sum_{m=1}^{+\infty}\sum_{\substack{n=-\infty \\ n\neq 0}}^{+\infty} \frac{J_n(m\pi M)}{m\pi}\sin(mx + ny - m\pi - n\pi/2)
\end{aligned}
\tag{B.32}
$$

or equivalently,

$$
\begin{aligned}
F_{ADs}(x,y) = {} & M\cos(y) \\
& + 2\sum_{m=1}^{\infty} \frac{1 - J_0(m\pi M)\cos(m\pi)}{m\pi}\sin(mx) \\
& - 2\sum_{m=1}^{\infty}\sum_{\substack{n=-\infty \\ n\neq 0}}^{+\infty} \frac{J_n(m\pi M)}{m\pi}\sin(mx + ny - m\pi - n\pi/2)
\end{aligned}
\tag{B.33}
$$

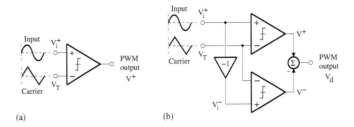

(a)   (b)

FIGURE B.5
Generation of (a) two-level and (b) three-level PWM waveforms.

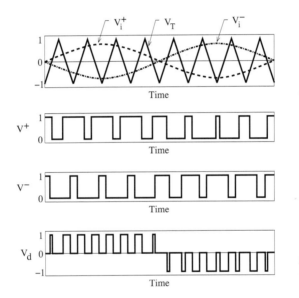

FIGURE B.6
Two-level and three-level PWM waveforms.

A double-sided modulation requires the use of a triangular carrier signal. Figure B.5 shows circuit structures that can be used for the generation of two-level and three-level PWM waveforms. Two-level and three-level PWM waveforms are represented in Figure B.6. The modulation switching method can be either of class AD (that is, the output can take two voltage levels) or class BD (meaning that the output can assume three voltage levels).

For a two-level PWM waveform, the Fourier series can be derived as follows

$$F_{AD_D}(t) = F_{AD_S}(t) - F_{AD_S}(-t) \tag{B.34}$$

Hence,

$$
\begin{aligned}
F_{AD_D}(x, y) = {}& M \cos(y) \\
&+ 2 \sum_{m=1}^{+\infty} \frac{J_0(m\pi M/2)}{m\pi/2} \sin(m\pi/2) \cos(mx) \\
&+ 2 \sum_{m=1}^{\infty} \sum_{\substack{n=-\infty \\ n\neq 0}}^{+\infty} \frac{J_n(m\pi M/2)}{m\pi/2} \sin\left(\frac{m+n}{2}\pi\right) \cos(mx + ny)
\end{aligned}
\tag{B.35}
$$

The output spectrum contains harmonics of the switching frequency, and inter-modulation components between the input and carrier frequencies that occur symmetrically around each of the carrier frequency harmonics.

In the case of a three-level PWM waveform, we have

$$
F_{BD_D}(t) = \frac{F_{AD_D}(t) - F_{AD_D,\pi}(t)}{2}
\tag{B.36}
$$

where

$$
\begin{aligned}
F_{AD_D,\pi}(x, y) = {}& M \cos(y - \pi) \\
&+ 2 \sum_{m=1}^{+\infty} \frac{J_0(m\pi M/2)}{m\pi/2} \sin(m\pi/2) \cos(mx) \\
&+ 2 \sum_{m=1}^{+\infty} \sum_{\substack{n=-\infty \\ n\neq 0}}^{+\infty} \frac{J_n(m\pi M/2)}{m\pi/2} \sin\left(\frac{m+n}{2}\pi\right) \cos(mx + n(y - \pi))
\end{aligned}
\tag{B.37}
$$

Thus,

$$
\begin{aligned}
F_{BD_D}(t) = {}& M \cos(y) \\
&- 4 \sum_{m=1}^{+\infty} \sum_{\substack{n=-\infty \\ n\neq 0}}^{+\infty} \frac{J_n(m\pi M/2)}{m\pi} \sin\left(\frac{m+n}{2}\pi\right) \sin\left(\frac{n}{2}\pi\right) \sin\left(mx + ny - \frac{n}{2}\pi\right)
\end{aligned}
\tag{B.38}
$$

Due to the fact that the output spectrum does not contain any harmonics of the carrier frequency, the intermodulation components are located symmetrically around the even harmonics of the carrier frequency.

B.3 Bibliography

[1] Z. Wang, "An analysis of charge-pump phase-locked loops," *IEEE Trans. on Circuits and Systems–I*, vol. 52, pp. 2128–2138, Oct. 2005.

[2] K. Nielsen, "A review and comparison of pulse width modulation (PWM) methods for analog and digital input switching power amplifiers," *Proc. of the 102nd AES Convention*, Munich, Germany, Preprint 4446 (G4), Mar. 1997.

Index

Printed and bound by CPI Group (UK) Ltd, Croydon, CR0 4YY

24/10/2024

01778281-0016